3/21/92

D1539140

Vita Mathematica
Volume 5

Edited
by Emil A. Fellmann

Norbert Wiener, 1945. Courtesy of Mrs. Margaret Wiener.

Norbert Wiener 1894–1964

by P. R. Masani

1990   Birkhäuser Verlag
Basel · Boston · Berlin

Author's Address:
P. R. Masani
Department of Mathematics
University of Pittsburgh
Pittsburgh, Pennsylvania 15260
USA

Library of Congress Cataloging-in-Publication Data

Masani, P. Rustom,
Norbert Wiener, 1894–1964 / by Pesi R. Masani.
(Vita mathematica: v. 5)
Bibliography: p. Includes index,
ISBN 0-8176-2246-2
1. Wiener, Norbert, 1894–1964. 2. Mathematicians—United States—
Biography. I. Title. II. Series.
QA29.W497M37 1990
510'.92'4—dc19

CIP-Titelaufnahme der Deutschen Bibliothek

Masani, P. R.:
Norbert Wiener: 1894–1964 / by Pesi R. Masani. – Basel;
Boston; Berlin: Birkhäuser, 1990
(Vita mathematica; Vol. 5)
ISBN 3-7643-2246-2 (Basel ...) Gb.
ISBN 0-8176-2246-2 (Boston) Gb.
NE: GT

© 1990 Birkhäuser Verlag Basel
Printed in Switzerland
Cover Design: Albert Gomm swb/asg, Basel
Layout: Silvia Leupin
ISBN 3-7643-2246-2
ISBN 0-8176-2246-2

This book attempts to trace the interaction
between mathematical genius and history that has led
to the conception of a stochastic cosmos.

God forever geometrizes.

Plato, c. 400 B.C.

The neglect of mathematics for thirty or forty years has nearly destroyed the entire learning of Latin Christendom. For he who does not know mathematics cannot know any of the other sciences; what is more, he cannot discover his own ignorance or find its proper remedies. So it is that the knowledge of this science prepares the mind and elevates it to a well-authenticated knowledge of all things. For without mathematics neither what is antecedent nor consequent to it can be known; they perfect and regulate the former, and dispose and prepare the way for that which succeeds.

From a letter of Roger Bacon, Franciscan friar and experimental philosopher, to Pope Clement IV, c. 1267.

I learnt to distrust all physical concepts as the basis for a theory. Instead one should put one's trust in a mathematical scheme, even if the scheme does not appear at first sight to be connected with physics. One should concentrate on getting an interesting mathematics.

From a 1977 lecture by P.A.M. Dirac, theoretical physicist and Nobel Laureate.

Contents

A) America's second Leibniz. B) The difficulties of a biographer. C) Scope and focus. D) Poetry, prose and mathematical symbolism. E) Policy as to mathematical usage.

Solomon and Leo Wiener. The gift of tongues. Tolstoyan conversion and arrival in New Orleans. Mutualistic communities in the 1880s. Leo's "Mark Twain" characteristics: "the Russian Irishman". Marriage and family. Professorships at Missouri and Harvard.

Norbert's Cambridge home. Early biological curiosities. Peabody School and Leo's tutelage. The "avenger of blood". Infantile maladroitness and threat of myopia. Old Mill Farms and Ayer High School. Love for the democracy of the New England town.

A) An eleven-year-old freshman at Tufts College. B) A semester in zoology. C) The "black year" at the Sage school of philosophy. D) The return to Harvard.

A) Wiener's emancipation. B) The doctrine of types. C) The Tractatus and logical empiricism. D) Wiener's philosophical activity. E) Columbia and its word-minded philosopher (1915). F) Harvard and the Docent Lectures (1915–1916). G) The unpleasant stint in Maine (1916). Note: The Peano postulate system.

The young war enthusiast. Futile efforts to enlist in the armed forces. Staff work with the Encyclopedia Americana. Computer at the U.S. Army Proving Grounds. Private in the U.S. Infantry. Fowlersque touch.

A) Wiener joins the Massachussetts Institute of Technology. B) The relation of space and geometry to experience [22a]. C) Postulate systems. D) The Brownian motion. E) Theory of the potential. Note 1: Boolean algebra. Note 2: Algebraic field. Note 3: The Brownian movement. Note 4: The "Borel" terminology. Note 5: Hausdorff dimension.

Slow promotion and resulting competitiveness. Mal-allocation of national income. Margaret Engemann and marriage. Search for positions abroad.

A) Harmonic analysis. B) Heaviside's operational calculus [26d, 29c]. C) From power to communications engineering. D) Generalized harmonic analysis. E) Tauberian theory. F) The intensity, coherence and polarization of light. Note 1: A proof of § 9A (9). Note 2: Historical remarks. Note 3: Spectral synthesis. Note 4: Conditional Banach spaces.

A) The Uncertainty Principle in classical physics. B) The collaboration with Max Born. C) The work with Struik and Vallarta on unified field theory. D) The philosophical interlude. Leibniz and Haldane. E) Quantum theory and the Brownian motion. Hidden-parameters. Note: The mapping M_μ.

A) Radiative equilibrium in the stars. B) Significance of the Hopf-Wiener equation; causality and analyticity. C) Paley and the Fourier transformation in the complex domain.

A) The need for ergodicity in Gibbsian statistical mechanics. B) Wiener's work in Ergodic Theory. C) The contingent cosmos, noise, and Gibbsian statistical mechanics. D) The Second Law of Thermodynamics. Entropy. E) The homogeneous chaos and the Wiener program in statistical mechanics. F) Anisotropic time and Bergsonian time. G) Information, negentropy and Maxwell's demon. Note 1: Definition of the Hamiltonian. Note 2: Thermodynamic entropy. Note 3: Boltzmann statistical entropy. Note 4: Proof of equality of entropy and information.

A) The analogue computer program at MIT. The Wiener integraph. B) The Lee-Wiener network. C) Further work with Lee. The refugee problem. Visit to the Orient. D) Wiener's 1940 memorandum on an electronic computer for partial differential equations. Note 1: The Wiener integraph. Note 2: Proof of the relation § 13B (3). Note 3: The Lee-Wiener network. Note 4: Scanning procedure to solve the PDE.

A) Project D.I.C. 5980: Its military and scientific significance. B) Idealization of the flight trajectory and resulting tasks. C) The mathematical theory. D) Operation of the anti-aircraft predictor. E) Resolution of the man-machine concatenation. F) The problem of transients. G) Secrecy, overwork and tension. H) Kolmogorov's paper and prediction theory. I) The Kolmogorov-Wiener concatenation. Note 1: Bigelow's design. Note 2: The equation § 14C (3); theory of instrumentation.

A) Rosenblueth, friend and philosopher. B) Teleology in the animal and the machine. C) Mexico, and the return to physiology. The heart. D) Muscle clonus. E) The spike potential of axons. F) The statistics of synaptic excitation. G) Voluntary, postural and homeostatic feedback. The moth-cum-bedbug. Note 1: The nervous system. Note 2: Spike potential of axons.

A) McCulloch, the poetic and mathematico-logical neurologist. B) The

Turing machine. C) The computing and pattern-recognizing abilities of the brain. D) Sensory and muscular-skeletal prosthesis. E) The computer as prosthesis for the brain. F) Brain waves and self-organization. Note: Structure of the Turing machine.

A) The Teleological Society. B) The parallel evolutions of von Neumann and Wiener. C) Von Neumann's letter on the direction of cybernetical research; molecular biology. D) Wiener's thoughts on molecular biology.

A) Cybernetics and its historical origins. B) The subject-matter of cybernetics. C) Is cybernetics a science? D) Soviet views on cybernetics. E) Ontogenetic learning. F) Phylogenetic learning and non-linear networks. G) The distinction between scientific inquiry, and stratagem or contest. Note 1: Comparison with life processes. Note 2: Mathematical representation of non-linear filters.

A) Automatization and the labor movement. B) The educational challenge. C) Anti-homeostatic aspects of the American economic system. D) Wiener's economic ideas. E) The control of conflict in civil society. F) Communal information. G) The grammar of the social sciences. H) Wiener and Marxism.

A) Wiener's thoughts on geopolitics. B) Wiener's letter to Walter Reuther on geopolitics, and von Clausewitz's principles on war. C) Type-classification in military science. D) The pitfalls in computerized atomic war games. Time scales. E) Synopsis of Wiener's military thought. F) Von Neumann's position; science and human welfare. G) The Black Mass, twentieth century. Note: Von Clausewitz's principles of war.

A) The problem of evil. B) Phylogenetic corruption, the Fall. Homo peccator. C) Redemption: Mythic perception and faith as integral parts of man's ascent. D) Redemption: Self-abnegation and duty. E) The long-time State. The mandate from heaven. F) Sin grows with doing good: Tragedy and catharsis. G) Rational Logos, or rational and altruistic Logos?

A) Wiener's literary work. B) Wiener's theatrical initiatives. C) The mathematician's credo. D) Why mathematics is a fine art. E) The variance between Wiener's and Halmos's views on mathematics and art. The scholastic position. F) Wiener's thought on machine creativity.

A) The lovable quirk. B) Wiener as public figure and teacher. C) Resignation from the National Academy of Sciences; muddled ideas on science academies. D) Unobjective attitude towards Harvard. E) Wiener and George David Birkhoff.

System of Referencing and Notation

In this book, a notation such as § 8C refers to Section C of Chapter 8. A notation such as § 8C (6) refers to formula (6) appearing in Section C of Chapter 8.

Two-digit numbers followed by a Roman letter in square brackets, e.g. [53h], refer to the Bibliography of Norbert Wiener at the end of this biography (see p. 377). The '53' in [53h] refers to the year of publication, 1953. Numbers preceded by MC in square brackets, e.g. [MC, 57], refer to the folders in the Manuscript Collection of Wiener (in the MIT Archives) in which the documents appear. Roman numerals in square brackets, e.g. [IV], refer to Defense Department Documents pertaining to Wiener, listed after the Bibliography.

Numbers in braces, e.g. {K3}, cite the references listed at the end of the book (see p. 392).

The following notation is used intermittently in the book.

$\mathbb{N}, \mathbb{R}, \mathbb{C}$ refer to the classes of natural, real and complex numbers, respectively.

S stands for the Wiener class of functions, defined on p. 102–.

S also refers to the thermodynamic entropy, cf. § 12D.

$:=$ means equal by definition.

1_A is the indicator function of the set A; $1_A(x) = 1$ or 0, accordingly as x is in or not in A.

G.H.A.   stands for generalized harmonic analysis.

Acknowledgment

This book is an enlargement of an article in the Encyclopedia of Computer Science and Technology[1], that has grown and grown as I became more and more aware of the enormity and profundity of Wiener's thought. Its scope and purpose are indicated in the Prologue (Chap. I, §§ A,B,C), and only the pleasant task of making acknowledgment remains.

Although the manuscript was seen by a few publishers, it might never have seen the light of day had Professor P.L. Butzer of the Technische Hochschule in Aachen not passed on a copy to Dr. E.A. Fellmann of the Euler Archiv in Basel. My sincere thanks go to him, and to Dr. Fellmann for his invitation to publish it in the series *Vita Mathematica*.

Let me rethank Professors J. Belzer and A. Kent and Professors W.T. Martin, G. Birkhoff and M.H. Stone (deceased) for their initial help in 1979–80. Second, I want to thank Ms. Helen Samuels of the Archive Section at MIT for placing Wiener's unpublished material for my use in May 1981, and Professor I.E. Segal of MIT for permission to use his list of Wiener's doctoral students. I am grateful to Mr. E.F. Beschler of Birkhäuser Boston, Ms. K. McLaughlin and Professor R.J. Parikh for reading and criticizing the manuscript in whole or part, to Professor J. Benedetto for providing me with inaccessible biographical data, and to my friend Mr. K.N. Karanjia of Bombay for sending me a less-known paper of Dirac from which I have quoted. Mr. Beschler was also very helpful in settling some copyright difficulties.

The following individuals were very generous and forthright in sending me photographs for the book: Mr. J. Bigelow, Ms. A. Bogovich, Dr. F. Perrin, Professor N.N. Rao, the late Mrs. Virginia Rosenblueth, Dr. W.A. Stanley, Dr. K. Thompson, Mr. T. Walley Williams and Mrs. Margaret Wiener. My sincere thanks go to them, and to Mr. M. Roper of our Fine Arts Department for his fine reproductions of some of the other photos. I also want to thank Mr. M. Yeates of the MIT Museum

---

1 Vol. 14 (1980), pp. 72–136, Editors J. Belzer, H. Holtzman & A. Kent, Marcel Dekker, New York.

and Dr. W. H. Jaco of the American Mathematical Society for waiving their usual user-fee for photographs.

For the typing of earlier versions of the manuscript I am indebted to Mrs. D. B. Berman, Miss D. Heron and Mrs. S. Rinni. Very special thanks, however, go to my present secretary Mrs. N. Rhodes. Without her personal interest in the project over the last two years and her discipline in typing and managing the overwhelming bulk of paper, there might have been not book but chaos!

Last but not least I wish to thank the staff of Birkhäuser Verlag for the meticulous execution of a long and complex manuscript.

University of Pittsburgh   P. R. Masani
June 1989

1   Prologue

A America's second Leibniz

A special January 1966 issue of the *Bulletin of the American Mathematical Society* bears the inscription:

> dedicated to the memory of Norbert Wiener in recognition of his towering stature in American and world mathematics, his remarkably many-sided genius, and the originality and depth of his pioneering contributions to science.

In 1963, Lyndon Johnson, the president of the United States, presented Wiener with the National Medal of Science. A mathematician on a par with the best in this century, Wiener was also a universal thinker of colossal proportions. If we think of Charles Sanders Peirce as being America's first Leibniz, then we must think of Wiener as the second.

Wiener's contributions embrace several areas of mathematics and mathematical philosophy, relativity and quantum mechanics, communication engineering, computer theory and physiology. He was the originator of an interscientific discipline called *cybernetics*, the general theory of communication and control in the animal and the machine, along lines that had been dimly perceived by Plato. Wiener's pioneering mind was always ahead of its time. In 1940 he had outlined the design of an ultra-high-speed electronic digital computer, and by the mid 1940s he had foreseen the coming age of automatization. More than any other person he conveyed to the public the philosophical, moral and social implications of automatization. He came to be known as the *Father of Automation*. This was his obituary caption in *The New York Times*, March 19, 1964, p. 1.

But Wiener's vision extended far beyond the arena of philosophy, mathematics, physics and technology. His far-reaching mind illuminated almost every phase of intellectual endeavor. In physiology, singly and in collaboration, he worked on rhythms within the human body and especially within the brain. He not only developed a theory of pattern recognition, but went on to see it applied to the prosthesis of artificial limbs and organs for the handicapped. He worked out a theory of learning and reproducing automata that shed light not only on the automatic factory, but also on the ontogenetic and phylogenetic learning of biological species. He showed how communication is the dynamic cement that sustains a community, and how sponsored misinformation is a danger. In his socio-economic writings he suggested ways to relieve the non-homeostatic aspects of the economy. On military matters he brought to bear a unique perspective based on his understanding of typology, and time scales of interacting mechanisms. This great thinker also reflected on the problem of war, and on the wider problems of evil and of human governance, shedding new light on old theological issues.

In addition, Wiener partook of humane endeavor. He conferred with the labor leader Walter Reuther on the circumvention of working-class adversity during the automatization of the United States economy. His correspondance reveals a feeling concern for suffering humanity, and a strong sense of social justice and social responsibility.

Although Wiener wrote more than twenty years ago, his writings bear vitally on our present predicament. Indeed, much contemporary analysis and commentary seems banal and naive when set beside the profundity of his thought. Had this thought received better attention forty years ago, we would not be facing so many acute problems today. We would be spared from even more acute problems in the future, were we to heed his thought today.

B The difficulties of a biographer

This enormous versatility is just one factor that precludes writing a simple biography of Wiener. A second is that we must deal with several individuals besides Wiener. For Wiener always depended on other minds for initial intellectual stimulation, and it is remarkable how each phase of his life bears the impress of one or more powerful minds.

A third and more agonizing difficulty appears when we turn from Wiener's mind to Wiener's temperament. Like Charles Peirce, he was no courtier caught in the net of monetary greed or establishmentarian wrangling. Proverbially absent-minded, amusingly quirkish and idiosyncratic, he was fundamentally a gentle and humane soul. He was, however, given to recurrent manifestations of petulance, egoism, emotional instability, irrational insecurity and anxiety. His moods could swing from euphoria to gloom or vice versa on slight provocation. He was an ex-prodigy, and his early formative years, firmly molded by his father, himself a genius, were not always happy. His early suffering, caused often by break-down of parental wisdom, left a scar on his adult personality. This one unfortunate aspect of his character was the source of his uneven relationships with colleagues, another complicating factor in writing about him.

Charles Sanders
Peirce, 1839–1914.

Courtesy of
Dr. William A. Stanley,
National Oceanic and
Atmospheric
Administration,
Washington, D.C.

America's first Leibniz

The biographer's predicament is, however, significantly lessened by the vivid account which Wiener himself has left us of his life until about 1954 in his two books [53h, 56g]. He has given an intimate account of the relationship between his creative work and his emotional, personal and social life. But the biographer must be wary of being swayed by these books, for, unfortunately, bits of the immature egoism we spoke of occasionally show up and detract from the otherwise very warm, lively, philosophical and historical narrative. There are deliberate omissions, and some passages are deluged with narcissistic terms such as "self-esteem", "recognition", and "acceptance". In describing his colleagues Wiener sometimes quoted gossip. Speaking of Edmund Landau, the distinguished mathematician whose course he attended in 1914, Wiener wrote:

> When people asked where to find his house in Göttingen he would say quite naïvely, "You will have no difficulty finding it. It is the finest house in the town". [56g, p. 24]

He also narrated what the mathematician P. Koebe is reported to have said while viewing the painting of *The Last Supper* by Leonardo da Vinci:

> How sad! This painting will pass away, while my theorem concerning the uniformization of analytic functions will endure forever! [56g, p. 159]

While some of these tales, on the foibles of big mathematicians, add to the amusement of the reader, Wiener did not always place them in a generous setting. Sentences in the book, such as

> I think that this was the time I met Koebe, a ponderous, pompous man — "the great expert in the theory of functions", as he was said to have been called by the passers-by in his native town in Brandenburg. [56g, p. 159]

while contributing little, have irritated many a friendly reader. His descriptions of some other colleagues were even harsher and non-factual. In short, Wiener's two books, though exceedingly valuable, are not uniformly authentic.

The MIT historian of science, G. de Santillana, said of Wiener, "In his reactions he was a child, in his judgments a philosopher". The transformation was striking indeed. It was like the great metamorphosis that occurred when the disorderly, unhappy and irritable Ludwig van Beethoven picked up his musical pen. With Wiener too, all traces of immaturity and eccentricity vanished when he picked up his scholarly pen. This writer has had the good fortune of studying all the 250-odd publications of Wiener. He can raise his right hand and say that in this corpus, apart from the lapses just noted in his historical writings [53h, 56g] and also [58f], he has found only three works, [47b, 48d, 49h], in which Wiener-noise damps out the Wiener-message. Wiener's Manuscript Collection in the MIT Archives comprises 900 folders, among which is Wiener's correspondence with about a thousand individuals, ranging from an Attica prisoner to leaders of industry and labor, and some of the world's great minds.[1] Among the few letters this writer has scanned, he came across only one (to Dr. Frank Jewett, in September 1941, in which Wiener tenders his resignation from the National Academy of Sciences) that was genuinely confusing (§ 23 C).

Wiener's life work, its enormous range notwithstanding, exhibits a coherence of thought from start to finish, reminiscent of a great work of art. To write a biography of a person of so towering a stature is an

---

1   Fortunately, this correspondence is now being studied by Dr. Albert C. Lewis of the Bertrand Russell Editorial Project, McMaster University, Ontario, Canada, and may see the light of day whithin a few years.

awesome undertaking. The difficulties cited add to the enormity of the task.

C Scope and focus

The basic proposition of cybernetics that

$$signal = message + noise,$$

and that the message, and not the noise, is the sensible term in communication, is applicable in all sorts of contexts with suitable reinterpretations of the three terms. It sheds some light on how the responsibility of writing about Wiener should be met. Wiener is the signal, and for us the Wiener-message, and not the Wiener-noise, must be of significance. Those who find glee in petty anecdote or petty rivalry or in instances of moral failure should turn elsewhere. In moments of emotional stress, Wiener was wont to make naive and even foolish utterances, which often contradicted his more mature judgments. We must be firm in demoting the importance of such instances of emotive Wiener-noise. There are, alas, situations in communication in which the noise damps out the message. In such cases one's policy ought to be clear: the noise must be dealt with, and its effects analyzed. When we come to such situations in Wiener's life, we shall regretfully discuss Wiener-noise.

Recent years have witnessed a drop in the standards of biographical and autobiographical style. Authors are drawn to portray aberration and to enhance the small, and often follow up with dubious conjecture linking frailty to scientific fruition. To cite an example from a recent life of Einstein:

> Newton—a fanatical puritanical man who seems to have died a virgin and, probably, had suppressed homosexual tendencies—could be vicious and petty, given to violent rages . . . {B9, p. 133}

No evidence is provided for what "seems" and what "probably" happened. Next comes the conjecture: Newton "was able to channel his intellectual and psychic energies with such intensity and even ferocity on scientific discovery", "may be even" because of "his psychological handicaps" {B9, p. 133}. All this is quite dubious, and sheds no clear light on Newton's creativity.

The same author continues: "He (Newton) spent at least as much time, perhaps more, on alchemy and Biblical chronology as on physics" {B9, p. 123}. Aside from the dubiety of the time-estimation, what is lacking is allusion to the fact that Newton was not unique in this regard, that Kepler dealt with astrology, Leibniz spent considerable time on princely genealogy, and Pascal was both scientist and mystic and believed in miracles. As with those who condemn the Pythagoreans for their interest in numerology, the distinction between scientific work hours and off hours is overlooked, as is the possibility that numerology, chronology, and even alchemy, may be worthier pastimes than watching certain television, or reading certain magazines, or overeating, or getting intoxicated.

To cite another instance where the small is enlarged, a distinguished scientist and autobiographer explains why he presents an assessment of his discovery that recreates his "first impressions" rather than one "which takes into account the many facts" he had "learned since the structure was found" {W2, p. 13}. It is because the latter

> ... would fail to convey the spirit of an adventure characterized both by youthful arrogance and by the belief that the truth, once found, would be simple as well as pretty. {W2, p. 13}

The deceptivity of "first impressions" does not set scientific creation apart from other activities. All of us stumble on our "first impressions". As for "youthful arrogance", it is not an essential part of scientific adventure, as the example of young Einstein in the Berne patent office amply testifies, cf. {F4}. Thus the discussion of these two items may confuse rather than enlighten many a lay reader as to what constitutes creativity in science, in this instance the creativity that revealed the structure of DNA.

To convey any kind of comprehension of Wiener's great and complex contribution, a biographer must filter out Wiener-noise. There is a danger, however, that he must guard against, viz. letting the pendulum swing to the other extreme, and so classifying as "noise", Wiener-signals that contravene or challenge conventional and well-accepted attitudes. Accordingly, in this book we have dwelt on the wisdom behind several of Wiener's iconoclastic views: the analogy between entropy and moral evil, the amoral and immoral aspects of the American market economy and of American education, the pseudo-scientific side of econometrics, the pitfalls of atomic diplomacy, the incompleteness of the Copenhagen interpretation of quantum mechanics, to mention a few.

This book is designed for the general reader interested in the history of ideas as well as for the scientist. Although we discuss Wiener's childhood and his personal life, the prime focus is on Wiener's place in the history of ideas. A great deal of detail has been left out on his extensive travels, his meetings with interesting people, his deep reflections on his father and on the problems of prodigies. These omissions are not serious in vew of their coverage in Wiener's own writings and those of others such as S.J. Heims { H6 }.

D Poetry, prose and mathematical symbolism

It is imperative for the reader to understand the crucial *role of mathematical symbolism* in the life of science and in the life of man, if he is to sense the spirit of Wiener's work. In this section we shall explain the role of this symbolism, rather than the symbolism itself. The reader does not have to understand the mathematical formulae in the book in order to read it, but he should be able to see the very crucial role they play.

The paramount place of language in the life of man testifies to the importance of symbolism. But mathematical symbolism transcends ordinary prose in the following profound ways:

1. *The use of mathematical symbolism eliminates the waste of mental energy on trivialities, and liberates this energy for deployment where it is needed, to wit, on the chaotic frontiers of theory and practice. It also facilitates reasoning where it is easy, and restrains it where it is complicated.*

This thought has been admirably stated by A. N. Whitehead, a profound thinker whose positive influence on Wiener will emerge in the course of this book:

> It is a profoundly erroneous truism, repeated by all copy-books and by eminent people when they are making speeches, that we should cultivate the habit of thinking of what we are doing. The precise opposite is the case. Civilization advances by extending the number of important operations which we can perform without thinking about them. Operations of thought are like cavalry charges in a battle — they are strictly limited in number, they require fresh horses, and must only be made at decisive moments. {W7, p. 61}[2]

Consider, for instance, the triviality that the product of the sum of two numbers with a third is equal to the sum of the product of the first and third and the product of the second and third. Mathematical symbolism allows us to dispose of this 32-word triviality by the simple formula

$$(a + b) \cdot c = a \cdot c + b \cdot c. \tag{1}$$

Gone is the long-winded verbosity, the hallmark of bureaucratic chicanery and fake labor; instead, in Whitehead's words,

> ... we can make transitions in reasoning almost mechanically by the eye, which otherwise would call into play the higher facilities of the brain. {W7, p. 61}

---

2 This little book of Whitehead for the beginner provides an excellent insight into the spirit and scope of mathematics.

Thus from the equality (1), it takes but a glance to see, for instance, that

$$(a + b) \cdot c - b \cdot c = a \cdot c.$$

Only mathematical symbolism has this power of enlisting both mind and eye in logical reasoning. The eye is able to discern new and interesting results, which would otherwise go undetected. Between 500 B.C. and 1850 A.D., mathematics made great strides, but the science of logic was dormant. This was due in large measure to its bondage to prose. Its great advances began in the 1850s when G. Boole and C.S. Peirce so symbolized the subject that algebraic operations similar to those for numbers could be performed, and the eye and mind could work together.

At the foundational level of a science, where it is easy to commit logical errors, the rigidity of mathematical symbolism plays a restraining role. B. Russell, one of Wiener's most important mentors, has explained how this happens:

> Now, in the beginnings, everything is self-evident; and it is very hard to see whether one self-evident proposition follows from another or not. Obviousness is always the enemy to correctness. Hence we invent some new and difficult symbolism, in which nothing seems obvious. Then we set up certain rules for operating on the symbols, and the whole thing becomes mechanical. In this way we find out what must be taken as premiss and what can be demonstrated or defined. For instance, the whole of Arithmetic and Algebra has been shown to require three indefinable notions and five indemonstrable propositions. But without a symbolism it would have been very hard to find this out.{R6, p. 77}

Thus, as Russell remarks, in "the beginnings" of a subject "symbolism is useful because it makes things difficult". The same symbolism makes things easy and quick at the advanced levels of the subject, where the risk of fundamental error is much less.

2. *In general the true propositions of mathematical science are visually more beautiful than those that are false. This holds for the mathematical propositions that claim to enunciate the laws of nature. Thus mathematical symbolism makes possible the employment of not just the intellect but also the aesthetic faculty in the quest for truth.*

The late Cambridge mathematician, G.H. Hardy, again one of Wiener's mentors, expressed this in the words, "There is no permanent place in the world for ugly mathematics" {H4, p. 25}. For instance, the formula

$$\iint_S \left[\left(\frac{\partial P_3}{\partial x_2} - \frac{\partial P_2}{\partial x_3} \right) \ell_1 + \left(\frac{\partial P_1}{\partial x_3} - \frac{\partial P_3}{\partial x_1} \right) \ell_2 + \left(\frac{\partial P_2}{\partial x_1} - \frac{\partial P_1}{\partial x_2} \right) \ell_3 \right] dS$$

$$= \int_C (P_1 dx_1 + P_2 dx_2 + P_3 dx_3), \quad (2)$$

viewed as a visual design, has considerable aesthetic charm.[3] But with the proper interpretation of the symbols x_1, x_2, x_3, P_1, P_2, P_3, ℓ_1, ℓ_2, ℓ_3, etc., it becomes an important mathematical truth. Indeed, so important is this truth (known as Stokes' Theorem) and the ingredients it involves, that it has become standard practice to write a single boldfaced **P** to represent the symbolic triad P_1, P_2, P_3, now called a *vector*, and to use the symbol curl **P** for the triad (or vector) formed by the coefficients of ℓ_1, ℓ_2, ℓ_3; thus

$$\mathbf{P} = (P_1, P_2, P_3), \quad x = (x_1, x_2, x_3)$$

$$\text{curl } \mathbf{P} = \left(\frac{\partial P_3}{\partial x_2} - \frac{\partial P_2}{\partial x_3}, \frac{\partial P_1}{\partial x_3} - \frac{\partial P_3}{\partial x_1}, \frac{\partial P_2}{\partial x_3} - \frac{\partial P_3}{\partial x_2} \right).$$

With this notation and with suitable interpretation of the multiplication of such vectors by numbers, the formula (2) condenses to

$$\iint_S (\text{curl } \mathbf{P}) \cdot dS = \int_C \mathbf{P} \cdot d\mathbf{x}. \tag{3}$$

The aesthetic aspect of mathematical formalism gives deep insights into the structure of the universe, and has therefore considerable heuristic value in scientific discovery. Of the many examples of this, one of the most inspiring is Maxwell's discovery of the electromagnetic theory of light. It involves the concept of the *curl* just mentioned.

The state attained by electromagnetic theory in the late 1850s can be summed up in four equations governing the electric and magnetic field vectors[4] **E** and **H**. Only two of these concern us, viz.

$$\frac{\partial \mathbf{H}}{\partial t} = -c \cdot \text{curl } \mathbf{E}, \quad 0 = \text{curl } \mathbf{H}. \tag{4}$$

---

3   Thus on the right hand side of (2) the subscripts 1, 2, 3 are simply arranged one after the other. On the left hand side the subscripts are mixed-up, but according to a pretty pattern. Thus, in the coefficient of ℓ_1, not 1 but 2 and 3 appear; in the coefficients of ℓ_2, not 2 but 3 and 1 appear; and similarly for the coefficients of ℓ_3. Also, in each coefficient the two subscripts appear once in the numerator and once in the denominator. Finally, notice that if we take 1, 2, 3 cyclically, i.e. think of 1 as succeeding 3, then the terms in coefficients having the minus sign are just those in which the subscript in the denominator succeeds that in the numerator.

4   The reader should think of a *vector* as a quantity possessing both a magnitude and a direction. An example is the *gravitational force* that the sun exerts on the earth. This has a definite direction at each moment, viz. that from the earth's center to the sun's center. But it also has a strength or magnitude depending on the distance between the centers. Much the same is true for the electric and magnetic forces surrounding a body. Another example of a vector is the *displacement*, such as that of a billiard ball on a table.

The new investigations of the subject by the Scottish physicist James
Clerk Maxwell, based on the fascinating ideas of the British experimental physicist Michael Faraday, led him, however, to the pair of equations

$$\frac{\partial \mathbf{H}}{\partial t} = - c \cdot \text{curl } \mathbf{E}, \quad \frac{\partial \mathbf{E}}{\partial t} = + c \cdot \text{curl } \mathbf{H}. \tag{5}$$

The second equation in (5) flew in the face of evidence, for it conflicted
with the second equation in (4). But this did not deter Maxwell. He had
a strong intuitive feeling that the mathematically more beautiful pair (5)
had to be right.[5] Assuming the equations (5), he proceeded to find out
what they entailed. The elimination of $\mathbf{H}$ from the two equations in (5)
yields the equation

$$\frac{\partial^2 \mathbf{E}}{\partial t^2} = c^2 \left(\frac{\partial^2}{\partial x_1^2} + \frac{\partial^2}{\partial x_2^2} + \frac{\partial^2}{\partial x_3^2} \right) \mathbf{E}. \tag{6}$$

There is a similar equation for $\mathbf{H}$, obtainable by the elimination of $\mathbf{E}$. It
had been known since 1750 that equations of the form (6) represent
waves propagating in space with the speed c. The computation of c from
experimental data showed that it closely approximated the known speed
of light. Thus the equation (6) and its associate for $\mathbf{H}$ assert that the
electric and magnetic vectors propagate through space with the speed of
light. This led Maxwell to propound his *electromagnetic theory of light*
to the effect that all radiation (heat, light, ultraviolet light, etc.) is
electromagnetic wave propagation. To test the theory, electromagnetic
waves had to be produced in the laboratory, but this was not technically
possible at that time. Nine years after Maxwell's death, the German
physicist H. Hertz was able to produce such waves in his laboratory. A
few years later, it became clear that Maxwell was right: all radiation,
heat, light, x-rays, etc. is electromagnetic and obeys his equations. Not
only did this bring about tremendous unification in physics, but the
whole field of radio was opened up. (What Hertz had produced were
radio waves).

In so proceeding on the basis of mathematical aesthetics, Maxwell was continuing a great tradition begun by Pythagoras in 500 B.C.,
strongly endorsed and admired by Plato, and practiced by every great

5 Notice the complete symmetry (but for the minus sign) in the pair of
   equations (5). According to (5), $\mathbf{E}$ depends on the fluctuations of $\mathbf{H}$ in exactly
   the same way as $\mathbf{H}$ depends on the fluctuations of $\mathbf{E}$. In (4) there is no such
   symmetry in the roles of $\mathbf{E}$ and $\mathbf{H}$. If we accept, as Hardy did, that "there is
   no permanent place in the world for ugly mathematics", then we must place
   our bet on the equations (5).

physicist from Kepler on. It is based on the faith that the universe is fundamentally harmonious, and that therefore the enunciation of its fundamental principles will be symbolically aesthethic. It is the faith which Einstein expressed in the words, "I believe in Spinoza's God who reveals himself in the harmony of what exists".

The large-scale production of radio waves, which Hertz had found and whose existence was predicted by Maxwell nine years earlier on the basis of mathematical harmony, has changed human technology in profound ways that has affected us all. Electronics has become a very important branch of engineering. These changes were stepping stones to the further changes that have brought on our age of automatization. In these later transitions too, the quest for mathematical harmony has played a vital role, as we shall see from Wiener's work. We would refer the reader who seeks to understand the place of mathematical harmony in the life of man to Whitehead's two essays, *Mathematics as an element in the history of thought* {W8, Ch. II} and *Mathematics and the Good* {W10}.

3. *Mathematical symbolism alone allows the creation of concepts by which seemingly opposing aspects of reality, that are revealed as science grows, can be effectively synthesized.*

In the infancy of the human race, each situation was perceived by the mind as a blurred organic whole. Man's earliest mode of perception was *mythic* rather than logical or analytic, and it is a profound saying that "poetry is the mother tongue of humanity" {C9, p. 35}. Indeed, human life would be intolerable without the mythic mode of perception: great art hinges on such perception.

It is not generally recognized, however, that as science advances it is forced to view reality from widely different perspectives, and that the integration of these divergent aspects into a unified whole calls for the same sort of maneuver that the poet undertakes. The only difference is that in science, this maneuver is undertaken first and foremost from an urge to express the truth rather than from an urge to portray beauty or narrate experience. It is in this undertaking that the role of mathematical symbolism is decisive.

To illustrate this we can do no better than start with the figure given in A.W. Watt's interesting book, *The Two Hands of God* {W3}, Fig. 1. This figure has a dual aspect: if we focus on the white we see a chalice, and if we focus on the black we see two faces in profile. Treating both types of attention equally, the only honest caption for the figure would be "Chalice, face-profiles".

Now the very physicist H. Hertz, whose experimental creation of radio waves led to the acceptance of the Maxwell theory that *all radia-*

tion (including light) consists of electromagnetic waves, also discovered that when light falls on thin metallic sheets, electrons are ejected. In 1905, Einstein showed that this so-called *photo-electric effect* can only be explained by the hypothesis that *light consists of discrete photons*, i.e., *atoms of energy*. Thus, as with the Fig. 1, light has a dual aspect, and a correct description for it would be *wave-particle*, or briefly, *wavicle*. In the early 1920s de Broglie predicted an analogous dual nature for matter itself—a prediction beautifully confirmed by subsequent experiments. See Fig. 2.

Obviously a good theory of the atom and of radiation must cover both their wave and their particle aspects, and must allow the graceful transition from one standpoint to the other. This can be and has been done by using ideas that can only be expressed in mathematical symbolism. In quantum mechanics the *state of the system* is described by vectors $(x_1, x_2, \ldots)$, which are like the triads (x_1, x_2, x_3) we mentioned earlier, except that instead of the three components x_1, x_2, x_3, there are now infinitely many components $x_1, x_2, x_3. \ldots$ The collection of all such infinite-dimensional vectors, in which an analogue of the Pythagorean theorem on right-angled triangles prevails, is called a *Hilbert space*.[6] Its beautiful theory belongs to pure mathematics, but this theory is just what is needed to deal with the (otherwise baffling) *wave-particle* or *wavicle* concept as a logical entity.

The seemingly opposing aspects of reality, which we found in the atomic world, also show up in the world of ordinary-sized bodies. The demonstration of this was one of Einstein's important contributions. The shapes, sizes, masses of ordinary objects, and the time-intervals between events depend on the *frame of reference* (FR) in which these quantities are measured. Physical theory has to integrate the different sets of numbers so obtained (one set for each FR) into a unified whole. The concepts needed for this are associated with the *4-dimensional space-time continuum*. The key to this is again the right mathematical symbolism. What is involved is best conveyed in the words of Sir Arthur Eddington (his "measure code" being synonymous with our "frames of reverence"):

> To grasp a condition of the world as completely as it is in our power to grasp it, we must have in our minds a symbol which comprehends at the same time its influence on the results of all possible kinds of operations. Or, what comes to the same thing, we must contemplate its measures according to all possible measure-codes—of course, without confusing the different codes. It might well seem impossible to realise so comprehensive an outlook; but we shall find that the mathematical calculus of *tensors* does represent and deal with

---

6   Named after the great German mathematician D. Hilbert (1862–1943).

Fig. 1   Chalice or face profiles?

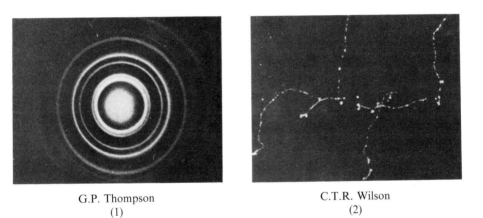

G.P. Thompson   C.T.R. Wilson
(1)   (2)

Fig. 2   The wave-particle or "wavicle"

Electrons behaving as (1) *Waves*, in passing through a thin metal film;
(2) *Particles*, in passing through a gas.

world-conditions precisely in this way. A tensor express simultaneously the whole group of measure-numbers associated with any world-condition; and machinery is provided for keeping the various codes distinct.

. . . .

And, just as in arithmetic we can deal freely with a billion objects without trying to visualise the enormous collection; so the tensor calculus enables us to deal with the world-condition in the totality of its apsects *without attempting to picture it.* {E1, p. 3} (emphasis added)

These remarks, it is hoped, have convinced the reader why it would be a folly to try to convey the intellectual contribution of Wiener without bringing in mathematical symbolism. If he or she can understand this symbolism, well and good. But if not, the reader is advised to skim over the symbolic portions of the book, but to keep an eye for the form of the symbolism, even though the content may be unintelligible.

E Policy as to mathematical usage

To add to the interest of the non-mathematical reader, we begin each section with an informal exposition of the mathematical ideas involved. Wherever possible, we quote Wiener's own popular or semi-popular exposition of these ideas.

Within each section, where mathematics is involved, we have a mathematical crescendo that starts from the informal exposition, and terminates towards the end of the section. In the later parts of such a section, we do state results familiar to mathematicians, without bothering to define the terms. Technical detail, and topics that might interest only a few readers are set as Notes at the end of the chapter.

The non-mathematical reader is requested to attend to the discussion at the start, and to adopt a more or less cavalier attitude towards the mathematics, which might appear as the reader advances. As such a reader skims over these later portions, it is hoped that an eye will be kept on the beauty of the symbolism.

2   Count Tolstoy and Wiener's Father

Grandfather Solomon Wiener hailed from a Jewish family of scholarly journalists living in Krotoschin in Russia. He moved from there to settle down in Byelostock in White Russia, where he married a girl from a Jewish family of tanners. Their son Leo, born in 1862, was a genius.

Solomon, who was discontented with Orthodox Judaism and had a penchant for literary German, put Leo into a Lutheran gymnasium in Minsk. There Leo learned not only German, but also Russian, the state language, French and Italian, the languages of the cultured, as well as Greek and Latin. At age 13 Leo began supporting himself by tutoring. He moved to a gymnasium in Warsaw and picked up Polish in the course of talking to people. After graduation and a brief stint in the Medical School of the University of Warsaw, he joined the Polytechium in Berlin to train as an engineer. There, having to work with a Serbian and a Greek draughtsman, he added Serbian and modern Greek to his linguistic repertoire.

Leo realized that an engineering career was not for him. Some wealthy relatives of the Wieners in the banking business in Berlin tried to lure him into a career in banking. But Leo, now 17, underwent a Tolstoyan conversion like many a 19th-century idealist. He renounced for good the use of tobacco, alcohol and meat, and in league with a student friend decided to start a vegetarian socialist community in Central America. His friend reneged, and the 18-year-old Leo landed in New Orleans in 1880 all by himself with fifty cents in his pocket, after an adventurous voyage during which he studied English and Spanish.

Like a true Mark Twain character, Leo worked in a factory and on the railway, and then turned to farming in Florida and Kansas. Had he arrived a few years earlier, he would have found viable mutualistic if not socialist communities in the very region over which he was roaming, such as the ones in Cedar Vale, Kansas, and in Bethel, Missouri. He might have fitted into the former, for it was made up of "Russian materialists" and "American spiritualists" living in harmony, and among its members were "a Russian sculptor of considerable fame, a Russian astronomer, and a very pretty and devoted and wonderfully industrious Russian woman", cf. {N4, pp. 324-, 353, 354}. By the 1880s, however, these communities had died, and Leo would have had to move to states such as Iowa or South Dakota to find ones that were still thriving, cf. {Al}. But Leo's commitment ot Tolstoyism was not that strong: he trampled on and wound up working in a department store in Kansas City. According to Wiener, Leo did stumble across a Fourierist colony. "It had gone to seed, and all the efficient poeple had left it, while all the rogues and footless, incompetent idealists remained" [53h, p. 19].

Seeing a sign "Gaelic lessons given" on a church wall in the city, Leo enlisted in the class and soon became its instructor, earning the nickname "the Russian Irishman". A bit tired of his jack-of-all-trades existence, Leo applied for a teaching position in the Kansas City school system, and proved to be a successful and popular teacher. Later, he moved to the modern languages department of the University of Missouri at Columbia.

To backtrack a little, the poet Robert Browning was the rage of the ladies of Kansas City, and the erudite Leo was invited to speak at their "Browning Clubs". There he met a Miss Bertha Kahn, daughter of a department store owner of German-Jewish descent. In 1893 in Columbia, Leo, then a professor, and Bertha were married. Browning's verse play *On a Balcony* has three characters, Norbert, Constance, and the Queen. So when a son was born to Leo and Bertha on November 26, 1894, they christened him "Norbert", and when in 1898 a daughter was born they called her "Constance". Who, if anyone was to be christened the Queen, we shall never know, for a daughter born in 1902 was named Bertha and a son born in 1906 was named Fritz. Unhappily, a son born in 1900 and another in 1911 died in infancy.

While Bertha's father Henry Kahn, was a Jew, her mother came from the Ellinger family, rooted in Missouri for at least two generations, that was only partially Jewish. As Wiener tells us, "The family was hovering in a state of unstable equilibrium between its Jewish background and an absorption into the general community" [53h, p. 23].

As for his mother Bertha, Wiener wrote:

> She had been brought up with the indulgence often extended to the belle of the family. I remember one photograph of her made when I was about four years old. She looked extremely handsome in the short sealskin jacket of that period. I had great pride in that picture and in her beauty. She was a small woman, healthy, vigorous, and vivacious, as she has indeed remained to this present day. She still carries herself like a woman in her prime.
>
> In the family of divided roots and Southern gentility into which she was born, etiquette played a perhaps disproportionately large part, and trespassed on much of the ground which might be claimed by principle. It is small wonder, then, that my mother had, and conceived that she had, a very hard task in reducing my brilliant and absent-minded father, with his enthusiasms and his hot temper, to an acceptable measure of social conformity. [53h, p. 27]

Bertha and
Leo Wiener,
parents of
Norbert Wiener

In 1895 the overgrown modern languages department of the University of Missouri was subdivided into separate departments. Due to some nepotism, Leo was denied the chair in German. He then headed for Boston, where, after holding jobs at Boston University, the New England Conservatory, the Public Library and other institutions, he joined the Department of Slavic Languages and Literature at Harvard. He eventually became a full professor and retired from Harvard in 1930.

Leo's writings include a history of Yiddish literature and a 24-volume English translation of Tolstoy's complete works, among a host of others. Leo had wide intellectual interests and interesting friends and acquaintances, among them liberal European scholars such as Thomas Masaryk. Leo could speak forty languages. Tolstoyan only to a degree, his religious attitudes were agnostic and humanistic. He had a broad knowledge of the natural sciences and mathematics. Only when he had completed college could Norbert surpass his father in matters mathematical.

Leo Wiener's political sympathies were democratic, Menshevik and anti-Zionist.

> My father had been desperately opposed to the Communists from the beginning. At least part of this was because his intimate connections with Russia were with such men as Milyukov, who was a Menshevik and associated with the unsuccessful Kerensky regime. . . . from the time of the revolution on, not only did my father's researches drift further and further away from Russia, but thread after thread of his personal contact with that country was broken. [53h, p. 235]

Abbott Lawrence Lowell, the president of Harvard, on the other hand, wanted the professors in the Slavic Department to take an interest in contemporaneous developments in the Soviet Union and its Western neighbors. Thus arose a difference of opinion between the two men. Another source of friction was Leo's open opposition to the policy adopted by Harvard after World War I on a Jewish "numerus clausus", i.e. a certain percentage restriction on the number of Jews admitted to Harvard. Both factors contributed to the denial of Leo's request to continue his Harvard connection beyond the normal age of retirement.

Leo Wiener died in 1939. Wiener describes how much languages and philology had been a part of his father by depicting a scene when his father was nearing death by apoplexy:

> . . . his fine intelligence no longer enabled him even to recognize the loved ones about him. And I remember that he spoke as though he had the gift of tongues, in English, in German, in French, in Russian, in Spanish. Confused as to what he saw about him, his languages nevertheless were clear, grammatical, and idiomatic. The pattern went through the fabric, and neither wear nor attrition could efface it. [53h, p. 47]

Wiener has left us with a vivid description of his father's personality, ideals and his disappointments from failure to attain them:

> . . . a small, vigorous man, of emotions both deep and quick, sudden in his movements and his gestures, ready to approve and to condemn, a scholar rather by nature than by any specific training. In him were joined the best traditions of German thought, Jewish intellect, and American spirit. He was given to overriding the wills of those about him by the sheer intensity of his emotion rather than by any particular desire to master other people. [56g, p. 18]
> . . . he was essentially a German liberal of type well known in the middle and later nineteenth century, fully in sympathy with the German intellectual tradition as it had come down from the time when Goethe had been a truer symbol of German aspirations than the Emperor Wilhelm II was ever to become. Separated as my father was from Germany, largely self-educated, and outside the orthodox German academic tradition, he still hoped for many years that by sheer intellectual strength and integrity he could win from Germany the recognition accorded to a great German scholar.

In this expectation Father had never been completely realistic. He was too innocently honest a man to be wordlywise. It took him many years to learn how the great German intellectual tradition had come to be subservient to a group of vested interests in learning. [56g, p. 46]

3 New England and Wiener's Early Training, 1901–1905

Wiener was fundamentally molded by the democracy of the New England of his childhood days. Later in life he came to value his sojourns away from New England because they let him see "the great lines of its spiritual character" from a distance [53h, p. 48].

For the rest, Wiener's childhood was shaped by the milieu around his father's home. His Cambridge, Massachusetts, neighbors in 1901 were Maxime Bocher, the mathematician, and Otto Folin, the physiological chemist, whose children became his playmates. The Wieners had a friend in the great physiologist Walter Cannon of the Harvard Medical School, and Norbert felt free to address his childish questions to him.[1] His early biological curiosities were encouraged by his father's mycological hobbies. He also met his father's interesting friends: Prince Kropotkin, geographer and anarchist, whom he saw in Cambridge and in London, and the British Zionist, Israel Zangwill, in London, with whom his father often had arguments.

At six Norbert could read freely, and by eight he had overstrained his eyes and was threatened with myopia. Apart from the *Treasure Island* type of fiction, he devoured books on science, especially natural history. He kept reading, although as he recalls "much of it was over my head" [53h, p. 65]. He "longed to be a naturalist as other boys longed to be policemen or locomotive engineers". But shades of the mature Wiener were already present in the child:

> Even in zoology and botany it was the diagrams of complicated structures and problems of growth and education which excited my interest as fully as the tales of adventure and discovery. [53h, p. 64]

His infantile maladroitness made school a difficult proposition for the six-year-old Norbert. His reading was far above average, but his handwriting was bad and his arithmetic weak. Soon after putting him into the

---

1 Cannon coined the word "homeostasis" which later became an important term in Wiener's lexicon. His writings were later to shape Wiener's ideas on cybernetics, and Cannon's assistant, Dr. Arturo Rosenblueth, was to become one of Wiener's most important collaborators, (cf. Ch. 15).

Peabody School in 1901, Leo realized that what the boy needed was not the cribbing of multiplication tables and the like, but something interesting and challenging. So he removed Norbert from school and placed him under his own tutelage with courses in algebra and classical languages. This continued for about two years, after which Norbert reentered school at age eight.

A critical idealist understands that effective action must accord with the laws of nature, but often his earnestness gets the better of his judgment and his temper. This was the case with Leo in his educational dealings with Norbert. Norbert recalls how his fumbling in algebra or grammar turned "a gentle and loving father" into "an avenger of blood", and describes how his father's harsh "What?"'s and "Do it again"'s terrified him and worsened his responses, only to enrage his father even further and to make him hurl insults in German. The lessons often ended in a family scene with a crying Norbert running into his mother's arms. Pages 67–68 of [53h] bear eloquent testimony to the destruction of the learning process caused by a teacher who equates discipline with a mixture of harshness and irony, and resorts to the public exposure of a child's stupidity, and to his harassment in the presence of others.

Unlike the fathers of the prodigies John Stuart Mill and Samuel Butler, Leo did not invoke any moral or religious sanction for his harsh instructorship. Rather, he justified it on (somewhat questionable) educational grounds. In a 1911 article on "New ideas in child training", he wrote:

> It is nonsense to say, as some people do, that Norbert and Constance and Bertha are unusually gifted children. They are nothing of the sort. If they know more than other children of their age, it is because they have been trained differently. [53h, p. 158]

What kept Norbert from falling into total despondency during this difficult period, 1901–1903, was his image of his father as not just a "taskmaster", but also a hero who was trying to attain the impossible ideal for his son.

But Norbert never completely recovered from the strains of this period. The differences he noticed between himself and other children, stemming both from his precociousness and from his poor eyesight and poor muscular coordination, and the wounds suffered by his ego during conflicts with his father, turned him into an ego-defender who sometimes treated the external world as a medium for ego-aggrandizement. At such moments, he lost his philosophical touch. His reaction, at age 18, to his father's statement (quoted above) is an instance. All he could see in it was the implication: "my failures were my own, but my

Norbert Wiener
at age seven

successes were my father's" [53h, p. 159]. To the much more interesting topic of Leo's article, whether or not precociousness can be induced by training, he does not respond, despite his fascination with "problems of growth and education" since 1903 [53h, p. 64].

For six months during this period, Norbert was barred from reading on doctor's orders, and had to cope with exclusively oral and manual instruction in languages and science. Here there was no conflict of egos, and Wiener, the philosopher, describes his compulsory ear-training during these months as a "most valuable" discipline,

> for it forced me to be able to do my mathematics in my head and to think of languages as they are spoken rather than as mere exercises in writing. [53h, p. 76]

Fortunately the ban on reading was lifted after six months, and Wiener was able to read freely until 1946, and then after an eye operation till the end of his life. But he always preferred working on a blackboard to working on a sheet of paper.

Even after he left the Peabody School, Norbert continued to play with his former mates on the school playground. From his own account he had his normal fill of childish fun and pranks. He also mingled with children at a Unitarian Sunday School and read in its library. Among his favorite books was the science fiction of H. G. Wells and Jules Verne. These pleasant relationships were unfortunately ruptured when the Wieners moved to Old Mills Farm near Harvard, Massachusetts, in 1903. (Leo had not lost his old love for farming, and wanted his children to share in this.) Norbert soon made new friends among the village children.

In the autumn of 1903, at age eight, Norbert resumed schooling at the Ayer High School near his new home. Because of his precocity he was placed in a class with boys seven years his senior. During play periods he had to go to an adjoining junior school to find compatible playmates, still two or three years older than him. "They (the other children) viewed me socially as an eccentric child, not as an underaged adolescent" [53h, p. 94]. These words were written in 1952, but obviously it is the boy of eight and not the philosopher who is speaking. What the philosopher had to say on the relationship between myopia, muscular incoordination, clumsiness and ineptness is worth noting:

> Muscular dexterity is neither a completely muscular nor a completely visual matter. It depends on the whole chain which starts in the eye, goes through to muscular action, and there continues in the scanning of the eye of the results of this muscular action. It is not only necessary for the muscular arc and the visual arc to be perfect, each by itself, but it is equally necessary that the relations between the two be precise and constant.
> Now a boy wearing thick glasses has the visual images displaced through a considerable angle by a small displacement in the position of the glasses upon the nose. This means that the relation between visual position and muscular position is subject to a continual readjustment, and anything like an absolute correlation between these is not possible...
> A further source of my awkwardness was psychic rather than physical. I was socially not yet adjusted to my environment and I was often inconsiderate, largely through an insufficient awareness of the exact consequences of my action...
> A further psychic hurdle which I had to overcome was impatience. This impatience was largely the result of a combination of my mental quickness and physical slowness. I would see the end to be accomplished long before I could labor through the manipulative stages that were to bring me there. When scientific work consists in meticulously careful and precise manipulation which is always to be accompanied by a neat record of progress, both written and graphical, impatience is a very real handicap. [53h, pp. 127, 128]

Even during the school term Norbert had to submit to his father's supervision of his studies. As he recounts, his father would listen with "half an ear" while continuing his own work. "But half an ear was fully adequate to catch any mistakes" that Norbert made and to provoke harsh reprobation [53h, p. 94].

Norbert graduated from the Ayer High School in the summer of 1906 at the age of eleven. During the three years he spent in its environs he had made many friends and had many a happy moment. Years later he expressed the meaning of this experience in his life as follows: "I had a chance to see the democracy of my country at its best in the form in which it is embodied in the small New England town" [53h, p. 101]. Describing his revisit about forty years later to these same environs, by then partially industrialized, he was to write:

> I have the impression that my friends in this small industrial town represent a sort of stability without snobbishness which is universal rather than provincial, and that the structure of their society compares well with the best that a similar place in Europe would have to offer. When I go back among them it is expected of me, and rightly expected of me, that I revert in some measure to my status as a boy among the elders of the family. And I do so gratefully, with a sense of roots and security which is beyond price for me. [53h, p. 101]

Wiener's love for the New England farm folk persisted until the end. He expresses "eternal thanks" to his father for the property he acquired in South Tamworth, New Hampshire, in 1909, [53h, p. 140], where Mrs. Margaret Wiener, his widow, still resides. Of the New Hampshire farmer he wrote: ". . . whether you can love him or not, and very often you can, you can and must respect him because he respects himself" [53h, p. 142].

4 Harvard and Wiener's University Training, 1906–1913

A An eleven-year-old freshman at Tufts College

Leo Wiener decided not to send his son to Harvard College in order to spare him the strain of taking the Harvard entrance examinations and of the subsequent embarrassment of being "an eleven-year-old at Harvard". But Harvard University was the center of Norbert's educational gyrations, and the dominant factor in his advanced training.

It was Tufts College in Boston that Wiener entered in 1906, and from which he received his B.A. in mathematics, *cum laude*, in 1909 at age 15. He speaks highly of his professors at Tufts. His first mathematics course on the theory of equations, including Galois theory, was in his own words "really over my head" [53h, p. 104]. He fared better in his other, more engineering-oriented, mathematics courses. A good deal of his study consisted of extracurricular experimentation and reading, especially in philosophy. The two philosophers who influenced him most were Spinoza and Leibniz:

> The pantheism of Spinoza and the pseudomathematical language of his ethics mask the fact that *his is one of the greatest religious books of history*; and if it is read consecutively instead of broken up into axioms and theorems, it represents a magnificent exaltation of style and an exertion of human dignity as well as of the dignity of the universe. As for Leibniz, I have never been able to reconcile my admiration for him as the last great universal genius of philosophy with my contempt for him as a courtier, a placeseeker, and a snob. [53h, p. 109] (emphasis added)

The next most influential philosophical figure for Wiener was William James, whose Lowell lectures on pragmatism he attended. But in spirit Wiener was in many ways closer to the then unknown Charles Sanders Peirce than he was to William James. Besides philosophy, Wiener's extracurricular activities also included the study of biology.

At this stage Wiener began to display an innate ability in practical engineering design. He and his friends designed an electromagnetic coherer for radio waves and an electrostatic transformer. He himself was clumsy in the handling of apparatus, but felt at home in a laboratory amid understanding colleagues who could deal manually with his ideas. He also participated in biological experiments, but here guilt feelings sometimes intervened. He recalls [53h, p. 112] the remorse he felt when, in the course of an unauthorized experiment, he and his friends accidentally killed a guinea pig.

After his graduation from Tufts, Wiener suffered a short depression. Apart from adolescent insecurities and anxieties about "success", endemic to our economic order, he was haunted by obsessions of death, of the world's desire to witness the collapse of ex-prodigies, and other illusions. The depression faded away, but not without leaving a scar: Wiener turned to the one thing he felt he had, viz. mathematical prowess, and considered using it for self-adulation and career wrangling. His own words are quite exaggerated and misleading, as we shall see in the sequel.

> My study of mathematics gave me a consciousness of strength in a difficult field, and this was one of its great attractions. My mathematical ability at this time was a sword with which I could storm the gates of success. This was not a pretty or a moral attitude, but it was real, and it was justifiable. [53h, pp. 121–122]

But this narcissistic passage, with its ignorant conclusion, cannot be entirely disregarded, for now and then Wiener's conduct was egocentric. It marred his relationship with colleagues, and one would imagine his mathematical creativity as well. Such narcissistic misuse of his intellectual powers, though infrequent, was a blot on his basic humaneness and generosity, and on his scientific greatness.

B A semester in zoology

In the fall of 1909 Wiener enrolled as a graduate student in the Harvard Zoology Department with the intention of doing a doctorate. This was his own decision, but nothing came of it. His bad eyesight and manual clumsiness made him a bad histologist, his drawings were poor, and his note-taking inadequate. After one semester his father decided to transfer him to the Sage School of Philosophy at Cornell University for the purpose of working for a doctorate in philosophy. Wiener resented this fresh demonstration of his father's authority.

Most would agree that Wiener's transfer from biology to philosophy was a good thing, but Wiener's own evaluation of the matter is worth stating, as it brings out an important prosthetic and practical aspect of the scientific methodology, of which most people are unaware:

> There are many ways of being a scientist. All science originates in observation and experiment, and it is true that no man can achieve success who does not understand the fundamental methods and mores of observation and experiment. But it is not

absolutely necessary to be a good observer with one's own eyes or to be a good experimenter with one's own hands. There is much more to observation and experiment than the mere collection of data. These data must be organized into a logical structure, and the experiments and observations by which they are obtained must be framed that they will represent an adequate way of questioning nature.

The ideal scientist is without doubt the man who can both frame the question and carry out the questioning. There is no scarcity of those who can carry out with the utmost efficiency a program of this sort, even though they may lack the perspicacity to frame it: there are more good hands in science than there are good brains to direct them. Thus, although the clumsy, careless scientist is not the type to do the greater part of the work of science, there is other work for him in science if he is a man of understanding and good judgment.

It is not very difficult to recognize the all-round scientist of whose calling there is no doubt. It is the mark of the good teacher to recognize both the laboratory man who may do splendid work carrying out the strategies of others, and the manually clumsy intellectual whose ideas may be a guide and help to the former. When I was a graduate student at Harvard, my teachers did not recognize that despite all my grievous faults, I might still have a contribution to make to biology. [53h, pp. 128–129]

Indeed, Wiener did make a contribution. In stressing the experimental, hypothesizing and creative side of science, the passage reveals the contributions of Roger and Francis Bacon, and Aquinas and Kant, respectively, to scientific methodology. The Kantian aspect is further elaborated in [22a], cf. § 7B.

C The "black year" at the Sage school of philosophy

Wiener's first days at Cornell University were marred by another emotional crisis. He heard Professor Thilly of Cornell remark to his father about the story that Rabbi Moses Maimonides was one of the distant ancestors of the Wieners. This was the first time that the 15-year-old Wiener realized that he was a Jew! This fact had been withheld from him by his parents. The sudden realization that his mother had often spoken derogatorily of Jews, that her maiden name, Kahn, was a variant of the Jewish Cohen, and that she had lied when she denied his cousin's earlier statement that they were Jewish were shattering:

> The wounds inflicted by the truth are likely to be clean cuts which heal easily, but the bludgeoned wounds of a lie draw and fester. [53h, p. 147]

"The black year of my life" was Wiener's epithet for the year 1911. With much justice Wiener wrote:

> Who was I, simply because I was the son of my mother and father, to take advantage of a license to pass myself off as a Gentile, which was not granted to other people whom I knew. . . . My protection may have been well intended, but it was a protection that I could not accept if I were to keep my integrity.
> If the maintenance of my identity as a Jew had not been forced on me as an act of integrity, and if the fact that I was of Jewish origin had been known to me, but surrounded by no family-imposed aura of emotional conflict, I could and would have accepted it as a normal fact of my existence, of no exceptional importance either to myself or to anyone else. [53h, p. 147]

But this was not to be the case:

> . . . an injudicious attempt to conceal from me my factual Jewish origin, combined with the wounds which I suffered from Jewish anti-Semitism within the family, contributed to make the Jewish issue more rather than less important in my life. [53h, p. 148]

Not only was this issue thrust into undue importance, but a crisis involving intense suffering was engendered. Wiener sadly recounts his immediate response to the crisis:

> . . . I alternated between a period of cowardly self-abasement and a phase of cowardly assertion, in which I was even more anti-Semitic than my mother. [53h, pp. 148–149]

It was after considerable groping that Wiener learned to cope with this issue. The agnosticism he had imbibed from his parents ruled out a return to Orthodox Judaism, and his honesty forbade the complete renunciation of his Jewishness. At last he saw the light, that anit-Semitism was just one manifestation of the much larger problems of human prejudice and human exploitation. The consequence was a shift towards enlightened liberalism [53h, p. 155].

During the 19th and early 20th centuries, many a Jewish family accepted Christianity either from conviction or from expediency. The Mendelsohn and Marx families in Germany and the von Neumann family in Hungary are known examples. A remarkable case was that of the French Jewish philosopher Henri Bergson, who wanted to convert to Roman Catholicism in the late 1930s, but hesitated to undergo baptism in view of the plight of European Jews under fascism. When Vichy France promulgated anti-semitic measures in 1940, the enfeebled man, close to death, stood in line to register as a Jew at his own request. Shortly thereafter, as per his wishes, a Catholic priest prayed at his funeral. (At that time the Church, in its supreme folly, had his books on its Index.)

The Wiener episode was markedly different. Here the parents, while inculcating in their son a reverence for integrity, practiced dishonesty by hiding from him his true heritage, and castigating it by slandering the Jews. Why had not Leo Wiener told Norbert about the Jews and about his Jewish descent when he was nine or ten? And what a difference it might have made, if Leo had told him of the great Jew of the 12th century (their reputed ancestor), the Rabbi Moses Maimonides, of how he often wrote in Arabic, the universal respect he enjoyed among scholastics, Christian, Islamic and Jewish, and of the unalloyed intellectualism represented in his memorable words:

> My son, so long as you are engaged in studying the mathematical sciences and logic, you belong to those who go round about the palace in search of the gate. When you understand physics, you have entered the hall; and when after completing the study of natural philosophy, you master metaphysics, you have entered the innermost court and are with the king in the same palace. { M2, p. 79 }

But there is no evidence that Leo, a half-baked Tolstoyan at best, ever understood the importance of transcendence in the solution of problems, human, aesthetic, or scientific. His injudiciousness was symptomatic of the paralysis that often accompanies incomprehension of the transcendent. It was Norbert Wiener, age twelve, who felt an instinctive attraction to Spinoza's thought.

At Cornell Wiener tackled Plato's *Republic* and attended courses on the 17th- and 18th-century British philosophers. Understandably, his performance was not good enough to secure a renewal of his Sage Fellowship. This setback led his father to remove Norbert from Cornell and shuttle him back to Harvard, this time to the Philosophy Department—a step which only aggravated Wiener's insecurity.

D The return to Harvard

Wiener reentered Harvard in September 1911 in a department with many great names. He took courses with George Santayana, Josiah Royce, G.H. Palmer and R.B. Perry, and also with E.V. Huntington of the Mathematics Department. He speaks glowingly of most of them, and singles out Huntington for kindness. Wiener decided to work for his Ph.D. with Royce, but the latter's health had broken down and Wiener's doctoral dissertation, "A comparison of the algebra of relatives of Schroeder and of Whitehead and Russell", was completed under the guidance of Professor K. Schmidt of Tufts College. It was a good dissertation, of which Wiener was later to say exaggeratedly, ". . . I had missed every issue of true philosophical significance" [53h, p. 171].[1]

---

1   He was referring to his neglect of the theory of types (cf. 5B). An interesting account of Wiener's thesis has been given in Dr. I. Grattan-Guiness' paper. {G 13}

Norbert Wiener in 1912 just before he received his doctorate from Harvard

Wiener received a Ph.D. from Harvard in June 1913. This happy event was accompanied by the good news that he had been awarded an overseas traveling fellowship by Harvard University, the John Thornton Kirkland Fellowship.

As a result of training from his father, Wiener had a Ph.D. at the age of 18, about six years below the average age. These valuable years were at his disposal. This was a compensation for the suffering that Wiener had to endure. The aggravation of the boy's difficulties by the confusion of discipline with harshness and intimidation, and by misinformation about his Jewish origin, were tragic events. But Leo Wiener's fundamental objective of keeping the boy on his intellectual toes, of putting him in contact with good minds, of giving him the run of libraries and laboratories, and a chance to love the countryside is something for which we must all be thankful.

5   Bertrand Russell and Wiener's Postdoctoral Years, 1914–1917

A   Wiener's emancipation

Wiener decided to utilize his Harvard traveling fellowship to study mathematical philosophy at Cambridge University under Bertrand Russell, on whose work his thesis had been based. Russell was Lecturer in the Principles of Mathematics, and had just completed the monumental and epoch-making *Principia Mathematica* { W11 } together with his former teacher, the philosopher Alfred North Whitehead. Wiener well describes this phase of his life as "emancipation": he was to be in a superb environment amid great minds, with the Atlantic Ocean acting as a potential barrier to paternal interference. But it was only with his father's intervention that this "emancipation" was procured.[1] Russell's autobiography {R8, Vol. I, pp. 345, 346} contains a letter he received from Leo Wiener. It is reproduced here in full, since it gives us a glimpse of Leo's interesting personality, and the tale it tells has a touch of humor. The emphasis we have added corroborates Norbert's complaints about his father's overbearing attitutes.

> 29 Sparks Street
> Cambridge, Mass.
> June 15, 1913

Hon. B.A.W. Russell
Trinity College
Cambridge, Eng.

Esteemed Colleague:

My son, Norbert Wiener, will this week receive his degree of Ph.D. at Harvard University, his thesis being "A Comparative Study of

---

1   Wiener's last sentence in [53h, p. 179], with the words "I then wrote to Russell", is thus somewhat misleading. Also misleading is the short account he gives of the first meeting he and his father had with Russell [53h, p. 183]. It transpires from Russell's letters that neither of the Wieners impressed him favorably.

the Algebra of Relatives of Schroeder and that of Whitehead and Russell." He had expected to be here next year and have the privilege of being your student in the second semester, but as he has received a travelling fellowship, he is obliged to pass the whole of the year in Europe, and so he wishes to enjoy the advantage of studying under you at Trinity during the first half of the academic year. He intended to write to you about this matter, but his great youth,—he is only eighteen years old and his consequent inexperience with what might be essential for him to know in his European sojourn, leads me to do this service for him and ask your advice.

Norbert graduated from College, receiving his A.B., at age of fourteen, *not as the result of premature development or of unusual precocity, but chiefly as the result of careful home training, free from useless waste, which I am applying to all of my children.* He is physically strong (weighing 170 lbs.), perfectly balanced morally and mentally, and shows no traits generally associated with early precocity. I mention all this to you that you may not assume that you are to deal with an exceptional or freakish boy, but with a normal student whose energies have not been mis-directed. Outside of a broad and liberal classical education, which includes Greek, Latin, and the modern languages, he has had a thorough course in the sciences, and in Mathematics has studied the Differential and integral Calculus, Differential Equation, the Galois Theory of Equations, and some branches of Modern Algebra (under Prof. Huntington). In philosophy he has pursued studies under Professors Royce, Perry, Palmer, Munsterberg, Schmidt, Holt, etc., at Harvard and Cornell Universities. His predilection is entirely for Modern Logic, and he wishes during his one or two years' stay in Europe to be benefited from those who have done distinguished work in that direction.

Will he be able to study under you, or be directed by you, if he comes to Cambridge in September or early October? What should he do in order to enjoy that privilege? I have before me The Student's Handbook to Cambridge for 1908, but I am unable to ascertain from it that any provisions are made for graduate students wishing to obtain such special instruction or advice. Nor am I able to find out anything about his residence there, whether he would have to matriculate in Trinity College or could take rooms in the city. This is rather an important point to him as he is anxious, as far as possible, to get along on his rather small stipend. For any such information, which would smooth his first appearance in a rather strange world to him I shall be extremely obliged to you.

I shall take great pleasure to thank you in person for any kindness that thus may be shown to my son, when, next year, you come to our American Cambridge to deliver lectures in the Department of Philosophy.

Sincerely yours,

Leo Wiener

Professor of Slavic Lan-
guages and Literatures at
Harvard University

Wiener entered Cambridge University in the autumn of 1913. Following Russell's advice not just to concentrate on the foundations of mathematics, but also to find out where its different branches were going and to watch as well the frontiers of physics, Wiener attended the mathematics courses given by Baker, Mercer, Littlewood and Hardy. By far the most influential among these was G.H. Hardy:

> In all my years of listening to lectures in mathematics, I have never heard the equal of Hardy for clarity, for interest, or for intellectual power. If I am to claim any man as my master in my mathematical training, it must be G.H. Hardy. [53h, p. 190]

It was while taking Hardy's course that Wiener wrote his maiden paper [13a] on transfinite ordinals. Wiener's warm relationship with Hardy lasted till the latter's death in 1947. He probably knew Hardy better than any other American mathematician, and he was invited to write Hardy's obituary in the *Bulletin of the American Mathematical Society* [49f].[2]

But Wiener's forte during this early period was in the foundations of mathematics rather than in its superstructure, and Russell rather than Hardy was his guide, friend and philosopher on all matters, except personal morality. Wiener did not approve of Russell's libertinism, and his witty remark on the philosophical libertine is worth recalling:

> There is a great deal in common between the libertine who feels the philosophical compulsion to grin and be polite while another libertine is making away with the affections of his wife and the Spartan boy who concealed the stolen fox under his cloak and had to keep a straight face when the fox was biting him. [53h, p. 192]

Wiener took two courses from Russell, which exceeded anything in his experience by their constructive use of Einstein's relativistic ideas and of the theory of types. He found the courses "extremely stimulating". To understand their influence we should note that in his thesis

2   This interesting obituary contained a few factual errors that were pointed out
     in a short communication to the *Bull. Amer. Math. Soc.* (vol. 55, 1949,
     p. 1082) by Hardy's colleagues Littlewood, Polya, Mordell, Titchmarsh and
     Davenport.

Wiener had dealt extensively with the *Principia Mathematica*, but strange as it may seem, he had leaned toward Schroeder's point of view and had missed the philosophical import of Russell's theory of types, cf. {G13}. When Wiener relearned this theory from Russell's lectures and realized what he had overlooked, it changed his entire outlook in a far-reaching way, affecting not only his early work on mathematical philosophy, but also his later ideas on communication, computers and war (cf. § § 19A, 20C below). The lectures were "masterpieces" [53h, p. 194]. "For the first time I became fully conscious of the logical theory of types and of the deep philosophical considerations which it represented." [53h, p. 191]

What made the Russellian doctrine of types so influential? A brief digression on this question is necessary.

B The doctrine of types

The edifice of pure mathematics is founded on the Cantorian idea of a *Menge* or *set*, and on the *diagonalization procedure* and principle of *transfinite induction* which Cantor had introduced. But, as logicians and mathematicians know, the informal logical employment of these ideas and principles results in the "diagonalization paradoxes" of Russell, Richard, Burali-Forti and Cantor, and thereby jeopardizes the very footing on which mathematics rests. In 1903 Russell's doctrine of types {R4, App. B} gave mankind its first clear glimpse of a method to safeguard mathematics from such paradoxes. With it, man's intellectual reach began moving forward.

To comprehend this profound intellectual development and its impact on Wiener, it is necessary to appreciate the scope of the underlying Cantorian idea of *set*, and to see how its informal consideration inevitably leads to contradictions.

First, a *set* or *Menge* (also called *class* or *aggregate*) is a grouping together of certain objects according to some (unambiguous but otherwise arbitrary) rule or method. The objects which are grouped together are called *members* or *elements* of the set. If A is a set, we write '$x \in A$' as shorthand for 'x is a member of A', and '$x \notin A$' as shorthand for 'x is not a member of A'. The symbol '$\in$' is called the membership-predicate; it corresponds to one sense of the traditional copula 'is'. Sets in this Cantorian sense are abstract objects. A physical entity such as a heap of stones is not a set of stones. The notation '$\{a_1, a_2, \ldots, a_n\}$' denotes the set whose members are precisely the entities $a_1, a_2, \ldots, a_n$. Thus {Paris} is the set with precisely one member, viz. Paris, and {Paris, London} is the set with precisely two members, viz. Paris and London.

The inconcistency inherent in an informal treatment of the set-concept is most easily seen by following Russell, and noting that while most sets are not members of themselves, some are. The set C of all cows is obviously not a cow. So C is not a member of C; briefly, $C \notin C$. Likewise, for the set F of all finite sets, we see that each of the sets

$$\{1\}, \quad \{1, 2\}, \quad \{1, 2, 3\}, \quad \text{etc.}$$

is finite and therefore a member of F. But there are infinitely many such members, and so F itself is not finite; thus $F \notin F$. On the other hand, the set I of all infinite sets is infinite. For each of the sets

$$\{1, 2, 3, \ldots\}, \quad \{2, 3, 4, \ldots\}, \quad \{3, 4, 5, \ldots\}, \quad \text{etc.}$$

is infinite and therefore a member of I, and there are infinitely many of them. Thus I is itself an infinite set, and so $I \in I$.

Now consider Russell's set R, which by definition has as members all and only those sets X which are not members of themselves; in brief,

$$X \in R \quad \text{if \& only if} \quad X \notin X. \tag{1}$$

Given a set A, the formula (1) enables us to give a direct answer to the question: Is A a member of R? It is, if $A \notin A$; it is not, if $A \in A$. Thus for the sets C, F and I considered above, we find that $C \in R$, $F \in R$ but $I \notin R$. If, however, we ask the pertinent question 'Is R a member of R?', we get into trouble, for the substitution of 'R' for 'X' in (1) yields the contradiction

$$R \in R \quad \text{if \& only if} \quad R \notin R. \tag{2}$$

In words, R is a member of R if and only if R is not a member of R. The contradiction (2) is called *Russell's paradox* or, more accurately, *Russell's antinomy*.

In 1902 Russell wrote to G. Frege in Germany informing him that his system of mathematical logic was inconsistent since the contradiction (2) was deducible from its axioms. Its effect on Frege is best described in the words of Carnap:

After years of laborious effort, Frege had established the sciences of logic and arithmetic on an entirely new basis. But he remained unknown and unacknowledged. The leading mathematicians of his time, whose mathematical foundations he attacked with unsparing criticism, ignored him. His books were not even reviewed. Only by means of the greatest personal sacrifices did he manage to get the first volume of his chief work [*Grundgesetze*] published, in the year 1893. The second volume followed after a long interval in 1903. At last there came an echo—not from the German mathematicians, much less the German philosophers, but from abroad: Russell in England attributed the greatest importance to Frege's work. In the case of certain problems Russell himself, many years after Frege, but still in ignorance of him, had hit upon the same or like solutions; in the case of some others, he was able to use Frege's results in his own system. But now, when the second volume of his work was also printed, Frege learned from Russell's letter that his concept of class led to a contradiction. Behind the dry statement of this fact which Frege gives in the Appendix to his second volume, one senses a deep emotion. But, at all events, he could comfort himself with the thought that the error which had been brought to light was not a peculiarity of his system; he only shared the fate of all who had hitherto occupied themselves with the problems of the extension of concepts, of classes, and of aggregates— amongst them both Dedekind and Cantor. {C3, p. 137}

It fell on Russell to vindicate the tradition of Cantor, Dedekind and Frege by divesting it of all contradictions. His solution, first outlined in 1903 {R4, App. B}, more definitively stated in a paper {R5} in 1908, and completed in the *Principia Mathematica* with Whitehead {W11}, was in a nutshell as follows. Every entity in the universe of discourse is assigned one of the types 1, 2, 3, ... The type 1 is assigned to each individual, i.e. to each entity (like a cow) which is not a set. The type 2 is assigned to any set whose members are of type 1, the type 3 is assigned to any set whose members are of type 2, and so on. Sets having members of more than one type are deemed "impure" and banned from the mathematical edifice, as are all expressions of the form '$x \in y$' and '$x \notin y$', unless y's type is one higher than x's type. In particular, all sentences of the form '$x \in x$' and '$x \notin x$' and with them their compounds such as (1) and (2) are proscribed. By this policy of enlightened censorship, all the theorems of Cantor, Dedekind and Frege are salvaged, while all contradictions are held at bay.

The *Principia Mathematica* thus marked the fulfillment of G. Frege's 1884 program for the reduction of real number arithmetic to logic. Thereby, it exposed once and for all the fallaciousness of the traditional view that the science of arithmetic deals with the laws of "quantity", whereas that of logic deals with the laws of "thought". Moreover, to use K. Gödel's words, the subject "was enriched by a new instrument, the abstract *theory of relations*" on which is based the theory of measurement {G9, p. 448}. By the soundness of its emphasis, the *Principia Mathematica* also thrust forward other valuable conceptions. For instance, its systematic use of recursive definitions brought to the forefront the idea of *recursion*, which, when subsequently set in a proper metamathematical[3] framework by K. Gödel {G8}, A.M. Turing {T8} and others, had revolutionary ramifications in mathematical philosophy, and via the work of C.E. Shannon, Wiener, von Neumann and

---

3 A metamathematical setting is one in which a clear distinction is made between the formal language and the language used in talking about it. The metalanguage is the one used in talking about a language. When we discuss German poetry in English, German is the language and English is the metalanguage.

Bertrand Russell
1872–1970

Photograph taken in
1916

others on automata theory, even in industrial technology. It ushered in
the age of automatization.

The technical complexities brought in by the theory of types were
overwhelming, and starting with Zermelo in 1908 simpler alternatives
were proposed to keep out the paradoxes. Roughly speaking, these
systems cut down the infinite hierarchy of types 1, 2, 3, . . . to just two:
"element and non-element", "set and class", "set formally expressible,
and set not so expressible". Among these was the system of John von
Neumann {V1} (1925), who in the 1930s and 1940s was to become an
influential figure in Wiener's life and work, cf. §§ 9F, 12B and Chs. 17, 20.
Based on two types, *set* and *class*,[4] it embodies several major simplifica-
tions. Many a worker, such as K. Gödel, who initially embraced the axi-

---

4   The von Neumann system, being a type theory, bears the impress of the 1903
    Russellian doctrine. We stress this point, for a reader of Professor Ulam's
    account of these developments {U1, p. 11} may come away with a contrary
    impression.

omatization of the *Principia Mathematica*, shifted to the von Neumann axiomatization in later work. The axiomatization of the *Principia Mathematica* itself underwent reform which made the system more manageable, cf. Quine {Q1, pp. 163–166}.

For Wiener, however, type theory was much more than a device to ward off the antinomies. Behind it lay the hierarchical classificatory attitude that Wiener found most stimulating. For instance, in later years he assigned types 1, 2, 3 to automatons A, B, C, in case B evaluated the performance of A, and C that of B. Likewise he assigned types 1, 2, 3, . . . to the military categories; tactics, strategy, general considerations to use in framing strategy, ... Thus for him (and possibly others), it was the full-fledged Russellian doctrine of types that was stimulating.

Wiener was also struck with Russell's paradox. He was wont to see this paradox behind many a situation where most of us might see none. Roughly, he saw it in any circumstance involving two entities x and y, in which trouble ensues when y becomes equal to x, or nearly equal to x. He detected it in the self-feedback maladies of engineering, e.g. the blare due to excessive proximity of microphone and loudspeaker, in many themes and plots in mythology, drama and fiction down to slapstick, as when an old lady, who forgivingly takes back her own radio set from her repentent robber, is promptly arrested for "accepting stolen property". Wiener took just as readily to the more Cantorian "paradoxes of the superlative", which yield self-contradictory concepts such as the set V of all sets. (On the one hand, the cardinal number a of V is obviously the largest cardinal; on the other by a theorem of Cantor a is smaller than the cardinal number 2^a of the set of all subsets of V—a contradiction.) Wiener's favorite self-contradictory concept was "the totally efficient slave". (Self-contradictory, because to be fully efficient one has to be free.)

To Wiener such paradoxical concepts were not just matters of fun. He used them tellingly to illustrate phenomena such as the Roman household in which the Greek philosopher-slave becomes the real master. On the other hand, Wiener would have been drowsed off by the minute dissection of paradoxes that many an Oxford don has carried out in the name of Wittgenstein. To him all paradoxes were birds of the same feather, Russellian in some (unspecified) way or other. While this cavalier attitude may be logically assailable, it served him admirably in putting the paradox to heuristic use in research, and in wielding it to drive home many a serious point in his persuasive discourse. But the non-metamathematical setting of the *Principia Mathematica* precluded Wiener, this love notwithstanding, from understanding and marshalling the full potency of the paradox in the disciplined and creative way in which K. Gödel (age 24) was able to do in 1931. For it was Gödel's remarkably original, judicious and rigorous use of both the so-called Richard paradox and the idea of recursion, in a proper metamathematical setting, that yielded the important methatheorem on the incompleteness of all axiomatic set-theories embracing Peano arithmetic, cf. H. Weyl {W5, p. 223}. This metatheorem says roughly that in any formal language in which Peano's postulates governing the integers 1, 2, 3, . . . can be written, there is a (meaningful) arithmetical statement P such that neither P nor its negation is provable from these and other given postulates. (See Note, p. 65. for the Peano postulates.)

Thus the *Principia Mathematica* was a watershed in the history of mathematical thought which affected many minds and set the philosophical tone of the age in which Wiener began his research. It offered strong shoulders on which giants could stand.

C The Tractatus and logical empiricism

A sequel to the *Principia Mathematica* appeared in 1919 in the *Tractatus logico-philosphicus*, by Ludwig Wittgenstein {W14}. This work, full of new ideas, in turn had tremendous repercussions. It led the philosophers of the Vienna Circle in the 1920s, 1930s and 1940s to formulate the main theses of *logical empiricism* on the scientific methodology:

1. the analyticity, or the lack of factual content, of all logical and mathematical statements;
2. the hypothetical character of all empirical statements;
3. the paramount importance of mathematical concepts in the formulation of general hypotheses of the sciences, and of logical and mathematical theorems in the transition from such hypotheses to verifiable experimental and observational statements.

The resulting faith in the intimacy of logic-mathematics on the one hand and the empirical sciences on the other, despite a clear-cut separation between the two, is best illustrated by three quotations from Einstein:

> Geometry (G) predicates nothing about the relations of real things, but only geometry together with the purport (P) of physical laws can do so. Using symbols, we may say only the sum of (G) + (P) is subject to the control of experience.
> As far as the laws of mathematics refer to reality, they are not certain; and as far as they are certain, they do not refer to reality.
> This view of axioms . . . purges mathematics of all extraneous elements, and thus dispels the mystic obscurity which formerly surrounded the principles of mathematics. But the presentation of its principles thus clarified makes it also evident that *mathematics as such cannot predicate anything about perceptual objects or real objects.* {E4, pp. 35, 28, 30–31} (emphasis added)

The principles expressed in these quotations are less circumscribed than the positivist principles laid down earlier by the scientist-philosopher

E. Mach. This widening of the empiricist principles was essential, since the earlier, less flexible, attitudes at times encouraged opposition to new viewpoints—witness the opposition of both Mach and Ostwald to the Boltzmann's kinetic theory of matter. A widening of the principles is also demanded by the criticism leveled by Professor Quine {Q2} and others against narrower interpretations, cf. Carnap {C7, pp. 257–274}.

From the logical empiricist view point, a logically-determinate (or analytic) statement such as "Napoleon is in France or Napoleon is not in France" is devoid of factual content. It is true, but its truth merely reflects our linguistic conventions governing the terms 'is', 'or' and 'not'. On the whereabouts of Napoleon it is about as "informative" as the answer "it is soporific" to the question, "Why does morphine put me to sleep?" Likewise, the arithmetical statement "3 + 1 = 4" is certainly true, but its truth reflects our axioms governing the signs ' = ', ' + ', '1', '3', '4'. By itself it is useless in the empirical realm. It becomes extremely useful, however, when the signs are given suitable logical and empirical interpretations. It then enables us to draw conclusions such as "I bought four apples, three are on the table, and one is on the floor; so none is lost". It is to its role as a usefully interpretable formalism that mathematics owes its omnipresence among the sciences. But apart from this, mathematics is also valuable for its intellectual and aesthetic qualities, as we already saw in § 1D on symbolism.

One positive contribution ensuring from the logical empiricist R. Carnap was the exorcising of the so-called "conflict" or "competition" between the formalist, logisticist and intuitionist philosophies of mathematics. Carnap showed that the controversy was an illusion. It arose from confusion of syntax with semantics, of pure calculus with interpreted calculus, and from insufficient awareness of the metamathematical concepts of *subcalculus* and of the *realizability* of one calculus in another. Shorn of its fanciful aura, the intuitionist calculus is a proper, and for many purposes very useful, subcalculus of the ordinary calculi, and mathematico-logical calculi with number-theoretic specific terms and Peano axioms are realizable in logistical calculi. Which calculus is appropriate for a particular purpose is an extra-logical question, which has to be settled by considerations of scope, convenience and taste, cf. Carnap {C4, pp. 48–51, 26–29}, {C3, § 84} and {C5}.

D Wiener's philosophical activity

The intellectual climate of the period from the mid-1910s to the 1930s was marked by the *Principia Mathematica*, the *Tractatus* and the writings of the Vienna Circle. Not all the scientists of this period had read these works, but all could breathe the clean air that came from them, and all were aware of the towering presence of Albert Einstein and of his practice of the great tradition of blending deep physical intuition with abstruse mathematics to secure very powerful hypotheses at the empirical level.

This atmosphere affected Wiener's studies in a concrete way. Russell urged him to approach mathematical philosophy from the broadest standpoint, to concentrate not just on the foundations but also to look at the frontiers of mathematics and theoretical physics. This advice not only brought Wiener into contact with G.H. Hardy, as we noted earlier, but it also exposed Wiener to Bohr's atomic theory, the work of J.W. Gibbs on statistical mechanics and the Einstein-Smoluchowski papers on the Brownian motion, which were to prove of lasting value, cf. § 7D below. One wonders whether Wiener would have received the wise counsel he obtained from Russell had he been a student in our present fragmented and watered-down universities.

The work of the logical school in Cambridge (c. 1914) was constricted by the absence of metamathematics. The only contribution to axiomatic set theory that came from Wiener's contacts with Russell was his theoretical definition of the ordered pair as a set [14a]. This simplification, by completing a line of thought of Charles S. Peirce, marked a new stage in the history of logic, in which relations and functions had no longer to be treated as primitive terms. In Quine's words:

> In the reduction of logic in turn to the three present primitives,[5] one essential step was Russell's discovery of how to define complex terms in context; a second was Sheffer's reduction of conjunction, alternation, and denial to joint denial; and a third was Wiener's definition of the ordered pair. {Q1, p. 126}

Wiener's definition also took away some of the complexity of the type theory in the *Principia Mathematica*, cf. {Q1, p. 163}.

This was no mean achievement for a lad of 19. But for reasons unknown, Russell did not appreciate Peirce's simplification of relational logic and still less that of Wiener. Russell has described Peirce as "one of the most original minds of the late nineteenth century, and certainly the greatest American thinker ever", cf. {R9, p. 276}. But the folder [MC, 468] contains a 1914 handwritten exchange in which Russell comments, "I do not think a relation ought to be regarded as a set of ordered couples", to which Wiener responds, "It seems to me that what is possible in mathematics is legitimate". Wiener also wrote a paper [15b] critical of the then prelavent view that a closed system of mathematical logic is possible. This accorded with the intuitive feelings he had had

---

5 viz. '↓', '∀', '∈' (joint-denial, universal quantification, membership).

since childhood, and the subsequent work of Gödel was to bear him out. But in this he could merely surmise and not deliver.

Wiener also maintained contact with the other two philosophers, besides Russell, belonging to the "Mad Tea Party of Trinity", viz. J.M.E. McTaggart, the Hegelian, and G.E. Moore, the analytic moral philosopher. (These were, respectively, the Mad Hatter, the Dormouse and the March Hare.) Wiener enjoyed the intellectual life of Cambridge University. He has written glowingly of the interesting people he met at Russell's parties, such as the young C.K. Ogden and I.A. Richards, just then embarking on their *The Meaning of Meaning*. He had also made friends with some of the senior students who were prospective dons, such as F.C. Bartlett, later to become Professor of Psychology at Cambridge. "Half-starved" though he was with the "raw carrots and inedible Brussels sprouts" he got from his landlady, and with his own "occasional penny bars of chocolate", he wrote "I was both happier and more of a man than I had been before" [53h, p. 197].

Early in 1914, after spending Christmas with his family in Munich, Wiener wrote his paper [14b] wherein he adapted some ideas of Whitehead to develop a relation of "complete succession" (which is irreflexive and transitive but not total). This relation is well-adapted to a theory of time and space in which entities such as *instants* and *points* are regarded as constructs derived from psychologically less remote entities such as temporal events having duration and spatial extension. The construction of idealized scientific concepts from the more familiar ones suggested by perception, by use of relation theory, is Whitehead's *principle of extensive abstraction*. Whitehead had by then left Cambridge for London, and Wiener went there to make his acquaintance. As Russell was to be away during the May term of 1914, Wiener followed his advice to study at Göttingen. Before his departure, however, he completed an essay [14c] entitled "The highest good", which won him a Bowdoin Prize from Harvard, and a paper [14d] on relativism.

Several ideas that were to loom in Wiener's cybernetics (1948) appear in [14c, d] in nascent form. In the more fundamental of these papers [14d] Wiener denies first the position taken by F.H. Bradley and G.E. Moore that self-sufficient knowledge and self-sufficient experience are possible, and second he questions the possibilities of a priori knowledge or purely derived knowledge.

Taking what today we would call a metamathematical standpoint, Wiener emphasises the "grammar" of geometry, i.e. the rules governing the formation of geometric sentences and the assignment of meanings to the geometric terms 'point', 'line', etc. (For instance, 'point' might mean a tiny dot, 'line' a stretched string.) This grammar has to be formulated in a language M_1. The grammar of M_1 will require for its formulation a language M_2, and so on, until we decide to terminate the process with *rules which are not written or spoken but only thought* [14d, p. 567]. The *spirit* of the geometric language is Wiener's term for these unspoken unwritten rules. Thus, a great deal more is involved in geometric deduction and in the veracity of what is deduced than meets the eye. In this approach we see an inclination on Wiener's part towards the concept of metalanguage and towards the thought that was to come with Wittgenstein in 1919, to wit, "What we cannot talk about we must pass over in silence" {W14, p. 151}.

Relativism is Wiener's term for the philosophical position that denies the possibility of certitude in human knowledge, but not the existence of certainty. It is not to be confused with pragmatism. Indeed Wiener questions the Protagorean formula "man is the measure of all things . . . " on the ground that to escape from the dilemma of a multitude of egos, the term "man" has to be carefully construed. But it is far from clear that such can be done without turning the dictum into a useless tautology. Also, Wiener questions the clarity of the pragmatic view that the "meaning" (or "value") of a conception lies in its "practical consequences". For whether or not a consequence is "practical" will depend on the "purpose" or "goal" we have in mind. But pragmatism ducks the crucial question "Just when is a mental state a purpose?" [14d, p. 567]. Much later cybernetics addressed this question, in terms of the feedback in teleological activity, cf. [43b] and § 15B

Turning to Bergson, Wiener discusses the unbridgeable "gulfs" that Bergson saw between

> homogeneous duration and mathematical time, between purposes and mechanism, between life and matter, between language and thought, between that intuitive thought which allows the mutual interpenetration of idea with idea, and intellectual thought—that thought which deals in absolutely hard-and-fast concepts and clearcut distinctions. [14d, p. 570]

On Bergson's belief that these "fundamental dichotomies" are irreducible and sharp, Wiener wrote:

> Now to suppose the existence of absolutely sharp distinctions runs directly counter to the spirit of relativism and, I believe, of Bergsonianism itself. [14d, p. 570]

It was precisely the "gulfs" listed in the first quotation that cybernetics bridged about 20 years later. Furthermore this bridging could be undertaken only after cyberneticists realized that the anisotropy of time (Eddington's arrow), demanded by the laws of thermodynamics, had more to it than what the physicists attributed: it also made room for indeterminism, for teleological phenomena, growth, learning, and freedom—the very things that Bergson cherished but felt were missing. Chapter 1 of Wiener's *Cybernetics* [61c] with the title "Newtonian and Bergsonian Time" stresses this, while affirming that the absence of strict determinism in the universe does not destroy its cosmic character.

Relativism also allows for the emphatic readings of absolutism, pragmatism and Bergsonianism. Thus Wiener concludes that "all philosophies are nascent relativisms" (p. 574). Wiener's remarks of what we owe to our heritage, his analogy between the history of theories and the history of tools (p. 576), his affirmation of the self-critical nature of the scientific enterprise (p. 577), and his affirmation of the secondary position of ignorance and noise—the physicist does not stop working just because he can only arrive at approximate laws and can only have instruments of imperfect accuracy (p. 575) — are in near-perfect harmony with the mature positions he was to take in later life.

In the essay [14c] Wiener rejects the view that gradation of ethical behavior demands an ethical *summum bonum*, "the highest good", for orderings can exist with no absolute maxima. " . . . our ideals grow with our attainments, . . . the better a man becomes, the broader are the vistas of righteousness that open out before him" [14c, p. 512]. Wiener then tries to separate sets of feelings that constitute *conscience* from those that constitute prejudices. He answers the question as to why individual consciences should concur, and bring forth a *social conscience*, by pointing to the biological impulse to survive, and to its fulfillment by gregariousness, cooperation and concerted action. So originates in us an instinctive urge to tolerate, if not respect, the consciences and prejudices of others. But individual consciences may antagonize one another, and the urge towards tolerance and towards a common good is often overwhelmed by the warlike conscience. Force thus becomes an arbiter between conflicting objective goods. Thus, *objective morality is the "end-product" of a two-fronted struggle, one within the individual to reorder his ethical priorities, and the other between groups of individuals. As this struggle is an ongoing one, objective morality is far from being fixed, and must evolve.*

The thoughts expressed in [14c] reappear in more cybernetical form in Wiener's theory of learning machines that was to come 25 years later, cf. §§ 18E, F. All learning machines are dependant on feedback with the environment, and to improve their future performance, they must put to use the record of their previous performances—in short, they must all learn from experience. So too for our species, "our ideals" must "grow with our attainments" (p. 512). The view that "there is one fixed immutable idea of moral conduct" (p. 512), or of any other kind of conduct, ends the learning process at some point. Any such view is obviously incompatible with Wiener's later thought, in which learning is a never-ending process. What is a bit unexpected is that even at age 20, Wiener was repelled by the idea of a terminus to learning.

That Wiener's own ethical susceptibilities and his appreciation of the good in life had changed for the better is apparent from the way in which he began to comport himself. He never gave up the vegetarianism he had acquired from his father, and like his father had to go hunting for vegetarian restaurants during his travels. The only vegetarian restaurant he found in Göttingen had excellent cooking, but it was owned and operated by a very anti-Jewish family. Wiener amusingly recounts how he and another young Jew[6] "defied the wrath" of the rabidly anti-Semitic literature surrounding them, and "ate, nay even enjoyed" their savory meals. His Jewishness had ceased to be the nightmare that it was in 1909.

Wiener's activity at Göttingen is best described by quoting from a letter he wrote to Russell in June or July of 1914:

> At present I am studying here in Göttingen, following your advice. I am hearing a course on the Theory of Groups with Landau, a course on Differential Equations with Hilbert (I know it has precious little to do with Philosophy but I wanted to hear Hilbert), and three courses with Husserl, one on Kant's ethical writings, one on the principles of Ethics, and the seminary on Phenomenology. I must confess that the intellectual contortions through which one must go before one finds oneself in the true Phenomenological attitude are utterly beyond me. The aplications of Phenomenology to Mathematics, and the claims of Husserl that no adequate account can be given of the foundations of Mathematics without starting out from Phenomenology seem to me absurd.
>
> Symbolic logic stands in little favor in Göttingen. As usual, the mathematicians will have nothing to do with anything so philosophical as logic, while the pilosophers will have nothing to do with anything so mathematical as symbols. For this reason, I have not done much original work this term: it is disheartening to try to do original work where you know that not a person with whom you talk about it will understand a word you say.
>
> During the Pfingsten holidays, I called up Frege up at Brunnshaupten in Mecklenburg, where he spends his holidays. I had several interesting talks with him about your work. {R8, Vol. 2, pp. 39, 40}

In this letter the young Wiener clearly reveals the cleavage between the tradition represented by Husserl, and we might add by mathematicians prejudiced against the actual infinite such as Gauss, Kronecker and Poincaré, and that of the bolder school led by Cantor, Frege, Peano, Whitehead and Russell.[7]

6 Hyman Levy, later to become a professor of mathematics at London University, and a vocal member of the British left-wing.

7 "I protest against the use of infinite magnitudes as something completed, which is never permissible in mathematics," Gauss, 1831.
   "Later generations will regard *Mengenlehre* as a disease from which one has recovered," Poincaré, 1908, cf. {B5, pp. 556, 558}.

Wiener's mind was far from idle during his Göttingen days, for soon after his letter to Russell he completed a paper [15a] on synthetic logic which was to serve as the basis for his Docent Lectures at Harvard the following year. He also tells us that here he had his first experience of "the concentrated passionate work necessary for new research" [53h, p. 211]. With his mind absorbed by matters foundational, the courses of Hilbert and Landau did not register an immediate impact on Wiener. But they had a delayed effect, and as his mathematical interests widened he began to appreciate what he had learned from them. Wiener also enjoyed the seminars and social life of the mathematical circle at Göttingen. He compared the situation there to that at Harvard by "a deep draught of Muncher" as against "near beer". Among his companions at the beer garden were the prospective mathematicians Otto Szasz and Felix Bernstein.

Wiener returned to the United States in August 1914, for a holiday. World War I had broken out, but his Harvard traveling fellowship had been extended for the year 1914–1915, and so he returned to wartime England. The atmosphere at Cambridge was charged by the fervor of war. Russell's pacifism had made him an unpopular figure, and the situation hindered undisturbed philosophical investigation.

An interesting interlude for Wiener during this gloomy period was a meeting with T.S. Eliot, great poet-to-be. Eliot had been a fellow in philosophy and Indic philology at Harvard during the years 1908–1910 and 1911–1914, respectively, and knew Wiener. He came under the influence of Bergson at the Sorbonne during 1910–1911, and was at Merton College, Oxford in 1915 studying the idealist philosophy of F.H. Bradley. He wrote to Wiener in October and December 1914 about a get-together. From the tone of these letters it is evident that he knew Wiener quite well: "I can't imagine what on earth you are doing with McTaggart unless you are reading Hegel or drinking whiskey" [MC, 10]. They finally met in London, and with little money in their pockets, had "a not too hilarious Christmas dinner together in one of the larger Lyons restaurants" [53h, p. 225]. They must have had a considerable philosophical exchange, for there is a six-page letter to Wiener from Eliot in London dated January 6, 1915, in which he speaks of Wiener's papers [14a, b, c, d] and [15a], and comments at length on the question of relativism in philosophy, broached by Wiener in [14d]. Eliot's interest is understandable in view of his own association with Bergson and Bradley. He also described his own embarkation on a critical thesis on Bradley's theory of judgment, *Knowledge and experience in the philosophy of F.H. Bradley*. In this letter Eliot wrote:

> For me as for Santayana philosophy is chiefly literary criticism and conversa-
> tion about life; and you have logic which seems to me of great value. [MC,
> 11]

(The entire letter is reprinted in the *Coll. Works*, IV, pp. 73–75.)

Although Eliot's description of philosophy as "chiefly literary criticism" did not fit with Wiener's, the letter shows a commonality between their flexible attitudes. Decades later there was to be a remarkable concordance between Wiener's views on the inevitability of the collapse of homeostasis and Eliot's view on the inevitability of the recurrence of sin: "Sin grows with doing good" {E6, p. 44}. One can only surmise what influence, if any, Wiener had on Eliot's poetry.

E Columbia and its word-minded philosopher (1915)

As submarines were starting to surface in the Atlantic, Wiener's father beckoned him to come home. Upon Russell's advice Wiener decided to spend the rest of his traveling assignment at Columbia University in New York, possibly under the philosopher John Dewey. This was perhaps the only unfortunate advice that Wiener got from Russell.

Wiener arrived in New York in January 1915 and soon regretted coming to Columbia University:

> I found the skyscraper dormitories of Columbia depressing after Cambridge
> and Göttingen. I also found the life of the place unsatisfactory in its lack of
> coherence and unity. Almost the only bond between the professors, who
> lived widely scattered in University Heights apartment houses or in suburban
> bungalows, was an almost universal antagonism toward Nicholas Murray
> Butler[8] and everything that he stood for. [53h, pp. 221–222]

In 1918 he found the Aberdeen Proving Grounds of the United States army academically more congenial and "closer to that cloistered but enthusiastic intellectual life" he had had at Cambridge than Columbia University [53h, p. 259]. As for the "word-minded" John Dewey, "his social dicta did not translate easily into precise scientific terms and mathematical symbolism", and Wiener learned nothing from him [53h, p. 222].

By the epithet "word-minded", Wiener was referring to a bondage to words that kills the expression of the human spirit. Here is a sample:

> Take, for an example, such an elementary proposition as "this is sweet". *This*,
> as has been shown, marks a selective-restriction, made for a definite purpose,
> within an inclusive qualitative problematic situation. The purpose is the final
> consequence of a resolved situation in attainment of which "this" has a

---

8 President of Columbia University, 1901–1945.

special function to perform. If the predicate "is sweet" is an *anticipation* of the resolved situation, it means "this" *will sweeten* something if that operation is performed which is required to generate definite perceptible consequence. Or, it may record the achieved result of the execution of the operation: "This *has* sweetened something." When the operation is completed, *this* is definitely qualified as sweet. This fact is manifest not in in a proposition (although a proposition may report it for purposes of record or communication of information) nor in symbols, but in a directly experienced existence. Henceforth, "this" is a sweet *somewhat*. The quality *sweet* does not stand alone but is definitely connected with other observed qualities. As thus characterized, it enters into further situations in which it incorporates into itself additional qualifications. It is a sweet, white, granular, more or less gritty thing or substance, say, *sugar*. (Author's italics) {D3, p. 128}

This dreary, humorless, long-winded, and confusing passage is typical of almost all the 535 pages of Dewey's *Logic, the Theory of Inquiry* {D3}. The human hen cackles and cackles over the egg of language, but cannot break its shell! William of Ockham (c. 1230) spoke well when he said, "it is vain to do with more what can be done with less". In 1905 Dewey attributed his position on logic to C. S. Peirce, cf. e. g. {D3, p. 5}, and this perhaps was why Russell advised Wiener to go to Columbia. Peirce, however, was a symbol-minded philosopher with a penetrating understanding of history, and he built on the solid ideas of Anaxagoras, Duns Scotus, Aquinas, Roger and Francis Bacon and Giambattista Vico. Peirce's important concept of *abduction* is almost exactly the vital conception that the scholastics referred to as "saving the appearances". His pragmatism was in essence Vico's: the understanding of theory is deepened by its use in concrete construction. Uneasy even with the mild dilution of his ideas by William James, a much clearer and more articulate philosopher than Dewey, Peirce announced in 1905 "the birth of the word 'pragmaticism', which is ugly enough to be safe from kidnappers". If there are ideas of Peirce in Dewey's pragmatism, they are ematiated beyond recognition.

It is commonly held that philosophical activity, good or bad, is largely inconsequential in practical terms. The opposite is true: when philosophers falter, humanity loses. We will encounter the social ill-effects of the influence of Dewey's philosophy when we come to Wiener's discussions with the labor leader Walter Reuther on industrial automatization and its educational challenges (§§ 19A, B).

Wiener describes his Columbian sojourn as probably the "low point" of his career between the European "summit" and later "gradual ascent" [53h, p. 226]. He spent his days there absorbed in the Docent Lectures he was planning for Harvard in the coming academic year. In the course of this, his mind was drawn towards analysis situs, the branch of mathematics we now call *combinatorial topology*, the first shots of which had been fired by Euler and Poincaré, but which really got underway only in the early 1920s with the work of Alexander and Veblen, both of whom Wiener was to meet a couple of years later at the Aberdeen Proving Grounds of the U.S. Army. Thus Wiener was five years ahead of the time in his ventures. These were not systematic, however, and no pioneering contribution resulted. Unfortunately for us Wiener lost his manuscript sometime before 1920.

Wiener, writing for the lay reader, has described combinatorial topology (analysis situs) as:

> ... that strange branch of mathematics dealing with knots and other geometric shapes whose fundamental relations are not changed even by a thorough kneading of space so long as nothing is cut and no two remote points of space are joined. Topology includes the study of such things as the familiar one-sided Möbius's sheet of paper, which you get when you take a long, flat strip, rotate one end of it through half a revolution and glue the ends. It makes an excellent parlor trick to ask a layman what will happen to such a strip if you start cutting it down the middle until the ends of the cut meet. If you try this, you will find that even after the cut is complete the strip will remain in one piece but will now make a full revolution instead of half a revolution as you proceed around it. [56g, pp. 26, 27]

F Harvard and the Docent Lectures (1915–1916)

After vacationing in New Hampshire, Wiener returned to Harvard University in the fall of 1915 to assume his duties as Assistant and Docent Lecturer in the Philosophy Department. Apart from teaching a section or two of introductory freshman philosophy, he gave his Docent Lectures on *Synthetic logic and measurement theory* based on his researches in [15a] and its sequel [21a]. *Synthetic logic* is the subject in which the ideal entities needed in the sciences (e.g. point, line, number, perfectly rigid body, etc.) are not postulated, but instead defined in terms of more familiar, less ideal entities, by going up the type-ladder. A simple example is Russell's definition of *number* as a set of equinumerous sets. Equinumerousness is a simpler notion than number. For instance, we can test that the sets of fingers on the left and right hands are equinumerous by matching the fingers; the underlying concept of "five" is more abstract.

In the paper [15a], with certain basic assumptions Wiener attempts to derive from a rather mildly restricted n-ary relation R, (i.e. a relation between n things), a new relation S which has the attributes of an ordering relation, and can serve as a basis of measurement of complex psychic sensations. (Imagine R to be an n-ary analog of the binary relation "seems louder than" between two tones.) Roughly speaking, [15a] does for sensation-intensities like 'loudness' what his earlier paper [14b] had done for extensive quantities such as length or duration (cf. § 5D above). The sequel [21a] to [15a] deals, in the same spirit, with the

more difficult issue of sensations, the intensities of which are "not
susceptive to increments of arbitrary magnitude", e.g. the intensity of
redness, cf. [21a, p. 182–]. To carry out such analysis Wiener had to
modify and extend the theory of measurement given in Vol. 3 of the
Principia Mathematica. Wiener's Docent Lectures also covered his
philosophy of geometry, which appeared in print only in [22a]. (For a
description of its contents, see § 7B, below.)

We now know, thanks to the research of Dr. Grattan-Guiness
{G13} of the high opinion in which Russell held Wiener's work in
synthetic logic. His referee's report on [21a] to the London Mathematical Society reads as follows:

> This is a paper of very considerable importance, since it establishes a com-
> pletely valid method for the numerical measurement of various kinds of
> quantity which have hitherto not been amenable to measurement except by
> very faulty methods.
>   Although Dr. Wiener's principles can be applied (as he shows in the
> later portions of his paper) to quantities of any kind, their chief importance
> is in respect of such things as intensities, which cannot be increased indefi-
> nitely. Much experimental work in psychology, especially in connection with
> Weber's Law, has been done with regard to intensities and their differences,
> but owing to lack of the required mathematical conceptions its results have
> often been needlessly vague and doubtful. So far as I am aware, Dr. Wiener
> is the first to consider, with the necessary apparatus of mathematical logic,
> the possibility of obtaining numerical measures of such quantities. His
> solution of the problem is, as far as I can see, complete and entirely satisfac-
> tory. His work displays abilities of high order, both technically and in general
> grasp of the problem; and I consider it in the highest degree desirable that
> it should be printed. {G13, p. 104}

In fundamental ways, Wiener was a Russellian, and wisely depreciated
his "many points of disagreement and even friction" with Russell, cf.
[53h, p. 193].

Among Wiener's audience in the Docent Lectures was G.D.
Birkhoff, Harvard's greatest mathematician. This was a tribute to
Wiener. But Birkhoff spotted certain lacunae and inaccuracies in Wiener's treatment. Thus began a dialectic relationship of mutual appreciation and conflict between these two giants of American mathematics,
that lasted until the very end.

G The unpleasant stint in Maine (1916)

Wiener's assignment at Harvard was temporary. While his accomplish-
ments were remarkable for a 21-year old, they were not always rigorous,
as we just saw, nor were they substantial enough for a Philosophy
Department of the stature of Harvard to consider him for a long-term
appointment. Nor was Wiener a good teacher of undergraduate routine,

and this may have precluded the short-term renewal of his appointment. Also, there is truth in Wiener's contention that "in those days work in mathematical logic was equally unhelpful in getting jobs in either mathematics or philosophy" [53h, p. 270]. Wiener had to seek a job elsewhere, and on his father's suggestion he tried to get one in a mathematics department. With the aid of a teacher's agency Wiener got an instructorship in mathematics at the University of Maine in Orono, Maine, and began his duties there in the fall of 1916.

Words such as 'dull routine', 'examination cribbing' and 'copying of homework' occur in Wiener's description of the academic environment at Maine. What made him 'feel alive' were contacts with the statistician Raymond Pearl and his group, and getting off campus to Bangor. He left the University of Maine at the end of the academic year.

Note: **The Peano postulate system**

This system for the set N of natural numbers 1, 2, 3, . . . has two undefined terms 1 and S, i.e. successor, ($S(n)$ is to mean "the successor of n", e.g. $S(7) = 8$). It has four postulates:

(1) $1 \in N$;
(2) if $n \in N$, then $S(n) \in N$;
(3) for $m, n \in N$, $S(m) = S(n)$ if and only if $m = n$;
(4) if $E \subseteq N$, $1 \in E$, and $n \in E$ implies $S(n) \in E$, then $S = N$.

These postulates, together of course with the rules of deductive logic, serve admirably in proving all the results in number theory. Bogus questions such as "What really is the number 6?" are bypassed. The postulates and the theorems deducible therefrom have no factual content (§ 5C). With suitable interpretations of the symbols '1' and 'S', however, factual assertions such as "The week has 7 days" become true, and arithmetic becomes usefully applicable.

6 World War I and Wiener's Military Yearnings

> It was embarrassing for me to meet the soldiers everywhere, in the movies, in the streets, and even in the classrooms of the university, and to think that as a foreigner I was immune to the universal sacrifice. Several times I thought of enlisting, but was deterred by the fact that after all it was not yet my war and that to go into it before my parents were ready to accept the situation would be in some sense a very serious disloyalty to them. Then too, with my poor eyesight, I was not exactly the best soldier material; nor did I desire to sacrifice my life for a cause concerning the merits of which I was not yet fully convinced. [53h, p. 229]

Wiener was speaking of Cambridge University in late 1914 and early 1915, during his second visit (cf. § 5D). As the United States got closer to the war, Wiener got more and more involved. Ardently patriotic, he wanted to join the Army in anticipation of the United States' participation on the side of the Allies, and his one big fear was flunking the entrance exams because of bad eyesight and manual clumsiness. He had a very firm conception of duty; witness his words written in 1960:

> I do not think that any honest man looks forward with any pleasure at the prospect not merely of dying in action but of having to face a situation in which the only honorable thing for him to do is to die in action. Nevertheless, in becoming a soldier, and especially in becoming an officer, he must accept the contemplation of this possibility. [60e, p. 26]

In 1916 he enlisted in the Harvard Regiment and during the summer went to Plattsburg, New York, to train for a commission in the Reserve Officers' Corps. He pursued this objective from Maine, and Professors William Ernest Hocking and Ralph Barton Perry and even President Abbot Lawrence Lowell wrote supporting letters on his behalf to the Commanding General of the Eastern Department in April 1917. But his marksmanship proved inadequate and his application was denied in May 1917, much to his disappointment. Wiener then tried to enter one of the other branches of the service, and after leaving Maine University he trained with the ROTC at Harvard in the summer of 1917. But he kept failing in one test or another. In an examination for a commission in the artillery he did rather well in mathematics, but flunked badly in horseback riding after falling off "an old nag which was as steady as a gymnasium horse" [53h, p. 244].

One has to read Wiener's letters during this period to realize that he had become something of a "warmonger". A "Dear Dad" letter from Orono, Maine dated March 1, 1917, about a month before the United States' entry into the war, is especially pungent:

> ...I have the misfortune to have a hostess who is both suffragette and pacifist—the yellow banner and the yellow streak. I wonder how long Wilson will continue to turn our other cheek? I am afraid that the West is suffering from hypertrophy of the pocket-book and atrophy of the conscience. It is a sad thing for a nation when *to epithmetikon* assumes control over *nous* and *thumos*, to use the language of the Republic.
>
> Have you run across any good anathemata among your medieval documents? I am getting a little short of cusswords for use against the Germans and their American allies. [MC, 15]

The letter ends with the wish to be "back in the good old Harvard Regiment:—as corporal of the 12th squad, company B". One gets a less exhilarating picture of Wiener's attitude to the war from his 1953 account of these events in [53h, p. 243].

Having failed in all his efforts to join the armed forces, Wiener searched for a job related to the war effort, and in the fall of 1917 got an apprenticeship in the General Electric Factory at Lynn, Massachusetts. He helped run some combustion tests, using a bit of mathematics in thermodynamic calculations. He felt like an "honest workman", "tired but happy". But his father, who knew the managing editor of the *Encyclopedia Americana*, then in Albany, New York, secured a job for him there as a staff editor, and Wiener abruptly terminated his services at GE. In 1953 he confessed to his spinelessness in this act:

> Though I felt morally obligated to stay on at the General Electric Company, I was too dependent on my father to dare to contravene his orders, and so I had to present my shame-faced resignation to the engineers who had given me my chance in Lynn. [53h, p. 248]

Wiener found Albany "a paradise of neatness, tradition and civilization" [53h, p. 252]. In a letter to his sister in October 1917, he described the "Americana" aspect of the *Encyclopedia* as follows:

> Our encyclopedia is called the Americana, because its staff consists of an ex-newspaper man of British Provenience and German extraction, one Irish-Canadian, one French ditto, one Canadian Baptist minister, one Russian Jewess, and my honored self. [MC, 22]

Wiener wrote a large number of articles for the Encyclopedia on subjects philosophical, mathematical and linguistic. Several of these were repeated in subsequent editions. (All are reproduced in the *Coll. Works* III and IV.) He even toyed with the idea of collecting some of them into a book. Wiener has described his experience in this phase of his life as follows:

> With all the shortcomings and unpleasant sides of hack writing, it was a wonderful training for me. I learned to write quickly, accurately, and with a minimum of effort, on any subject of which I had a modicum of knowledge. [53h, p. 251]

But he realized that despite the enjoyable and educative aspects of such a position, it could not be the terminus of his intellectual activity. He looked for a way out, preferably for one that led to the war effort. The way opened when he received a letter from Professor (then Major) Oswald Veblen, dated July 5, 1918, offering him the position of computor at the U.S. Army Proving Grounds in Aberdeen, Maryland, at a salary of $1200 per year with 30% for overtime. Wiener readily accepted after resigning his position at Albany, with high hopes of making himself useful in the war effort.

At the Proving Grounds Wiener joined in the construction of range tables for the new types of artillery and ammunition that were being designed. The demands for speed and accuracy necessitated the use of computing machines. As Wiener put it:

> It was a period in which all the armies of the world were making the transition between the rough old formal ballistics to the point-by-point solution of differential equations, and we Americans were behind neither our enemies nor our allies. In fact, in the matter of interpolation and the computation of the corrections of the primary ballistics tables, Professor Bliss of Chicago made a brilliant use of the new theory of functionals. Thus the public became aware for the first time that we mathematicians had a function to perform in the world. [53h, p. 256]

Wiener felt that much significant mathematical activity during the post-war years emanated from those disciplined at Aberdeen: he cites Veblen, Bliss and Alexander, among others. He speaks glowingly of his contacts there with Hubert Bray and Philip Franklin (later to become his brother-in-law and colleague) and others. He wrote: "Whatever we did, we always talked mathematics", and

> ...I am sure that this opportunity to live for a protracted period with mathematics and mathematicians greatly contributed to the devotion of all of us to our science. [53h, p. 258]

But even this active involvement with war work was not enough for Wiener, for in October 1918, he enlisted as a private in the U.S. Army. Once in he began to regret it, for the war was nearing its end and

Norbert Wiener (far right) in army uniform at Aberdeen Proving Grounds, 1918

little purposeful activity remained. He describes his transfer to the Fort Slocum Depot as a "sentence to the penitentiary". In the "penitentiary" he surprisingly encountered Dr. Harry Wolfson of Harvard, later to become a world-renowned scholar of the Semitic philosophy of the prescholastic period and one of Harvard's great professors. They had a good time together. Wiener summed up his army experience as follows:

> Even with a temperament not suited for a regimented life and a more than average desire to see what I was doing and to know what it meant, I had found a few months of army life a haven at a time when I had been very tired for years from making my own decisions. It has been said many times that the motives of the soldier and the monk are curiously similar. [53h, p. 261]

Wiener's stint as an army private had its painful moments, but also its funny side. During sentinel duty, armed with a rifle and bayonet, he "found it hard not to drowse off and to keep sufficiently alert to challenge the Officer of the Day" [53h, p. 260].

Wiener's attitude to World War I and his heroically futile efforts to get into combat, despite his obvious unsuitability for it, bring to mind the misadventures of the much older Englishman, H. W. Fowler, author of The *King's English* and the well-known *Dictionary of Modern English Usage*, who enlisted in England in 1915 at age 57 (after mis-stating it as 44), trained for nine months and was sent to the front only to be found unfit for combat, returned to base and assigned to kitchen duties. Both individuals were ardently patriotic and amusingly idiosyncratic, and both, we might add, had a penchant for excellent prose. Wiener was lucky in finally finding a congenial position as computor; Fowler was not. But while the Englishman was at peace with himself, Wiener had fears of German militarism that sometimes bordered on the irrational, as some of his unpublished correspondence, and especially an unpublished piece "Royalism in Germany" [MC, 495] that he wrote in 1919 after the kaiser's dethronement, would suggest.

7 From Postulate Systems to the Brownian Motion and Potential Theory

A Wiener joins the Massachusetts Institute of Technology

After his discharge from the U.S. Army in January 1919, Wiener took a job with the *Boston Herald* as a reporter, to tide him until his return to academic life in the fall. He was asked to cover the strike in the textile industry in Lawrence, Massachusetts, and Wiener gives an interesting account of how his sympathies swung toward labor as he learned more and more of the facts [53h, pp. 266–268]. The *Herald* published Wiener's reports on the strike, but the editors did not like his handling of his next assignment, on the candidacy of General C.B. Edwards for president of the United States, and he was fired. Wiener's explanation of this seems plausible: "... I had not learned to write with enthusiasm of a cause in which I did not believe" [53h, p. 268]. Wiener now had spare time at his disposal, which he spent on studying mathematics and writing papers.

Upon the advice of Professor W.F. Osgood of Harvard, Wiener applied for an opening at the Massachusetts Institute of Technology for the fall of 1919. The Mathematics Department there was not oriented towards research. Largely on Osgood's recommendation, he was offered a one-year instructorship, even though he had no post-graduate degree in mathematics. Wiener accepted, and with help from colleagues, notably H.B. Phillips and Philip Franklin, Wiener became a part of MIT.

During this period Wiener's intellectual interests began to move away from the philosophical foundations of mathematics towards its superstructure. He began by finalizing a lengthy paper on his 1915 Harvard Philosophy Docent Lectures on geometry and experience. But concurrently his attention began to focus on the postulates of specific systems then engaging the curiosity of mathematicians. The second, and most dramatic, stage in the transition brought him to the frontiers of probabilistic analysis (the Brownian motion), and a third stage to the frontiers of potential theory.

B The relation of space and geometry to experience [22a]

In [22a], a long paper in the *Monist*, Wiener tries to show that Kant's particular *synthetic a priori* view of geometry fails to explain why geometry, "which seems to scorn experience as its basis, yet furnishes results of the utmost empirical application and value". He agrees with Kant that geometry is a synthetic *a priori* science, but departs from him by claiming that it has an *empirical* subject matter. By a penetrating analogy with statistical tables, Wiener illustrates how this is possible. The data in the table are empirical, but the idea of a statistical table, its parameters, etc. and its organizing principles (e.g. to divide the sum x_1 + x_2 + x_3 by 3 and not by 7) are a priori, i.e. not contributed by the data. Likewise in geometry we may regard a *point* "not as a direct object of experience but as a certain arrangement or collection of objects of experience" (p. 20). Continuing in the footsteps of Whitehead and Russell, Wiener then goes on to demonstrate "the point as a tabulation of solids" (pp. 31–45), defining a point as a class of convex bodies. He picks convex bodies because convexity is so easy to perceive: " . . . hollowness can roughly be judged by the eye and the finger without reference to straight lines, convexity may also be determined by a more or less direct reference to experience" (p. 34). In short, Wiener brought type-based synthetic logic to bear on Kant's doctrine of space.

The discovery of non-Euclidean geometries, in which the proposition "the sum of the angles of a triangle is two right angles" is denied, shows that this Euclidean proposition is *analytic* (not *synthetic*) *a priori*, i.e. its validity is logical. But the proposition may also be viewed as *synthetic a posteriori* and belonging to physics, in which case (as Einstein affirmed, cf. § 5C) it is neither necessary nor certain. Thus Wiener's contention that geometry has an empirical subject-matter is valid only with the addendum "viewed as a branch of physics". From the perspective of our current wisdom, to wit that there are no synthetic a priori propositions and that pure geometry and physical geometry must be distinguished, much of Wiener's formulation is outmoded. But when the paper appeared, it aroused the interest of the British philosopher C.D. Broad.

The wisdom in Wiener's paper lay rather in his pointing out that mathematical concepts such as "mean", "variance", "point", "line", etc. which are extracted from empirical data, are contributed by the human mind, and to this extent they are *a priori*, i.e. not empirical. This applies to all scientific renderings of pre-scientific notions, what Carnap has called *explication* {C6, Ch. 1}, e.g. the explication of musical pitch by "wavelength" or "frequency", of hot versus cold by "temperature", of quantity of motion by "momentum", and so forth. The tremendous philosophical insight and ability of Wiener's young pioneering mind stand out, when we remember that much of our current understanding of these issues was articulated long after his paper was written. Einstein's own views were expressed in 1922. Wiener was only 20 when he presented these incisive thoughts at Harvard in 1915.

C Postulate systems

It is worthwhile to hear Wiener's own semi-popular description of the subject of postulation:

> The geometry of the Greeks went back to certain initial assumptions, known variously as axioms or postulates, which were conceived to be unbreakable rules of logical and geometrical thought. Some of these were of a predominantly formal and logical character, such as the axiom that quantities equal to the same quantity must be equal to each other. Another, with a more purely spatial content, was that known as the parallel axiom, which asserts that if we have a plane containing a line ℓ and a point P not on that line, then through P and in that plane one and only only one line can be drawn that will not intersect ℓ. [56g, p. 51]

The efforts to prove that (non-Euclidean) geometries exist in which the last postulate is violated, bore fruit at the hands of Lobatchevsky and Bolyai in the 1830s:

> From that time on it came to be more and more clearly understood that the so-called postulates of geometry, and indeed of other mathematical sciences, were not undeniable truths. They came to be regarded as assumptions which we could make or refuse to make concerning the particular mathematical systems which we wished to study further.
>
> This tentative attitude in mathematics, in which the postulates became suppositions made for the sake of further work rather than fundamental principles of thought, gradually began to be the standard point of view of mathematicians in all countries. [56g, p. 52]

As an example of what Wiener meant, we may cite the *Peano postulates* for the system $\mathbb{N}$ of natural numbers 1, 2, 3, . . . given in the Note, p. 65. These postulates serve admirably in proving all the results of number theory, and one need not bother as to what the number 1 "really" is. In the same way there are postulate systems covering more complex systems, for instance, for two operations $\oplus$ and $\circ$ corresponding to addition and multiplication.

Wiener's own activity in postulate systems came from his amiable contacts with and high regard for E. V. Huntington of the Harvard Mathematics Department who worked extensively in this field, and whose course he had attended in 1912. His first paper in this area, on Boolean algebras [17a], marks a clear departure from synthetic logic, and bears the impress of Huntington.

The postulates of a Boolean algebra govern the operations, such as the union $\cup$, intersection $\cap$, complementation $'$, and inclusion $\subseteq$, that exist between the subsets A, B, C, etc. of a set V (often denoted by 1). The subsets include the void set $\emptyset$ (often denoted by 0). These operations are easy to understand. For instance, if V is the set of all human beings,

A the set of Asians, B the set of boys, and C the set of Chinese, then $A \cup B$ is the set comprising all Asians and all boys, $A \cap B$ the set of all Asian boys, and A' the set of all non-Asian people. Also since all Chinese are Asians, $C \subseteq A$. In the postulational treatment, we forget all about sets, and instead lay down a few postulates such as $(A \cup B)' = A' \cap B'$ (which are fulfilled by sets) from which all the other propositions can be deduced.

In 1904 Huntington had given a set of nine such postulates for a Boolean algebra $\mathbb{A}$, governing $\cup$, 0, 1, $'$, which were the only specific primitive terms. He could then define the other operations such as $\cap$, $\subseteq$, etc. Wiener asked if operations other than union could satisfy the same nine postulates. Towards this end, Wiener used Boole's result that every binary operation $\oplus$ for $\mathbb{A}$ has the form

$$x \oplus y = (a \cap x \cap y) \cup (b \cap x \cap y') \cup (c \cap x' \cap y) \cup (d \cap x' \cap y'),$$

$x, y, \in \mathbb{A}$, where a, b, c, d, are fixed elements in $\mathbb{A}$. In [17a] working with this expression, Wiener settled the question and effectively summed up the invariant theory of Boolean algebras. (For details, see Note 1 on p. 89.)

This was in 1916. In 1919, Wiener's postulational interests shifted to the more classical algebraic concept of a *field*, i.e. a system allowing the four operations of addition, subtraction, multiplication and division. (A familiar example of a field is the set of all integers and fractions, positive, negative, such as 0, -3, $1/(-3)$, $5/2$, $-11/29$.) His objective in the papers [20a–d] was to make a complete study of the field at the postulational level, to wit:

> the study of *all* the sets of postulates in terms of which the system may be determined, and of the manner in which the system may be determined in terms of them. [20d, p. 176] (emphasis added)

Now Wiener had been strongly impressed by I. M. Sheffer's 1913 discovery (actually re-discovery of an 1880 unpublished result of C. S. Peirce) that a single connective $\downarrow$ ('neither-nor') suffices for truth-functional sentential logic. For instance, the sentence 'p or q' can be rendered '$(p \downarrow q) \downarrow (p \downarrow q)$', and similarly for 'and', 'not', 'only if', etc. In [20b], in the same spirit, Wiener introduced a single binary operation (which we shall write '$*$') subject to 7 postulates, and showed the equipollence of this system having just one connective $*$ to the usual one for a field $\mathbb{F}$. Wiener's $x*y$ is our familiar $1 - (x/y)$; e.g. $3*5 = 2/5$, $5*3 = -2/3$. But $3*0$ has no meaning, and thus $*$ is not strictly a binary operation for $\mathbb{F}$. (For details, see Note 2, p. 89.)

Wiener's spare time reading in 1919 included the works of the distinguished French mathematician Maurice Frechet, and soon his postulational inclinations started veering towards topology and the newly discovered metric vector spaces. Contacts with Frechet during the Strasbourg International Congress in the summer of 1920 strengthened this trend. Recall that in 1915 Wiener's attempts at a synthetic-logical development of topology had failed (§ 5E). He now tried his hands at a postulational approach.

The usual postulates for the topology of a space X center on the concept of *proximity* (or closeness) or the cognate concept of *closure-point* of a subset E of X, i.e. of a point c in X, which is in E or is such that points of E come arbitrarily close to c. (For instance, in the space X of all numbers, $1/5$ is a closure-point of the infinite set $\{1, 1/2, 1/3, \ldots\}$, but so is the number 0, which is not in the set.) Next, a *topological transformation* (or *homeomorphism*) f of X is defined to be a one-one transformation that preserves the closure concept, i.e. is such that if f transforms E into E' and c into c', and c is a closure point of E, then c' must be a closure-point of E'. Topology then becomes the study of the other invariants of the class Σ of homeomorphisms of X.

In [22c], guided by some work of Frechet, Wiener decided to start with X and Σ as primitive terms, and to subject them to postulates that would make X a topological space and make Σ the set of all its homeomorphisms. This forced him to define the *closure-point* of a subset E of X as a point c for which $f(c) = c$, whenever f is in Σ and $f(x) = x$ for each x in E. Closed sets, were then defined as usual. His objective thereafter was to ensure that Σ is a group of one-one functions on X onto X which carry closed sets into closed sets. Wiener investigated various circumstances in which this requirement is met, citing the specific spaces due to Frechet, F. Riesz and others, then in vogue.

In [22b], evidently written after [22c], Wiener went on to characterize the linear continuum (i.e. the ordinary indefinite straight line viewed as a topological space X) in these terms, departing thereby from the earlier characterizations given by Huntington, Veblen, and R.L. Moore. He laid down 9 axioms on (X, Σ) and proved that this X can be put in one-one correspondence with the real number system $\mathbb{R}$ so that Σ becomes the group of homeomorphisms of $\mathbb{R}$.

Wiener's postulational interests turned next to vector spaces. The important concept of a vector is well explained by Wiener:

> ... an ordinary space of three dimensions contains in itself directed quantities like arrows, which can be added to one another, as for example by traversing the step indicated by one arrow and then the subsequent step indicated by another and by considering this double step as if it were a single step. ... it has long been known that similar geometries exist in spaces of more than three dimensions and, in fact, in spaces of an infinity of dimensions. [56g, p. 55]

Such steps or arrows are called *vectors*. Note that they can be of different lengths. The length of a vector x, also called its *norm*, is denoted by $|x|$. Wiener's postulational work pertained to spaces (i.e. sets) V of normed vectors.

To see the significance of the norm, we must go back to 1905, when Frechet revealed the profundity behind the age-old idea of distance. He introduced *metric spaces*, i.e. spaces X for which it makes sense to speak of a *numerical distance* $d(x, y)$ between any points x, y in X. Frechet showed that the theory of limits, worked out during the 19th century to rigorize the calculus of Newton and Leibniz, extends to metric spaces. The distance (also called *metric*) of course serves as a measure of proximity: the smaller the distance $d(x, y)$, the "closer" are the points x and y. Thus metric spaces are topological spaces, of an attractive sort in which proximity is numerically measurable. Now the Pythagoreans knew that many numbers, such as $\sqrt{2}$ and π are "irrational", i.e. are not equal to fractions m/n, where m, n are integers. Cantor's way of incorporating these irrationals into the number system was to postulate that any numerical sequence $(x_1, x_2, \ldots x_n, \ldots)$ for which $x_m - x_n$ tends to zero, as m, n tend to infinity, has a limit ℓ. Frechet called a metric space X *complete*, when it fulfilled the anologous Cantorian postulate, i.e. when each sequence $(x_1, x_2, \ldots x_n, \ldots)$ in X, for which $d(x_m, x_n)$ tends to zero, has a limit ℓ in X.

Now notice that in a system V of normed vectors, we can introduce the distance d by the formula $d(x, y) = |x - y|$. Thus *normed vector spaces are metric vector spaces*.

What Wiener did in the papers [20e] and [22c] was to consider *affine normed vector spaces* X: to each pair x, y in X corresponds a vector xy in V, where V is a normed vector space of arbitrary dimension.[1] In these papers he gave postulates for V and then for X. Such vector spaces were arousing a lot of curiosity at that time. Only in his paper [23g], however, did Wiener impose the requirement that the metric defined by his norm be complete. S. Banach, a Polish mathematician, later to gain great stature, had, however, anticipated Wiener by a few months in his own paper, which Wiener cited in the paper [23g]. The spaces are therefore called *Banach spaces* (or "espaces du type B"). Wiener dealt with Banach-space valued functions of a complex variable and extended classical results, but he established no deep results of the sort that Banach was proving. His explanation as to why he did not pursue the

---

1   In [22c, § 7] Wiener also defined a local ("im kleinen") affine space, i.e. in current terms, a *locally pre-Banachian topological manifold*, but he did nothing with it beyond citing the sphere and the torus as examples. In [22b] he also introduced "angles" in the space ℓ_2, but did not pursue this idea.

field of abstract spaces, on the importance of which Frechet had assured him, reads (cf. also [56g, pp. 60–64]):

> . . . I left it completely for Banach to open up, as its degree of abstractness struck me as rendering it rather remote from that tighter texture of mathematics which I had found to give me the highest esthetic satisfaction. I do not regret having followed my own judgment in the matter, as there is only a certain amount of work that a mathematician can do in a given time, however he distributes his efforts. It is best for him to do this work in a field that will give him the greatest inner satisfaction. [53h, p. 281]

So let us not start speaking of "espaces du type BW", but just note in passing Professor R.L. Wilder's words:

> The technical and conceptual competence exhibited in these papers leaves no doubt . . . that he (Wiener) would have been highly successful had he chosen to continue. {*Coll. Works*, I, p. 319}

D The Brownian motion

We now come to a very important turning point in Wiener's researches. His primary thrust now shifted from the postulational treatment of mathematical superstructure to the superstructure itself, specifically to the then embryonic area of probabilistic functional analysis. Several factors seem to have been instrumental in affecting this shift which led to much of Wiener's finest work.

Preoccupied as he was with the foundations, his reading in the field of mathematical analysis had been slipshod. But during his spare time in early 1919 he studied the standard treatises of important analysts such as F.W. Osgood, V. Volterra, M. Frechet and H. Lebesgue. These books had come his way accidentally: they were bequests from his sister's fiancé, Gabriel Marcus Green of the Harvard Mathematics faculty, a promising mathematician, but fatal victim to a postwar epidemic. The books were eye-openers to Wiener, and no doubt encouraged his swing toward analysis. Another stimulus in the direction of functional analysis came from his meeting with Maurice Frechet at the Strasbourg Congress in the summer of 1920.

What Wiener needed was a push in a probabilistic direction, and this came from Dr. I.A. Barnett of Cincinnati, a student of E.H. Moore, the American pioneer in abstract analysis. Barnett referred to the growing interest in probabilistic questions in which the events are curves or functions, e.g. the paths traced by a swarm of flying bees or by an assembly of drunkards on the walk, and suggested this as something that Wiener might try. Wiener began trying in earnest. Although at that time the notion that events are measurable sets and probabilities are measures thereon was unknown, Wiener began experimenting with the

Lebesgue integral. His 1913 course with G.H. Hardy, his third great mentor, at last became relevant to his research.

Wiener has described the *Lebesgue integral* and its uses in lay terms as follows:

> ... It is easy enough to measure the length of an interval along a line or the area inside a circle or other smooth, closed curve. Yet when one tries to measure sets of points which are scattered over an infinity of segments or curve-bound areas, or sets of points so irregularly distributed that even this complicated description is not adequate for them, the very simplest notions of area and volume demand *high-grade thinking* for their definition. The Lebesgue integral is a tool for measuring such complex phenomena.
>   · · · · · · · ·
>   The theory of the Lebesgue integral leads the student from the measure of intervals to the measure of more complex phenomena obtained by combining sequences of intervals, and then to sets which can be approached by such sequences, while the sets of points excluded from them can be approached in a similar manner. ... It enabled Lebesgue to extend the notion of length or measure from the single interval to the extreme significant limits at which measure is possible. [56g, pp. 22–23] (emphasis added)

Wiener's intent was to extend Lebesgue's ideas so that they could apply to "intervals" each "point" of which is an erratic curve, such as the orbit of a flying bee. This was too difficult to do directly.

Fortunately, Wiener was aware of the 1919 paper of the British mathematician P.J. Daniell {D1}. Daniell had observed that in essence Lebesgue began with an integral $\int$ defined over a class $\mathscr{L}$ of conveniently chosen "elementary functions" (e.g. the step-functions or the continuous functions), and with these ingredients then built a much larger and mathematically more beautiful class $\mathscr{\bar{L}}$ of functions, and extended the integral, so that the extended integral $\bar{\int}$ over $\mathscr{\bar{L}}$ had richer properties. Daniell gave postulates that exemplified the Lebesgue procedure in the abstract. He considered a system $(\mathscr{L}, M)$, where $\mathscr{L}$ is a vector space of real-valued functions f, g, etc. over a set X, i.e. $\mathscr{L}$ is such that.

$$f, g, \in \mathscr{L} \ \& \ a, b \in \mathbb{R} \quad \text{implies} \quad af + bg \in \mathscr{L},$$

$\mathbb{R}$ being the set of real numbers, and where M is a function on $\mathscr{L}$ to $\mathbb{R}$, having the properties of the elementary integral $\int$. Daniell subjected $(\mathscr{L}, M)$ to a few simple but ingeniously chosen postulates, and proved the existence of an extension $\mathscr{\bar{L}}, \bar{M}$), in which $\mathscr{\bar{L}}$ and $\bar{M}$ share the deep attributes of full-fledged Lebesgue integration.

In [20f] Wiener made a beautiful adaptation of Daniell's work that was to stand him in good stead in his ventures. He started with a sequence π_n, $n = 1, 2, \ldots$, of finite partitions of the set X and a sequence of weightings w_n, $n = 1, 2, \ldots$, such that the w_n assigned a "weight" (i.e. positive number) to each cell of the partition π_n. In terms of these ingredients, which he subjected to simple postulates, Wiener defined both $\mathscr{L}$ and M so that $(\mathscr{L}, M)$ satisfied Daniell's postulates. This gave Wiener the Daniell extension $(\bar{\mathscr{L}}, \bar{M})$, which was able to deliver the goods. (For details on [20f], see Note 3 on p. 90.)

How could this purely mathematical work be brought to bear on the precipitous terrain that Wiener wished to explore? Nothing would have come out of it had not Wiener been able to enlist significantly his physical ideas and intuitions. He had developed such ideas and intuitions, for (cf. § 5D) he had accepted Russell's wise counsel to keep in touch with the frontiers of physics, and had read the work of J. W. Gibbs on statistical mechanisms, the Einstein-Smoluchowski papers on Brownian motion, and J. Perrin's book *Les Atomes*, cf. {P3}. The philosophical tone of the period, set by the *Principia Mathematica* and the work of Einstein, silenced hocus-pocus of the kind heard today about the "barriers" between "pure" and "applied" mathematics. Good physicists freely extolled the virtues of pure mathematics. Thus, what the French physicist and Nobel prize winner J. Perrin wrote in 1913 was music to Wiener's ears:

> Those who hear of curves without tangents or of functions without derivatives often think at first that Nature presents no such complications nor even suggests them. The contrary, however, is true and the logic of the mathematicians has kept them nearer to reality than the practical representations employed by physicists. {P3}

The immediate stimulus for Wiener's work apparently came from his reading Sir Geoffrey Taylor's now classic 1920 paper on *turbulence* {T3}, a step that few pure analysts would have undertaken. Taylor had characterized turbulence by certain averages dependent on the entire movement of the turbulent particle, and Wiener at once saw in this a possible opening for the ideas he had just introduced in [20f]. Wiener's attempts to handle turbulence in this way failed, but being aware of Brownian motion, a phenomenon vaguely akin to turbulence, he began trying his methods on the former. Here, thanks to his mastery of the field and its appropriateness to the mathematics he had devised, he made rapid progress. In a note [21c] submitted to the National Academy of Sciences in 1921 Wiener clearly demarcated his concept of the Brownian motion stochastic process and outlined its beautiful theory.

What is the Brownian movement, and why is it so important? Wiener describes it as follows:

> . . . let us imagine a pushball in a field in which a crowd is milling around. Various people in the crowd will run into the pushball and will move it about.

A. Einstein
(c. 1910)
1879–1955

Some will push in one direction and some in another, and the balance of pushes is likely to be tolerably even. Nevertheless, notwithstanding these balanced pushes, the fact remains that they are pushes by individual people and that their balance will be only approximate. Thus, in the course of time, the ball will wander about the field like the drunken man . . . and we shall have a certain irregular motion in which what happens in the future will have very little to do with what has happened in the past.

Now consider the molecules of a fluid, whether gas or liquid. These molecules will not be at rest but will have a random irregular motion like that of the people in the crowd. This motion will become more active as the temperature goes up. Let us suppose that we have introduced into this fluid a small sphere which can be pushed about by the molecules in much the way that the pushball is agitated by the crowd. If this sphere is extremely small we cannot see it, and if it is extremely large and suspended in a fluid, the collisions of the particles of the fluid with the sphere will average out sufficiently well so that no motion is observable. There is an intermediate range in which the sphere is large enough to be visible and small enough to appear under the microscope in a constant irregular motion. This agitation, which indicates the irregular movement of the molecules is known as the Brownian motion.

J.B. Perrin
1850–1942
Courtesy of
Dr. Francis Perrin

It had first been observed by the microscopists[2] of the eighteenth century as a universal agitation of all sufficiently small particles in the microscopic field. [56g, pp. 37, 38]

The Brownian movement occupies a very important place in the history of science, for it was the study of this movement, primarily by Einstein and the subsequent experimental verification of his theory by Perrin, that transformed the molecular-statistical theory of the world, ensuing from the work of Maxwell and Boltzmann, from a tenuous hypothesis into a well-confirmed one.

Surprisingly, Einstein was unaware that Brown had revealed this movement; he predicted its existence on purely theoretical grounds in his

2   The most careful among these microscopists was the British botanist Robert Brown, after whom the motion is named. In 1828 he observed the irregular "swarming" motion of pollen grains immersed in water. So persistent was this motion that he first thought that the grains were "live matter". He even assigned to them the male sex. He abandoned this view when he discovered that non-organic particles exhibit the same kind of movement.

1905 paper, stating that "if the movement discussed here can actually be observed . . . an exact determination of atomic dimensions is then possible" {E2, pp. 1, 2}. From the atomic hypothesis and the principle of equipartition of energy, Einstein deduced that at temperature T the mean-square displacement, $\bar{d}_t$, during a time-interval of length t, of a colloidal particle of radius a (such as Brown's pollen grain) is given by

$$\bar{d}_t^2 = \frac{RT}{3\pi a\mu N} \cdot t, \quad \{\text{E2, p. 81, (31)}\} \tag{1}$$

where R is the constant of the gas equation ($pv = RT$), μ is the viscosity of the liquid and N is Avogadro's number.[3] From this formula, N is expressible in terms of experimentally observable quantities and known constants. The relevant experiments were made by several physicists, among them J. Perrin. These yielded a value of N which matched its value obtained earlier by the entirely different method of diffusion of ions in a gas due to an electric field. This was a brilliant confirmation not only of the soundness of Einstein's research, but of the molecular-statistical theory itself.

This illustrious history may have encouraged Wiener to plunge into the study of the Brownian movement. But with his philosophical training, concern with sensation-intensities and their measurement (witness his papers [15a, 21a]) he must also have sensed the ubiquitous character of the Brownian movement. Not just the pollen grains get kicked around by the ceaseless molecular motion, so too do the electrons, whose flow in electrical apparatus control the transmission of messages, and the colloids, whose flow in biological organisms control life and mind. This realization of a randomness in the very texture of Nature profoundly affected Wiener's life work, as we shall see in this book.

Einstein had assumed that there is a positive number τ such that a time-interval of length τ is, in his words

> . . . . very small compared to the observed interval of time, but nevertheless, . . . such . . . that the movement executed by a particle in consecutive intervals of time τ are . . . mutually independent phenomena. {E2, p. 13}

From this premise he derived the result that the displacements in disjoint intervals are normally distributed, that "the mean (square) displacement

---

3   N is defined as the number of atoms in 16 grams of oxygen (or in w grams of any element, where w is its atomic weight). $N = 6.023 \times 10^{23}$ approx.

is ... proportional to the square root of the time" {E2, p. 17}, being given by equation (1).

Wiener's concern, unlike Einstein's, was with "the mathematical properties of the curve followed by a single particle" [56g, p. 38]. He therefore made the idealization that Einstein's conditions prevail for all positive lengths τ. To this idealized Brownian motion, "an excellent surrogate for the cruder properties of the true Brownian motion" [56g, p. 39], Wiener was able to apply the theorem proved in [20f]. In his first and remarkable paper [23d] Wiener followed a different approach, but the nexus with [20f] is clear from § § 3, 4 of the paper [24d]. Here Wiener defined the sequence $(\pi_n, w_n)_{n=1}^{\infty}$ with $X = [0, 1]$, so that it not only fulfills the premises of the [20f] theorem, but the extensions $(\mathscr{L}, \bar{M})$ of the resulting $(\mathscr{L}, M)$ also have the following additional properties. Write

$$\left\{ \begin{array}{l} \mathscr{X} = \{x: x \text{ is continuous on } [0,1] \text{ to } \mathbb{R} \ \& \ x(0) = 0\} \\ \mathscr{B}(\mathscr{X}) = \{B: B \text{ is a Borel subset of } \mathscr{X}\}. \end{array} \right. \quad (2)$$

Then $\mathscr{L}$ is the class of $\mathscr{B}(\mathscr{X})$ measurable functions on $\mathscr{X}$ to $\mathbb{R}$, and writing $\mu(B) = \bar{M}(1_B)$, where 1_B is the function with value 1 on B and 0 outside, μ has all the properties of Lebesgue measure. Thus, indeed, Wiener achieved his goal of assigning a measure μ to Borel sets B of functions or curves, just as Lebesgue had earlier assigned a length to Borel subsets of points of the line $\mathbb{R}$, and an area to Borel subsets of points of the plane $\mathbb{R}^2$. This μ is called *Wiener measure*. For more on [24d], and the terms "Borel" and "measurable", see Note 4, p. 90.

Excellent comments on the papers [21c, 21d, 23d, 24d] on the Brownian motion by K. Ito (*Coll. Works*, I), M. Kac {K1} and J. Doob {D6} are available, cf. also E. Nelson {N2}. However, so overwhelming have been the effects of this work on the development of analysis and probability theory, and later on of communication theory, that a little more must be said.

It was only in the 1930s that Wiener was able to unearth, and convey rigorously what is buried in these papers. By mapping the space $\mathscr{X}$, (cf. (2)) onto the interval [0, 1] so that the Lebesgue measure of the set corresponding to B equals the Wiener measure of B, Wiener characterized the Brownian motion as the stochastic process $\{x(t, \alpha): t \in [0, 1], \alpha \in [0, 1]\}$, governed by the conditions: (1) the increments $x(b, \cdot) - x(a, \cdot)$ are normally distributed random variables with mean zero and variances $\sigma^2 (b - a)$, i.e.

$$\int_0^1 |x(b, \alpha) - x(a, \alpha)|^2 d\alpha = \sigma^2(b - a), \quad \sigma = \text{const.}, \quad (3)$$

the abstract formulation of Einstein's equation (1); (ii) for non-overlapping intervals $[a, b]$, $[c, d]$ the increments $x(a, \cdot) - x(b, \cdot)$ and $x(c, \cdot) - x(d, \cdot)$ are stochastically independent. Equivalently we may characterize it as the process for which $x(t, \cdot)$ is normally distributed with zero mean and such that

$$\int_0^1 x(s, \alpha)x(t, \alpha)d\alpha = s \wedge t, \quad s, t \in [0, 1], \tag{4}$$

where $s \wedge t$ means the minimum of s and t.

Wiener showed that for almost all α in $[0, 1]$, the trajectories $x(\cdot, \alpha)$ are continuous everywhere but differentiable nowhere, thus confirming what Perrin's microscopic observations had suggested. Although the functions $x(\cdot, \alpha)$ are of "extreme sinusoidity", and non-rectifiable (i.e. even tiny pieces have infinite length), Wiener was able to define for any f in $L_2[a, b]$, a "Stieltjes" type integral

$$g(\cdot) = \int_a^b f(t)dx(t, \cdot) \quad \text{on } [0, 1], \tag{5}$$

and to enunciate the beautiful properties of the new random variables $g(\cdot)$ so obtained. Thus Wiener opened up the whole area of probability theory we nowadays call *stochastic integration*. These developments occur in the later works [33a, 34a, 34d] done in collaboration with Paley and Zygmund.

In the theory of stochastic processes the idealized Brownian motion occupies a central position, as the efforts of several mathematicians over the last fifty years have shown. Wiener may have surmised this, but he did *not* realize that his Brownian motion also penetrates deeply into the non-stochastic parts of mathematical analysis. A very interesting example of this is afforded by the initial value problem of the (one-dimensional) heat equation with potential term V:

$$u(\cdot, 0) = \varphi(\cdot) \quad \text{on } (-\infty, \infty),$$

$$\frac{\partial u}{\partial t} = \frac{1}{2}\frac{\partial^2 u}{\partial t^2} - V(x,t)u, \quad t \geq 0.$$

The solution $u(\cdot, \cdot)$ of this is given by

$$u(x,t) = \int_{C[0,\infty)} e^{-\int_0^t V\{x + f(\tau), t - \tau\}d\tau} \cdot \varphi\{x + f(t)\}w(df),$$

where $w(\cdot)$ is Wiener measure. (The dummy variable f refers to the path of a Brownian particle, now over the time-interval $[0, \infty)$, not $[0, 1]$.) This was discovered in 1948 by Professor Mark Kac under the stimulus of R.P. Feynman's cognate result in non-relativistic quantum theory, and has led to widespread use of functional integration in both mathematical analysis and field physics.

From a topological standpoint the graph of a function f in the class $\mathscr{X}$, being a continuous curve, is a one-dimensional manifold. But the graph of a typical f is extremely crinky: even a tiny segment has infinite length. In 1919, the German topologist F. Hausdorff introduced a concept of "dimension of a set" that reflects its irregularity. In the 1970s it was proved that this "Hausdorff dimension" of the typical path of the Brownian movement is 3/2, and so exceeds its ordinary dimension 1. (See Note 5 on p. 91.)

Since 1950 erratic sets akin to graphs of the Brownian functions have been encountered repeatedly by Dr. Benoit Mandelbrot of IBM and other workers in their researches into phenomena marked by intrinsic irregularities that persist even as we improve the accuracy of the scale of observation. Such are the jagged lines of cracks in rock filaments, for instance. To deal with such irregularities, Dr. Mandelbrot has singled out sets whose Hausdorff dimension exceeds the topological dimension, calling them *fractals* {M3, M4}, but it seems clear that this definition is too restrictive. Roughly speaking, fractals emerge after an infinite number of iterations of a step which involves breaking up a set as well as changing the scale. Postulates embodying these operations, with "fractal" as a primitive term, would seem to offer a good approach to this subject. Some significant hints as to such postulation were given in Professor James Cannon's 1982 lectures {C2}, unfortunately still unpublished. Cannon found cognate ideas in the theory of topological manifolds, especially of the so-called "hyperbolic type", that are quite useful. Thus Wiener's idealized Brownian motion has turned out to be the progenitor of a growing variety of fractals encountered in physics as well as in pure mathematics.

In the mid 1920s Wiener became conscious of the then discovered *shot effect* (Schrotteffekt) in electrical amplification, cf. [26b, § 6]. Einstein had surmised that thermal agitation of electrons in a conductor would give rise to random fluctuations of the voltage difference between the terminals, and in 1918 W. Schottky had worked out the effect this would produce in amplifiers. But this effect ("tube noise") was too small to be detected by the instruments then available. The predicted shot effect was experimentally verified only around 1927 when valve amplification became possible. Wiener saw in the shot effect another application of his work:

> This shot effect not only was similar in its origin to the Brownian motion, for it was a result of the discreteness of the universe, but had essentially the same mathematical theory. [56g, p. 40]

Despite their depth and wide scope, Wiener's papers on Brownian motion did not create any stir in the mathematical world. The leaders of American mathematics at that time were G.D. Birkhoff of Harvard, a Poincaré analyst, and Oswald Veblen of Princeton, a topologist. They were interested in physics, but primarily in the then exciting areas of general relativity and quantum theory; their horizons did not include the Brownian motion, still less Wiener's idealized Brownian motion. Another delimiting factor lay in the unclear presentation in the papers themselves. Professor Mark Kac has recounted his "depressing lack of success" in understanding them in his student days [K1, p. 55]. Only Paul Lévy in France was able to digest them because of his own active interest in the subject, and so perhaps was A. Khinchine in the Soviet Union.

Only after the dissemination of the pioneering work of R. von Mises and A. N. Kolmogorov on the foundations of probability theory in the 1930s was the importance of Wiener's work on the idealized Brownian motion appreciated. But Kolmogorov's epoch-making systematization of 1933 {K5} was itself a part of the harvest ensuing from Wiener's pioneering effort ten years earlier. In his preface to his monograph {K5} Kolmogorov calls attention to certain novel points of his presentation, citing among these *probability distributions in infinite dimensional spaces*, and emphasizing that these new problems "arose, of necessity, from some perfectly concrete physical problems". His allusion to his own paper with Leontovich entitled "Zur Berechnung der mittleren Brownschen Fläche" suggests that Wiener's work played a part in shaping Kolmogorov's train of thought. Moreover, the Wiener measure over the infinite dimensional space C[0, 1], is not a mere special case of the general Kolmogorov measure over such spaces, for latent in it was an idea that went beyond Kolmogorov's formulation. As Professor K. Ito has pointed out:

> Wiener's success in his rigorous definition of Brownian motion lies mainly in his remark that Wiener measure is concentrated in a compact family of continuous functions with arbitrarily high percentage. This idea, now called *tightness*, has become a basis of the modern theory of probability measures on function space . . . {*Coll. Works*, I, p. 518}

This tightness idea, buried in Wiener's work, was formally introduced in a general setting by the Soviet mathematician Yu. Prokhorov only in the 1950s.

E　Theory of the potential

There was another interesting and important phase of Wiener's activity during the early 1920s. He often consulted O. D. Kellogg, the Harvard authority on potential theory. We do not know what provoked Wiener's interest in this subject, which falls in the area of partial differential equations in which he had little background. But within a very short time his conversations with Kellogg brought him to the frontiers of this subject. Wiener then wrote six papers within a space of three years which revolutionized the field. In the words of the French authority M. Brelot, he "initiated a new period for the Dirichlet problem and potential theory" {B19, p.41}.

The concept of potential came into vogue a generation or two after Newton, when the idea of *energy* began to loom on the scientific horizon. Lagrange explicitly introduced the concept of *potential energy* into Newtonian mechanics, and demonstrated its considerable vitality both for dynamics and for the theory of gravitational attraction. It was soon realized that the cognate ideas of *temperature, pressure* and *voltage* play analogous roles in the field theories of heat, fluid mechanics and electricity. This led to the development of a general theory of the *potential* by the mid-19th century at the hands of Laplace, Poisson, Gauss and Green. Thereafter the subject has attracted the attention of some of the best minds.

A central problem in the subject is the so-called *Dirichlet problem*. Imagine a room or chamber R bounded by a very thin wall, i.e. by a material surface S. Different points s on the wall S are maintained at fixed temperatures $\varphi(s)$ by continuous thermal contact with different heat reservoirs at these fixed temperatures, but there are no heaters, refrigerators or reservoirs inside the room R itself. The temperature differences on S will cause a flow of heat along the wall and across the room. After a while, this flow will become *steady*, i.e. the temperature $u(r)$ at any point r in R will not change with the passage of time. The Dirichlet problem is to determine this steady-state temperature distribution $u(\cdot)$ in the room R, given its distribution $\varphi(\cdot)$ on the wall S. The problem is of obvious interest in thermal engineering, as are the cognate problems for voltage and pressure in electrical engineering and hydraulics. From the laws governing these fields it is known that $u(\cdot)$ must satisfy Laplace's equation

$$\Delta u = \frac{\partial^2 u}{\partial x^2} + \frac{\partial^2 u}{\partial y^2} + \frac{\partial^2 u}{\partial z^2} = 0 \tag{1}$$

at each point $r = (x, y, z)$ in R. Thus the Dirichlet problem, formulated mathematically, is: given $\varphi(\cdot)$ on S, find the $u(\cdot)$, satisfying (1) on R, and such that for any s in S, $u(r)$ tends to $\varphi(s)$ as r tends to s.

The solution to the problem, which depends on both the shape of the boundary S and the boundary-distribution φ, was known for smooth surfaces S and continuous functions φ. But, as Wiener learned from Kellogg, the issues are extremely complicated when the surface S has sharp dents and corners, and had remained unresolved. This mathematical complexity is reflected in the physical instabilities which occur when chambers have such crooked surfaces, as Wiener noted. Speaking of electrostatics, he reminds us:

Around such a sharp point the air is continually breaking down as an insulator, and if the field is large, a distinct corona effect will be seen in the dark. Many sailors have observed the curious effect known as the corposant, where nails and other pointed objects glow with a ghastly light in the electrified atmosphere of a thunderstorm. It is through something like this corona effect that a lightning rod relaxes the potential gradient of the charged atmosphere about it, by a gradual and unspectacular process, before the tensions build up to such a point that they may cause a disastrous stroke. [56g, pp. 81]

Wiener's great contribution was to show that no matter how rough the surface S, the Dirichlet problem has a "solution" in a genuine but non-classical sense, and to introduce several ideas of lasting value to accomplish this. As this work is very technical and has been commented on in the special *Bulletin* {A2} and in the *Coll. Works*, I, it will suffice to say just the following. In his papers Wiener introduced the now central and crucial concepts of *capacity* for arbitrary sets and a *generalized solution* for the Dirichlet problem as well as the criterion of regularity of the solution. To solve the Dirichlet problem, Wiener and H. B. Phillips considerably advanced the finite-difference technique of solving partial differential equations that has become standard forty years later with the advent of computers. Indeed the technique is central to the very method of electronic computation that Wiener was to propose, cf. § 13D.

An important link beween Wiener's generalized solution of the Dirichlet problem and his (earlier) idealized Brownian movement, which Wiener completely missed, was discovered by S. Kakutuni in 1944 {K2}. Let R be a bounded (open) region in the plane. Given (x, y) in R. take two independent Brownian motions $X(t, \alpha)$, $Y(t, \alpha)$, $t \geq 0$ and α in $[0, 1]$, starting at (x, y):

$$X(0, \alpha) = x, \quad Y(0, \alpha) = y, \quad \alpha \in [0, 1],$$

and let $\tau(x, y, \alpha)$ be the first time that this planar Brownian curve $(X(\cdot, \alpha), Y(\cdot, \alpha))$ leaves R. Then the generalized solution of the Dirichlet problem

$$\Delta u = 0 \text{ on } R, \quad u = \varphi \text{ on } B$$

where φ is continuous on the boundary B of R is given by

$$u(x, y) = \int_0^1 \varphi[X\{\tau(x, y, \alpha), \alpha\}, Y\{\tau(x, y, \alpha), \alpha\}]d\alpha.$$

This discovery has initiated a whole new approach to potential theory, which is being actively pursued.

It would seem from this far reaching work of Wiener and his followers that the omnipresent character of the Brownian movement in nature is reflected in the ubiquity of the idealized Brownian movement in the purely mathematical realms of analysis and probability.

Note 1: **Boolean algebra** (cf. § 7C).

In [17a] Wiener showed that the general binary operation ⊕ will satisfy Huntington's very axioms, if and only if in Boole's expression $d = 0$, $c = b$ and $a = 1$; with such a choice for a, b, c, d, the expression for ⊕ reduces to

$$x \oplus y = (x \cap y) \cup \{b \cap (x \cup y)\}.$$

For this ⊕, treated as the union, the universal elements 1, 0 become b, $\bar{b}$ and the dual operation ⊙ (corresponding to ∩) is given by

$$x \odot y = (x \cap y) \cup \{\bar{b} \cap (x \cup y)\},$$

but the new complementation is still $^-$. Wiener then showed that the one-one correspondence

$$x \to x' = (a \cap \bar{x}) \cup (\bar{a} \cap x)$$

is an isomorphism of $\mathbb{A}$, if and only if (with an obvious notation)

$$x \oplus' y = (x \cap y) \cup \{\bar{a} \cap (x \cup y)\},$$

so that $1'$, $0'$ are now $\bar{a}$, a. In this way Wiener was able to sum up the invariant theory of Boolean algebras.

Note 2: **Algebraic field** [20b] (cf. § 7C).

With his postulates on the sole binary operation∗, Wiener showed that 0 and 1 are uniquely defined in $\mathbb{F}$ by $x*x$ and $0*y$, $y \neq 0$. He defined multiplication for (non-zero) x, y, by

$$x \cdot y = \left(1*\{[\{(1*y)*1\}*x]*1\}\right)*1.$$

Next came subtraction defined by $x - y = x \cdot (y*x)$, and lastly addition: $x + y$ was defined to be the unique element s such that $s - y = x$. From the results of his earlier paper [20a], it follows that his ∗ is only one of a family of connectives ∗ of the form

$$x*y = \frac{A + Bx + Cy + Dxy}{E + Fx + Gy + Hxy}$$

with rational number coefficients, that do the same job. In [20c,d, 21b], Wiener defined two connectives ∗ and # to be *equivalent*, if each could be had by an iterated application of the other, and he studied conditions for equivalence, including in this the *complex field*. This last field, denoted by $\mathbb{C}$, comprises all numbers of the form $x + iy$, where x, y are ordinary real numbers, and $i = \sqrt{(-1)}$.

Note 3: **The Brownian movement** (cf. § 7D).

As stated earlier, in [20f] Wiener started with a sequence $(\pi_n, w_n)_{n=1}^{\infty}$, where π_n is a finite partition of a fixed set X, and w_n is a function on π_n to $\mathbb{R}_+$. He then took the class

$$\mathscr{L} = \{F : F \in \mathbb{R}^X \ \& \ \exists n \geq 1 \ni F \text{ is } \pi_n\text{-simple}\},$$

and for any F in $\mathscr{L}$, defined its mean-value $M(F)$ by

$$M(F) = \{ \sum_{t \in \Delta \in \pi_n} F(t) w_n(\Delta)\} / \{ \sum_{\Delta \in \pi_n} w_n(\Delta)\}.$$

Wiener assumed that π_{n+1} is a refinement of π_n, and w is finitely additive. Then $M(F)$ becomes independent of n, and is unique. Wiener subjected the π_n, w_n to further conditions (Kolmogorov's marginal conditions in essence) and showed that $(\mathscr{L}, M)$ then fulfills Daniell's conditions. Consequently, there is an extension $(\bar{\mathscr{L}}, \bar{M})$, $\mathscr{L} \subseteq \bar{\mathscr{L}}$, $M \subseteq \bar{M}$, $\bar{\mathscr{L}}$ being the class of "summable" functions and $\bar{M}$ the "Daniell" integral. Wiener showed that F is in $\bar{\mathscr{L}}$ if F is "uniformly continuous", i.e. inf.$_{n \geq 1}$sup$_{\Delta \in \pi_n}$ Osc(F, Δ) = 0. This work [20f] provided a firm mathematical footing for Wiener's ventures into physics.

Note 4: **The "Borel" terminology** (cf. § 7D).

First, as to the "Borel" terminology, for a topological $\mathscr{X}$, the class $\mathscr{B}(\mathscr{X})$ of *Borel subsets* of $\mathscr{X}$ is defined to be the smallest of the classes $\mathscr{F}$ such that

  (i) every closed subset of $\mathscr{X}$ is in $\mathscr{F}$;
 (ii) if $A \in \mathscr{F}$, then the complement $A' \in \mathscr{F}$;
(iii) if $A_1, A_2, \ldots, A_n, \ldots \in \mathscr{F}$, then the infinite union
       $A_1 \cup A_2 \cup \ldots \cup A_n \cup \ldots$ is in $\mathscr{F}$.

A function F on $\mathscr{X}$ to $\mathbb{R}$ is called $\mathscr{B}(\mathscr{X})$ *measurable* (or *Borel measurable*) if for $a, b \in \mathbb{R}$ with $a \leq b$, we have

$$\{x : x \in \mathscr{X} \ \& \ a \leq f(x) \leq b\} \in \mathscr{B}(\mathscr{X}).$$

Practically every function encountered in ordinary mathematical work is Borel measurable.

Wiener's ingenious definitions of π_n and w_n in [24d] are complicated. For this book the following general description should suffice. Each π_n is a partition of the function-space $\mathscr{X}$ in (2). Each π_n has $(2^n)^{2^n}$ cells. Each cell Δ in π_n is characterized by 2^n points $t_1, t_2, \ldots, t_{2^n}$ in [0, 1] and 2^n intervals $I_1, I_2, \ldots, I_{2^n}$ scattered over $\mathbb{R}$; the members of Δ are those f in $\mathscr{X}$ for which $f(t_k)$ lies in I_k for $k = 1, 2, \ldots, 2^n$. To this cell Δ, the weighting w_n assigns a weight based on the normal probability distribution found by Einstein.

Note 5: **Hausdorff dimension** (cf. § 7D).

The result that the Hausdorff dimension of the graph of a typical Wiener Brownian function f is $3/2$ is due to S. Orey in 1970. This result has been significantly generalized.

First, the Wiener Brownian motion can be defined for parameter-domains in $\mathbb{R}^p$, and taking values in $\mathbb{R}^q$. For $p > 1$ and $q = 1$, one has only to replace the RHS of (4) by $\Pi_1^p(s_i \wedge t_i)$ For $q > 1$, one then defines the motion as a vector of q stocastically independent $\mathbb{R}$-valued Brownian motions. In recent years the irregularities of these Brownian curves and surfaces have been subjected to very fine analysis. Write $X_{pq}(\alpha)\,(t) = x\,(t,\alpha)$, when $t \in \mathbb{R}^p$ & $x(t,\alpha) \in \mathbb{R}^q$. L. Tran {T6, T7} (1977) has proved that for almost all α,

$$\dim(\text{range } X_{p,\,q}(\alpha)) = \min\{2p,\,q\}$$

$$\dim(\text{graph } X_{p,\,q}(\alpha)) = \min\{2p,\,q+\frac{1}{2}\}$$

where dim is the Hausdorff dimension.

8 The Allotment of National Income, Marriage, and Wiener's Academic Career

In his work on potential theory, unlike that in Brownian motion, there were other persons pursuing the same ends as Wiener. There were Kellogg and his students, Lebesgue and his pupil Bouligand in France, and many others. At certain moments the egoist in Wiener took over from the mathematician and the philosopher. An unfortunate conflict occurred that made Wiener feel "thoroughly sick and discredited", and suffer an attack of bronchial pneumonia. A somewhat unobjective account of this episode is to be found in [56g, pp. 83–85]. The following brief and accurate statement of Professor Levinson indicates how the egoist in Wiener made matters worse by exaggerating the importance of the episode:

> Unfortunately there arose a problem of timing his publications so as not to compromise the theses of two of Kellogg's students. Actually the matter resolved itself satisfactorily although this is not made clear in Wiener's account. Wiener unfortunately felt that Kellogg had been unfair to him. Thus instead of feeling indebted to Kellogg he again felt that Harvard was mistreating him. {L8, p. 15}

Wiener's competitiveness, as revealed by this incident, had something to do with his slow promotion in the Mathematics Department at MIT. Between 1919 and 1924 Wiener did some of the best mathematical work of this century. During this period he received considerable human help from his MIT colleagues, but all he got officially from MIT, in 1924, was an assistant professorship without tenure. One might explain this circumstance by saying that it happened in an era of generally slow promotions, before the advent of President Karl Compton with his enlightened policy on the status of science at MIT, and that Wiener's pioneering work on Brownian movement was about ten years ahead of its time and few understood it.

To say this, however, is to say little, for in reality tension over promotion looms large because the scientists have to share a tiny fragment of the national income, consequent to its large-scale diversion in support of inefficient bureaucracies, substandard and questionable manufacture, and pseudointellectual culture. Unfortunately, Wiener did not analyze the problem objectively, and see it as the mal-allocation of national income. He harped on his not receiving offers from other institutions such as Princeton and Chicago. This last fact he attributed rather dubiously, on hearsay evidence, to lack of support from Harvard professors, in particular from G.D. Birkhoff.

When not personally involved, Wiener was capable of analyzing similar situations objectively, and often with penetrating wit. For example, speaking of the dangers which scientists face when the public becomes aware of the social utility of their work, he wrote:

> Emerson did not tell the whole truth about the fate of the man who devises a better mousetrap. Not only does the public beat a path to his door, but one day there arrives in his desecrated front yard a prosperous representative of Mousetraps, Inc., who buys him out for a sum that enables him to retire from the mousetrap business, and then proceeds to put on the market a standardized mousetrap, perhaps embodying some of the inventor's improvements, but in the cheapest and most perfunctory form which the public will swallow. Again, the individual and often delicious product of the old small cheese factory is now sold to the great cheese manufacturers, who proceed to grind it up with the products of a hundred other factories into an unpleasant sort of vulcanized protein plastic. [53h, p. 256]

But despite his tremendous imagination, Wiener was singularly incapable of relating the injustice he suffered to injustice in-the-large. This lover of *Alice in Wonderland* could not visualize his slow promotion and small emoluments as another facet of the very process that turned tasty cheese into "protein plastic". He also seemed to forget that subjection to injustice put him in the company of Socrates, Mozart, Cantor, Peirce, and others. Perhaps Russell never told him about the economics of the publication of the *Principia Mathematica* that set many like Wiener off on their researches. Be it noted that the resolution of the theoretical difficulties required seven years of intense effort (1900–1907) and the final writing three more years (1907–1910). What happened when Russell and Whitehead finally delivered the enormous manuscript to the Cambridge University Press is best recounted in Russell's words:

> The University Press estimated that there would be a loss of £600 on the book, and while the syndics were willing to bear a loss of £300, they did not feel that they could go above this figure. The Royal Society very generously contributed £200, and the remaining £100 we had to find ourselves. We thus earned minus £50 each by ten years' work. This beats the record of *Paradise Lost*. {R8, Vol. 1, p. 229}

Thus the intellectual rockets that uplifted Wiener's research and uplifted English poetry were products of injustice. The authors of Edwardian, Elizabethan and Tudor gossip reaped rich rewards as do our pornography queens. "Values are for the birds," said the devil with a chuckle.

While Wiener's confusion is not understandable, his quest for economic security certainly is. It was motivated to a large extent by his growing emotional attachment to Margaret Engemann, a young lady of German descent. She was a student in Leo Wiener's course on Russian literature. Later she became an assistant professor of modern languages at Juniata College in Huntingdon, Pennsylvania, and Wiener's courtship was intermittent and lasted for a few years. This was partly due to economic reasons. In 1926 at last, as Wiener wrote,

> ... the recognition I was receiving from Germany together with an improved economic status at MIT consequent upon it, now for the first time made it possible for me to look the responsibilities of marriage in the face. [56g, p. 110]

Norbert and Margaret were married in Philaldelphia in the spring of 1926, but had to part soon as he had to proceed to Göttingen and she to Huntingdon. Margaret was only able to join him in Göttingen in the summer. The deep need he had felt over the years for enduring feminine companionship, to which he has alluded in [56g, pp. 86, 87] and [53h, pp. 283–], was met at last. Margaret's stabilizing influence on Wiener both in his relationship with his parents and colleagues, and in his emotional life can hardly be exaggerated. His dedication in [53h] reads "To my wife, under whose gentle tutelage I first knew freedom". His "new personality as a married man" eased his social life considerably. [56g, p. 128]

Soon thereafter Wiener became "a very clumsy pupil in the art of baby sitting" and a family man [56g, p. 127]. His daughter Barbara was born in 1928 and his daughter Peggy in 1929. The two children, now Mrs. Barbara Raisbeck and Mrs. Peggy Kennedy, have families of their own. Raisbeck, an engineer, is an executive with A.D. Little Co., and a consultant. Kennedy is an official in the New York State Government. Wiener was a devoted and affectionate father. His autobiography [56g] describes many an intimate contact with his children. They shared much of his extensive travels and were exposed to schooling in several countries and cultures. Peggy even "collaborated" with her father, but only in a piece of mystery fiction. (See the letter to Alfred Hitchcock of Hollywood, page 339.)

The "improved economic status" that Wiener had anticipated did not materialize. In 1927, still an assistant professor, he decided to answer advertisements in the British magazine, *Nature*, for professorial positions at London and in Australia. His application for the Chair in Mathematics at the University of Melbourne in 1928 was backed by strong supporting letters from several of the world's leading mathematicians, among them C. Carathéodory, G.H. Hardy, D. Hilbert and

Marguerite Engemann
in 1926, the year of
her marriage to
Norbert Wiener

O. Veblen, but nothing came of it. Finally, in 1929 MIT promoted him
to an associate professorship. He might have remained in this intermediate category for another eight to ten years, were it not for the fact
that his forthcoming researches, although inspired by the needs of
engineering, put him in command of a well-established purely mathematical area, head and shoulders above those inside the establishment.
This led to his promotion to a full professorship in 1932.

In this last research Wiener's scholarly pen reigned supreme:
nothing was forced and egoism was gone. We must now turn to this
phase of Wiener's life, in which he again became a true practitioner in
the Pythagorean-Platonic tradition.

9 From Communications Engineering to Generalized Harmonic Analysis and Tauberian Theory

A Harmonic analysis

Harmonic analysis, for all its modern ramifications, has a history going back to Pythagoras and his interest in music and the vibrations of strings in the lyre. [56g, p. 105]

Wiener's mathematical ideas came from the free impingement on his mind of the scientific and technological changes that occurred around him. Wiener needed congenial intellectual contacts with individuals outside mathematics. In this regard he was very fortunate in finding at MIT a forward-looking electrical engineering department. The leaders of this department were the professors Dougald C. Jackson in the early 1920s and Vannevar Bush later on. Wiener, who as we saw in Chapter 3, had an early flair for things electrical, got along splendidly with the engineers who often sought his advice on mathematical methodology.

Wiener's fruitful interaction with the engineers came from his ability to express his deeply felt intellectual predilections in engineering terms. Central to his scientific thought were the notions of *pure tone* and *orthogonal expansion*, which constitute the core of the subject known as *harmonic analysis*. A general appreciation of harmonic analysis is essential to understanding Wiener's work with the engineers, and to most of his other work as well.

The analysis of a complex phenomenon or structure in terms of its basic constituents, and the synthesis of the former from the latter are endeavors common to the sciences, to the arts and to practical life. Central to them is the idea that *organization involves the recurrence of certain fundamental patterns*. We have *harmonic analysis and synthesis* in situations in which the basic constituents or fundamental patterns are *pure tones*. As Wiener suggests, such analysis and synthesis goes back to the Pythagorean analysis of musical chords and of the relation of musical pitch to the length of vibrating strings. But the subject really got underway with the acoustical advances of the 18th century, when the pure tones were identified with the sinusoidal functions $\cos \lambda t$, $\sin \lambda t$, of different frequencies λ, their very good acoustical realizations being given by the sounds of tuning-forks.

Here is how Wiener explains in lay terms the notion of the sinusoidal curve and that of harmonic analysis:

> ... let us suppose that we have a drum of smoked paper turning around and let us further suppose that we have a tuning fork vibrating parallel to the axis of the drum, and that to the end of this tuning fork is attached a straw which will make a white mark on the smoked paper. As the drum revolves at constant speed, the straw will leave an extended mark which we call a sinusoid.
>
> Let us now consider more complicated curves, made up by adding sinusoids. It is possible to add curves to one another by adding their displacements, that is to say, by combining two tuning forks of different rates of oscillation so that they both act on the same straw as it traces its path along a drum of smoked paper. In this motion we can observe two or more rates of oscillation in the same curve at the same time. The study of how to break up various sorts of curves into such sums of sinusoids is called harmonic analysis. [56g, p. 76]

Landmark contributions to harmonic analysis came from the great French mathematician Jean-Baptiste Fourier in the 1820s. A fundamental theorem of his asserts that periodic curves "can be broken into a number of separate sinusoids which repeat themselves at different rates", to use Wiener's words. [56g, p. 76]. A function f on $\mathbb{R}$ such that for each t, the values $f(t)$ and $f(t + T)$ are equal is called *periodic with period T*. The sinusoids $\cos t$, $\sin t$ have the period 2π, and obviously $\cos(\pi t/T)$, $\sin(\pi t/T)$ have the period $2T$. Fourier's theorem tells us that if f is a smooth real-valued periodic function on $\mathbb{R}$ with period $2T$, then it has an infinite series expansion

$$f(t) = \frac{a_o}{2} + \sum_{k=1}^{\infty} \left\{ a_k \cdot \cos\left(\frac{k\pi t}{T}\right) + b_k \cdot \sin\left(\frac{k\pi t}{T}\right) \right\}. \tag{1}$$

Knowing f we can find the coefficients a_k, b_k from the formulae

$$a_k = \frac{1}{T} \int_{-T}^{T} f(t) \cos \frac{k\pi t}{T} \, dt, \qquad b_k = \frac{1}{T} \int_{-T}^{T} f(t) \sin \frac{k\pi t}{T} \, dt. \tag{2}$$

In many an electrical application the values $f(t)$ are complex numbers $x + iy$, where x, y are in $\mathbb{R}$ and $i = \sqrt{(-1)}$, cf. § 7C. It is very convenient to replace the two real numbers $\cos \lambda t$, $\sin \lambda t$, by the single complex number

$$\cos \lambda t + i \sin \lambda t, \qquad \text{briefly } e_\lambda(t) \text{ or } e^{i\lambda t}. ^{[1]} \tag{3}$$

---

1 This last expression $e^{i\lambda t}$ is justified by the connection that Euler found in the 18th century between the sinusoids and the transcendental real number $e = 2.71828\ldots$ with a non-terminating decimal expansion.

For each real λ, the function e_λ is called a *complex sinusoid* or a *character* of $\mathbb{R}$.[2] In terms of the e_λ, the equations (1), (2) in Fourier's theorem become

$$f(t) = \sum_{k=-\infty}^{\infty} c_k e^{ik(\pi/T)t} \quad \& \quad c_k = \frac{1}{2T} \int_{-T}^{T} f(t) e^{-ik(\pi/T)t} dt. \quad (4)$$

It is also very convenient to denote the coefficient c_k by $\hat{f}(k)$, in view of its obvious dependence on f. The equations (4) then take the form

$$f(t) = \sum_{k=-\infty}^{\infty} \hat{f}(k) e^{ik(\pi/T)t}, \quad \hat{f}(k) = \frac{1}{2T} \int_{-T}^{T} f(t) e^{-ik(\pi/T)t} dt. \quad (5)$$

In some situations Fourier series prove inadequate, and we need instead Fourier integrals. As Wiener puts it:

> Fourier's name is also connected with other ways of adding together sinusoids in which the number of sinusoids to be added is too great to be represented by a first curve, a second curve, a third curve and so on. We may indeed have to add a mass of sinusoids which is entirely too dense to be arranged in one-two-three order. [56g, pp. 76, 71]

To put Wiener's words into symbols, it will suffice to give only the complex number formulation. The equations (5) now give way to

$$f(t) = \frac{1}{\sqrt{(2\pi)}} \int_{-\infty}^{\infty} \hat{f}(\lambda) e^{i\lambda t} d\lambda, \quad \hat{f}(\lambda) = \frac{1}{\sqrt{(2\pi)}} \int_{-\infty}^{\infty} f(t) e^{-i\lambda t} dt. \quad (6)$$

These equations are symmetrically related. $\hat{f}$ is called the *direct Fourier transform* of f, and f is called the *indirect Fourier transform of $\hat{f}$.* (Often we write $\check{g}$ to denote the indirect Fourier transform of g.)

The kind of functions f for which the Fourier transform is useful is well described by Wiener:

> The standard form of the theory of the Fourier integral, as developed by Plancherel and others, concerns curves which are small in the remote past and are destined to become small in the remote future. In other words, the standard theory of the Fourier integral deals with phenomena which in some sense or other both begin and end, and do not keep running indefinitely at about the same scale. [56g, p. 77]

Now a bounded increasing function $F(\lambda)$ "keeps running indefinitely" along the entire λ axis. Even so, it has a kind of Fourier transform, viz.

---

2 A *character* of $\mathbb{R}$ is defined as a non-zero continuous function f on $\mathbb{R}$ to $\mathbb{C}$ such that for all s, t in $\mathbb{R}$, we have $f(s + t) = f(s)f(t)$. A basic theorem asserts that the class of all characters of $\mathbb{R}$ is precisely the set $\{e_\lambda : \lambda \in \mathbb{R}\}$.

$$\check{F}(t) = \int_{-\infty}^{\infty} e^{it\lambda} dF(\lambda), \tag{7}$$

the last being a Lebesgue integral not with respect to the length measure, but with respect to F, which we may think of as the mass distribution along a stick laid out on the λ axis. By decomposition, such an integral can be defined when F is the difference of two increasing functions and even when it has an imaginary part of the same sort. $\check{F}$ is called the (indirect) *Fourier-Stieltjes transform* of F.

Finally, all the previous remarks extend to higher dimensional spaces. Thus, for ordinary 3-dimensional space $\mathbb{R}^3$, we take two points λ and x, $\lambda = (\lambda_1, \lambda_2, \lambda_3)$, $x = (x_1, x_2, x_3)$ and define the character e_λ by

$$e_\lambda(x) = e^{i(\lambda_1 x_1 + \lambda_2 x_2 + \lambda_3 x_3)}, \quad \text{cf. (3)}. \tag{8}$$

In terms of these characters we can again introduce (triple) Fourier series and Fourier integrals, and Fourier-Stieltjes integrals.

The question still remains as to why we should attribute so universal a significance to these sinusoids (real or complex) in preference to other families of simple functions that come to mind, such as the powers $f(t) = t^n$, or the indicators $1_{[a, b]}$ of intervals, for which $1_{[a, b]}(t) = 1$, for $a \leq t \leq b$ and $1_{[a, b]}(t) = 0$, for other t. Wiener's answer, gleaned from several of his writings, may be rendered as follows.

First, under the field conception introduced by Lagrange, Faraday and Maxwell, physical phenomena are explained in terms of the contiguous propagation of stresses and strains through the medium rather than by instantaneous "action at a distance". Together with the principle of classical determinism it entails that the basic natural laws are stateable in the form of partial differential equations (PDE's) in space-time. Second, science rests on faith in the uniformity of nature. A mathematical expression of this faith is the principle that the laws of nature are invariant under time and space translations. (Roughly, a law that prevails here and now, prevails at all places and all times.) For such invariant laws, it follows that the PDE's have coefficients not depending on the time. Third, these PDE's with time-independent coefficients can be treated as *linear*, when the phenomena studied stem from small initial disturbances of the medium from its equilibrium position.

Now it is a purely mathematical theorem that if at an instant the (small) disturbance of the medium is given by a function f (of three variables) on the space $\mathbb{R}^3$, then its disturbance t seconds later is given by $T(t)(f)$, where the *linear operator*[3] $T(t)$ is such that its action on the

---

3 The operator S is *linear* if $S(af + bg) = aS(f) + bS(g)$, where a, b are scalars, and f, g are vectors.

characters e_λ of $\mathbb{R}^3$, cf. (7), is especially simple: mere multiplication by a number, i.e.

$$T(t)\,(e_\lambda) = a(t,\lambda)e_\lambda, \quad \lambda \in \mathbb{R}^3, \tag{9}$$

where $a(t,\lambda)$ is a number, real or complex. (For a proof, see Note 1, p. 112.) If the initial (say $t = 0$) disturbance f of the medium can be represented as a combination of the characters, $f = \Sigma c_\lambda e_\lambda$, then form the linearity of $T(t)$, it follows at once that the disturbance at instance t is given by

$$T(t)(f) = \Sigma c_\lambda a(t,\lambda)e_\lambda.$$

This shows that the characters are absolutely intrinsic to the study of the phenomena governed by time-invariant linear systems, and the problems of expressing functions as combinations of characters, and finding the "Fourier coefficients" c_λ for a given f, in short, *harmonic analysis and synthesis*, are exceedingly important.

B Heaviside's operational calculus [26c, 29c]

The first task that Wiener undertook at the behest of his engineering colleagues was the rigorization of the 1893 operational calculus of Oliver Heaviside. As Wiener wrote: "The brilliant work of Heaviside is purely heuristic, devoid even of the pretense to mathematical rigor. Its operators apply to electric voltages and currents, which may be discontinuous and certainly need not be analytic" [26c, p. 558]. The thought underlying Wiener's rigorization is that "when applied to the function e^{nit}, the operator $f(d/dt)$ is equivalent to multiplication by $f(ni)$" [26c, p. 560]. Given an arbitrary function f, he dissected it into a number of frequency ranges, and applied to each range that expansion of $f(d/dt)$ which converged on this range.

By making these moves, Wiener was in effect groping towards the *theory of distributions* that was to come twenty-five years later with Laurent Schwartz. In § 8 of his paper, which deals with the operational solution of second order linear partial differential equations in two variables, Wiener wrote

> . . . there are cases where u must be regarded as a solution of our differential equation in a general sense without possessing all the orders of derivatives indicated in the equation, and indeed without being differentiable at all. It is a matter of some interest, therefore, to render precise the manner in which a non-differentiable function may satisfy in a generalized sense a differential equation. [26c, p. 582]

In this he not only anticipated Professor Schwartz, but as the latter tells us, in 1926 Wiener had seen farther than what all others were to see until 1946:

Il est amusant de remarquer que c'est exactement cette idée qui m'a poussé moi-même à introduire les distributions!* Elle a tourmenté de nombreux mathématiciens, comme le montrent ces quelques pages. Or Wiener donne une très bonne définition d'une solution généralisée; j'en avais, dans mon livre sur les Distributions, attribué les premières définitions à Leray (1934), Sobolev (1936), Friedrichs (1939), Bochner (1946), la définition la plus générale étant celle de Bochner; or la définition de Wiener est la même que celle de Bochner, et date donc de ce mémoire, c'est-à-dire de 1926, elle est antérieure à toutes les autres. {Coll. Works, II, p. 427}

In [26c] and its sequel [29c], an appendix to Vannevar Bush's book on *Operational Circuit Theory* {B22}, Wiener also took the important step of introducing the concept of *retrospective* or *causal operator*, thus initiating what is called the theory of *causality and analyticity*: the study of how one-sided dependence in the time domain leads to holomorphism in the spectral domain. In essence he defined an operator T on the space of signals on $\mathbb{R}$ to be *causal*, if and only if for each t,

$$f_1 = f_2 \text{ on } (-\infty, t) \quad \text{implies} \quad T(f_1) = T(f_2) \text{ on } (-\infty, t).$$

It follows easily that the transfer operator of a time-invariant linear filter with convolution weighting W, i.e. the filter which yields for input f the output g:

$$g(t) = (W*f)(t) = \int_{-\infty}^{\infty} W(t-s)f(s)ds, \tag{10}$$

will be causal, if and only if $W = 0$ on $(-\infty, 0]$. Thus in the causal case, (10) gets amended to

$$g(t) = (W*f)(t) = \int_{-\infty}^{t} W(t-s)f(s)ds, \quad t \text{ real.} \tag{11}$$

Equivalently, the Fourier transform $\hat{W}$ is holomorphic on the lower half-plane.

The work on the Heaviside calculus indicates how readily and profoundly Wiener could transform his engineering interests and insights into good pure mathematics. His objective in [26c], was to help engineers, and the paper ends with the sentence, "We thus have completely solved the telegrapher's equation for a semi-infinite line with a given impressed voltage at the end".

C From power to communications engineering

Wiener discerned a certain parallel between the developments of electrical engineering and of mechanics. In the latter the consideration of uniform motion gave way to that of simple harmonic motion and then

---

* Voir l'introduction de mon livre sur les distributions p. 4.

periodic motion, notably planetary motion, which, with the rise of statistical mechanics, in turn gave way to the study of highly random movements such as the Brownian motion. In electrical engineering there was first the direct current ("uniform level"), and then came the alternating current of one or several frequencies ("periodic level"). This corresponded to the stage of *power engineering*, the study of generators, motors and transformers, in which the central concept is *energy*. For this study classical mathematics (primarily advanced calculus, with complex numbers, differential equations and Fourier series) sufficed. But with the advent of the telephone and radio came *communications engineering* in which the central entity is the irregularly fluctuating current and voltage, which carries the *message* and which is neither periodic nor pulse-like (i.e. in L_2). As Wiener put it:

> A telephone line carries at the same time frequencies of something like twenty per second and frequencies of three thousand. It is precisely this variability and multiplicity of frequency which makes the telephone line an effective vehicle of information. The line must be able to carry everything from a groan to a squeak. [56g, p. 75]

Thus Wiener found in the voltage curve of a busy telephone line the same kind of local irregularity and overall persistence that he had encountered in the Brownian motion, and he began to associate the communication phase of electrical engineering with the statistical phase of mechanics. For these phases, new and more difficult mathematical methods were required. He set about to find them, spurred on by his engineering colleagues.

D Generalized harmonic analysis

Wiener felt that signals of wide varieties should be harmonically analyzable, and that for this, the wider class of irregular and persisting curves must be properly demarcated by use of new averaging operations. In this research the earlier work of professional mathematicians on the rigorous study of various classes of functions having convergent Fourier series and integrals was of little use. Wiener therefore had to seek his ideas from non-establishment "radicals" such as Lord Kelvin, Lord Rayleigh, Sir Oliver Heaviside, Sir Arthur Schuster and Sir Geoffrey Taylor, who had been interested in the harmonic analysis of allied random phenomena in acoustics, optics, and fluid mechanics. Another radical on whom Wiener could have leaned was Albert Einstein, but his 1914 work in harmonic analysis {E3} came to light only in 1985 (see Note 2, p. 112).

Building on the ideas of G. I. Taylor, Wiener introduced the class S of measurable functions f on the real axis $\mathbb{R}$ for which the *covariance function* φ:

Lord Kelvin
1824–1907

Lord Rayleigh
1842–1919

Sir Oliver Heaviside
1850–1925

Sir Arthur Schuster
1851–1934

Sir Geoffrey Taylor
1886–1975

British radical analysts ("mathophysicists" in Halmos's terminology) from whom Wiener derived some of his most important mathematical ("mathological") ideas. See § 22E.

$$\varphi(t) = \lim_{T \to \infty} \frac{1}{2T} \int_{-T}^{T} f(s + t)\overline{f(s)}ds \qquad (12)$$

exists and is continuous on $\mathbb{R}$. Many persistent irregular signals f (such as brain wave encephalograms) belong to this very large class S, which includes in particular the almost periodic functions of H. Bohr and A. S. Besicovitch and the periodic functions studied by Fourier.

The first bit of generalized harmonic analysis (G.H.A.) that Wiener did was to show that the covariance function φ of a function f in S is the Fourier-Stieltjes transform of a bounded, non-negative, non-decreasing function F on $\mathbb{R}$:

$$\varphi(t) = \int_{-\infty}^{\infty} e^{it\lambda}dF(\lambda), \quad \text{cf. } \S 9A(7). \qquad (13)$$

F is called the *spectral distribution* of f. In the special case in which φ has a finite Lebesgue integral, (13) gives way to

$$\varphi(t) = \int_{-\infty}^{\infty} e^{it\lambda}F'(\lambda)d\lambda = \sqrt{(2\pi)} \ \check{F}'(t), \qquad (14)$$

i.e. F' is just the Fourier transform of φ. The first to use a function such as F' was Schuster. At the turn of the nineteenth century, he used the size of $F'(\lambda)$ to locate the "periodicities" "hidden" in the erratic function f. Schuster called F' the *periodogram* of f. For the justification of (13) and the historical antecedents of (12), (13) and (14), see Note 2, p. 112.

Since the function φ averages out the irregularities of f, and is rather smooth, it is to us not surprising that φ has a Fourier-Stieltjes transform; indeed, Schuster himself had an inkling of this development. But Wiener did not stop here. He went on to define a Fourier associate s of the extremly irregular function f itself by the complicated formula:

$$s(\lambda) = \underset{A \to \infty}{\text{l.i.m.}} \frac{1}{\sqrt{(2\pi)}} \left(\int_{-A}^{-1} + \int_{1}^{A} \right) \frac{f(t)e^{-it\lambda}}{-it} dt$$
$$+ \frac{1}{\sqrt{(2\pi)}} \int_{-1}^{1} f(t) \frac{e^{it\lambda}-1}{-it} dt. \qquad (15)$$

This s is called the *generalized Fourier transform* of f. This function had no historical antecedents, and was exclusively Wiener's creation. He got both F and s in his memoir [30a], after much groping extending back to [25c]. Now a well-known result in classical harmonic analysis is the Bessel identity between a function f and its Fourier transform $\hat{f}$, viz.

$$\int_{-\infty}^{\infty} |f(t)|^2 dt = \int_{-\infty}^{\infty} |\hat{f}(\lambda)|^2 d\lambda. \tag{16}$$

Wiener's intuition told him that an analogous identity had to hold between his erratic f and its generalized Fourier transform s. He surmised that this identity had the form:

$$\lim_{T\to\infty} \frac{1}{2T} \int_{-T}^{T} |f(t)|^2 dt = \lim_{h\to 0} \frac{1}{4\pi h} \int_{-\infty}^{\infty} |s(\lambda + h) - s(\lambda - h)|^2 d\lambda, \tag{17}$$

and he showed that its correctness hinged on that of the simpler identity for non-negative g:

$$\lim_{T\to\infty} \frac{1}{T} \int_{0}^{T} g(t) dt = \lim_{h\to 0} \frac{2}{\pi h} \int_{0}^{\infty} g(t) \frac{\sin^2 ht}{t^2} dt. \tag{18}$$

But Wiener had a hard time proving the latter, and here matters stood in 1926.

E Tauberian theory

The stalemate broke in late 1926 when Wiener was in Europe on a Guggenheim Fellowship for 1926–1927. He had planned to spend the year at the Mathematics Institute in Göttingen, where he was scheduled to give a series of lectures.[4] But some inept publicity on his part and its unforeseen consequences strained his relations with Professor Richard Courant, the administrative head of the Institute. Instead of trying to heal the breach, Wiener exaggerated its importance and got more entangled by other inept acts. As a result, he was depressed in Göttingen, and his lectures were below par. He cut short his Göttingen stay and spent the remainder of his term at Copenhagen with Harald Bohr.

But the European trip was by no means a disaster. By happenstance Wiener made two contacts that proved to be very conducive to his efforts. At Göttingen he met an acquaintance, the British number-theorist A.E. Ingham, and learned from him that the identity (18) was, as they say, "Tauberian" in nature, and that his mentors Hardy and Littlewood were authorities on such matters. This was news to Wiener. Another important contact was with the Tauberian theorist Dr. Robert

---

4 The invitation to lecture had come from interest in the quantum mechanical aspect of another lecture he had delivered in Göttingen in the summer of 1925 (cf. Ch. 10).

G.H. Hardy
1877–1947

Photograph taken
around 1910

Schmidt of Kiel, whom Wiener met at a German mathematical meeting at Düsseldorf after the summer of 1926.

What happened in the next year or so was dramatic. Following Inham's advice, Wiener studied the work of Hardy and Littlewood, and noticed that they too were concerned with the equality of two means. Then suddenly he saw that a simple exponential change of variables

$$x = e^{-s}, \quad y = e^{-t}, \quad dx = -e^{-s}ds,$$

would reduce the integrals occurring in Schmidt's theorems to convolution integrals with which he was familiar from his electrical studies, cf. § 9B (10). Thus letting $f(u) = \varphi(e^u)$, $g(u) = e^{-u}\psi(e^{-u})$, we have

$$\int_0^\infty \varphi(x/y)\, \psi(x)\, dx = \int_{-\infty}^\infty f(t-s)g(s)ds = (f*g)\,(t).$$

The limit on the left as $y \to 0_+$, which Schmidt took, becomes $\lim_{t\to\infty}(f*g)\,(t)$. Thus the real problem was to get a theorem for the limit of convolution integrals. Wiener then went on to find such a theorem.

In [28b], after acknowledging that Hardy and Littlewood "had already succeeded in casting a large group of Tauberian theorems into the form of an equivalence of two means of positive quantities" [28b, p. 184], Wiener presented his new approach. This involved a hard attack, relying not on any classical Tauberian theorem, but on a number of novel lemmas. Perhaps the most remarkable of these lemmas, a theorem in its own right, asserts that if f on $[0, 2\pi]$ vanishes nowhere and has an absolutely convergent Fourier series, then $1/f$ also has an absolutely convergent Fourier series.

Wiener's final result in [32a] has a beautifully simple enunciation. Write

$$(W*f)(t) = \int_{-\infty}^{\infty} W(t - s) f(s) ds, \quad t \text{ real.} \tag{19}$$

Here $W \in L_1$ and $f \in L_\infty$. We may, cf. (10) above, regard $g = W*f$ as the response of an ideal convolution filter $W*$ with weighting W subject to the input signal f. When will the output g tend to a limit as $t \to \infty$? Wiener's Tauberian theorem gives the answer: *feed the signal f into another filter W_0* with a nowhere vanishing frequency response, i.e. where the weighting W_0 in L_1 is such that its Fourier transform W_0 vanishes nowhere. If the resulting response g_0 has a limit as $t \to \infty$, so will the original response g.* Wiener also proved a "Stieltjes" form of this theorem. It may be looked upon as its analogue when the input is a real or complex valued measure μ over $\mathbb{R}$ and the filter performs the convolution

$$(W*\mu)(t) = \int_{-\infty}^{\infty} W(t - s) \mu(ds). \tag{20}$$

All the classical Tauberian theorems, even the very deep, can be recovered from Wiener's two theorems: one has only to pick W, W_0, f and μ intelligently and change variables. This applies also to (18) and therefore to the Bessel-type identity (17) on which rests the appropriateness of Wiener's generalized Fourier transformation. Thus generalized harmonic analysis is put on a sound footing. But many other theorems are also uncovered. Among these, perhaps the most interesting is the one on the Lambert series, which bears on the analytic theory of prime numbers. It led Wiener and his former student from Japan, Professor S. Ikehara, to a simpler proof of the celebrated Prime Number Theorem, to wit,

$$\lim_{n \to \infty} \frac{\pi(n) \cdot \log n}{n} = 1,$$

where $\pi(n)$ is the number of primes not exceeding n. For this proof, all one had to know about the Riemann zeta function is that it has no zeros on the line $x = 1$ in the complex plane, and the traditional appeal to contour integration was avoided.

Wiener's friend, J.D. Tamarkin of Brown University, induced Wiener to gather his scattered results and present them systematially in two memoirs. In 1929 Wiener wrote the first memoir [30a] on G.H.A. In 1932 appeared his great memoir [32a] on Tauberian theorems. In the final organization of both memoirs Wiener received considerable assistance from both Professor Tamarkin and his colleague Professor Einar Hille. With the appearance of these publications Wiener was recognized as the master in these fields, and was awarded the Bocher prize of the American Mathematical Society in 1933.[5] In April 1934, he was elected a Fellow of the National Academy of Sciences of the USA.

The mathematical significance of the memoir [32a] is commented on extensively in the Wiener *Coll. Works*, II. Here only three remarks should suffice. (1) In 1938 H.R. Pitt from England proved a beautiful theorem involving the so-called "slowly oscillating functions" of J. Kamarta, a Yugoslav mathematician, from which both of Wiener's general Tauberian theorems can be deduced. (2) The deep significance of the methods employed by Wiener to establish his Tauberian theorems came to light only eight to ten years later when the eminent Soviet mathematician, I.M. Gelfand, developed the theory of Banach algebras. (3) The further pursuit of Wiener's ideas at the abstract level gave birth to an important branch of analysis called *spectral synthesis*. (Regarding (2) and (3), see Note 3, p. 113).

While Wiener's memoir [32a] had been fully assimilated and appreciated by the mathematical community, the same cannot be said of his memoir [30a] on G.H.A. As late as 1970, the authorities E. Hewitt and K. Ross remarked that it "has yet to be interpreted and assimilated in the standard repertory of abstract harmonic analysis" {H7, vol. II, p. 547}. Since this issue is thoroughly discussed in the *Coll. Works*, II, it will suffice to say that this writer has recently shown how [30a] can be interpreted in terms of current functional analysis by means of *helices* in Hilbert space. Helices, which are like the grooves on a cylindrical screw, are familiar curves in elementary differential geometry, but the first helix in an infinite dimensional Hilbert space appeared in Wiener's 1923 work on Brownian motion. Thus this recent development affects an even tighter integration of Wiener's G.H.A. and his theory of the Brownian movement, and exemplifies the fundamental unity in this thought. (See Note 4, p. 114.)

---

5 Named after the same Bocher with whose children he had played as a six-year-old. Sharing the prize with him was Marston Morse, a former student of G.D. Birkhoff, and another great mathematician-to-be.

F The intensity, coherence and polarization of light

In the memoire [30a], Wiener also demonstrated the propaedeutic role of G. H. A. in the field of optics. In Maxwell's theory of light, the flux of electromagnetic energy at a fixed point P in a medium traversed by light, through a small surface at P perpendicular to the direction of propagation, is proportional to $|\mathbf{E}(t)|^2$, where $\mathbf{E}(t)$ is the electric vector at P at the instant t in question. This led Wiener to regard $|f(t)|^2$ as the energy of instant t of the signal f in the class S and to regard

$$\varphi(0) = \lim_{T \to \infty} \frac{1}{2T} \int_{-T}^{T} |f(t)|^2 d\tau, \quad \text{cf. (3),} \tag{21}$$

as the *total-mean-power* or "brightness" or "intensity" of the signal f, since power = energy/time.

Central to Wiener's clarification of optical ideas was his tacit interpretation of the *photometer* as an instrument which, when impinged with a light signal f in S gives the reading $\varphi(0)$. The justification of this interpretation came in 1932 when J. von Neumann {V3} noted that no instrument reads instantaneously, but each has a response-time (or latency) $T > 0$. (Think of an ordinary clinical thermometer.) The reading of the instrument at instant t thus incorporates the values $f(\tau)$ of the incoming signal during the time-interval $[t - T, t]$, and gives, according to von Neumann, the time-average

$$\bar{f}_t = \frac{1}{T} \int_{t-T}^{t} f(\tau) d\tau.$$

In the case of the photometer exposed to incoming light, our $f(\cdot)$ is the electromagnetic energy, which is proportional to the function $|E(\cdot)|^2$, and consequently the observed reading on the photometer is

$$\bar{I}_t = \frac{1}{T} \int_{t-T}^{t} |E(\tau)|^2 d\tau. \,^6$$

Now a candle flame consists of an exceedingly large number of radiating centers, viz. thermally excited atoms. Each atom radiates for approximately 10^{-10} seconds, the starting moment being quite unrelated to the activities of its neighbors. Even if T were only a microsecond, billions of atomic events would be involved in the reading $\bar{I}_t$. The replacement of T by $2T$ or even $1000T$ will hardly affect $\bar{I}_t$. In short, our observed intensity $\bar{I}_t$ is an exceedingly good fit to the hypothetic mean

---

6 Here and below we assume rectilinear propagation, and write $\mathbf{E}(t) = E(t)\mathbf{u}$, where $\mathbf{u}$ is the unit vector in the direction of propagation.

$$\lim_{T\to\infty} \frac{1}{T} \int_{t-T}^{t} |E(\tau)|^2 d\tau.$$

Wiener tacitly hypothesized that the signal $E(\cdot)$ is in his class S, and he tacitly replaced the last limit by

$$\lim_{T\to\infty} \frac{1}{2T} \int_{-T}^{T} |E(\tau)|^2 d\tau = \varphi(0),$$

where $\varphi(\cdot)$ is the covariance function of $E(\cdot)$, thus reaching the conclusion that the intensity $\bar{I}_t$ recorded by the photometer is $\varphi(0)$. Wiener's replacement is valid if, in view of the chaotic origins of light, we make the reasonable hypothesis (as Wiener did later) that $E(\cdot)$ *is a trajectory of a stationary stochastic process*. For then the Birkhoff Ergodic Theorem (see § 12A) tells us that the last two time-averages are indeed equal in general.

Next, Wiener considered the *Michelson interferometer*, equipped at the output end with a photometer: camera, eye, etc. A simple calculation shows that with the incoming light, represented by the electric vector $E(\cdot)$, the observed intensity at the output end is

$$\varphi(0) + \varphi(\Delta\ell/c),$$

where $\Delta\ell$ is the difference between the lengths of the two arms, c is the speed of light, and φ is the covariance of the signal $E(\cdot)$. By turning the screws, i.e. changing $\Delta\ell$, we can find $\varphi(x)$, for any given (not too large) x. Thus Wiener saw in the Michelson interferometer an *analogue computer for the covariance function of light signals*. It was his favorite optical instrument, and he enjoyed exposing its principles to willing (and even unwilling) listeners.

Wiener's explication of the notion of *coherence of signals* came from his extension of G.H.A. to vector-valued signals $f = (f_1, \ldots, f_q)$ on $\mathbb{R}$. Wiener assumed that for each pair f_i, f_j, a *cross-covariance function* φ_{ij}:

$$\varphi_{ij}(t) = \lim_{T\to\infty} \frac{1}{2T} \int_{-T}^{T} f_i(t+\tau)\overline{f_j(\tau)} d\tau \tag{22}$$

exists and is continuous on $\mathbb{R}$. He then got a $q \times q$ matrical covariance function $\Phi(\cdot) = [\varphi_{ij}(\cdot)]$, and corresponding to (13), the equation:

$$\Phi(t) = \int_{-\infty}^{\infty} e^{it\lambda} dF(\lambda), \tag{23}$$

where now $F(\cdot) = [F_{ij}(\cdot)]$ is a $q \times q$ matrical spectral distribution, which Wiener called the *coherency matrix* of f. Wiener called the signals $f_1, \ldots,$

f_q *incoherent* when $\varphi_{ij} = 0$ for $i \neq j$, i.e. when $\Phi(\,\cdot\,)$ and $F(\,\cdot\,)$ are diagonal matrices.

It is well known that we cannot make light from *two* sources interfere. This is because the maintenance of phase relationships, such as are required for interference, is impossible under the atomic chaos in which light originates. The fact, just as familiar, that two adjacently placed equal candles are twice as bright as one, cannot, however, be given a satisfactory explanation except in Wienerian terms. Briefly, since the light signals f_1 and f_2 from the two candles are incoherent, the covariance function of the combined signal $f_1 + f_2$ is easily seen to be $\varphi_1 + \varphi_2$. Hence the observed intensity is $\varphi_1(0) + \varphi_2(0)$. When the two candles are equal, we get $2\varphi(0)$, i.e. twice the intensity of a single candle.

The coherency matrix also clarifies the notion of *polarization of light*. Take a pair of orthogonal axes in the plane π perpendicular to the direction of propagation, supposed fixed. Light is *transverse* wave propagation: the optical vector, which is identified with $E(t)$, lies in π. Consequently, it is representable by a $\mathbb{R}^2$-valued vector $(f_1(t), f_2(t))$. As t changes, this vector changes in direction (not merely in magnitude), and this fact accounts for the phenomenon of polarization. Roughly speaking, the light is "unpolarized" when this vector keeps swinging "erratically". Wiener made this idea precise. He called the light (completely) *unpolarized*, if the 2×2 matrical spectral distribution $F(\,\cdot\,)$ of this vector signal has the form $f(\,\cdot\,)I$, i.e. is a scalar matrix. He called the light *elliptically polarized*, if this $F(\,\cdot\,)$ is unitarily equivalent to a diagonal matrix with only one non-zero entry.

Thus with the reasonable hypothesis that (macroscopically observed) light signals are trajectories of an ergodic stochastic process, Wiener was able to justify what physicists such as Schuster and Rayleigh knew intuitively but could not formulate rigorously. His ideas have found a place in the standard repertory in optics, e.g. in the treatise by Born and Wolf {B18}. But the full significance of his work on coherent light, begun in 1928, emerged only with the advent of lasers, masers and holograms. Sir Dennis Gabor, the father of holography and Nobel Prize-winner, has described Wiener's coherency matrix as a "philosophically important" idea, which "was entirely ignored in optics until it was reinvented . . . by Dennis Gabor in England in 1955 and Hideya Gamo in Japan in 1956". Gabor points out that the matrix theory of light propagation was initiated actually by Max von Laue in 1907, and covers the transmission of information, and that "the entropy in optical information is a particularly fine illustration of the role of entropy in the Shannon-Wiener theory of communication" (*Coll. Works*, III, pp. 490–491).

Note 1: **A proof of § 9A(9).**

Fix t in $\mathbb{R}$ and write $A = T(t)$. Let τ_h be translated by h in $\mathbb{R}^3$, i.e. for h in $\mathbb{R}^3$

$$[\tau_h\{f\}](s) = f(s + h), \quad s \in \mathbb{R}^3.$$

Let $s \in \mathbb{R}^3$. Then by the stipulated commutativity of A and τ_h, we have

$$[A(e_\lambda)](s + h) = [\tau_h\{A(e_\lambda)\}](s) = [A\{\tau_h(e_\lambda)\}](s). \tag{24}$$

But since $e^{a+b} = e^a \cdot e^b$, $\tau_h(e_\lambda) = e_\lambda(h) \cdot e_\lambda$. Hence by (24), and the homogeneity of A,

$$[A(e_\lambda)](s + h) = [A\{e_\lambda(h) \cdot e_\lambda\}](s) = e_\lambda(h) \cdot [A(e_\lambda)](s).$$

Setting $s = 0$, we get, $[A(e_\lambda)](h) = [A(e_\lambda)](0) \cdot e_\lambda(h)$. More fully,

$$[T(t)(e_\lambda)](h) = [T(t)(e_\lambda)](0) \cdot e_\lambda(h).$$

Since $h \in \mathbb{R}^3$ is arbitrary,

$$T(t)(e_\lambda) = a(\mathrm{t}, \lambda) \cdot e_\lambda, \quad a(t, \lambda):\, = [T(t)(e_\lambda)](0).$$

Note 2: **Historical remarks** (cf. § 9D).

The justification of (13) rests on the observation that φ is a continuous non-negative definite function on $\mathbb{R}$, and therefore by Bochner's theorem, it is the Fourier-Stieltjes transformation of a non-negative mass distribution F. Wiener's proof of (13) in [30a] is much more complicated. It readily follows from (13) that when φ is in $L_1(\mathbb{R})$, we have (14). To show that functions f in S exist for which F is absolutely continuous, Wiener very ingeniously used the stochastic integral

$$f(\mathrm{t},\alpha) = \int_{-\infty}^{\infty} W(t\text{-}\tau)d_\tau x(\tau,\alpha), \quad \alpha \in [0,1], \quad W \in L_2(R),$$

of the Brownian motion $x(\cdot,\cdot)$, and showed that for almost all α, $f(\cdot,\alpha)$ is in S and that for $f(\cdot,\alpha)$ the spectral distribution F is absolutely continuous with $F'(\lambda) = \sqrt{(2\pi)} |\, \check{W}(\cdot)\,|^2$, a.e. where $\check{W}$ is the indirect Fourier-Plancherel transform of W, [30a, § 13].

The concepts of covariance φ of a signal f and its spectral distribution F, and their interconnection have an interesting history. The first one to introduce F via its density F' was Schuster $\{S5\}$ in 1899. But he did so not by (14) but by the formula

$$F'(\lambda) = \lim_{T \to \infty} \frac{1}{2T} \left| \frac{1}{\sqrt{(2\pi)}} \int_{-T}^{T} f(t)e^{-it\lambda}dt \right|^2 \tag{25}$$

which, even for absolutely continuous F, is valid only in very special cases. Nevertheless Schuster used it successfully to find the "hidden periodicities" in the irregular functions f that he encountered in empirical time-series. Schuster bypassed the covariance function φ, and was probably unaware of its existence.

The fact that Einstein was a participant in this history was revealed only in 1985, when a remarkable two-page heuristic note of his, written in 1914, dealing with f, φ, F', and carrying a version of (13), came to light {E3}. Einstein was unaware of the work of Schuster, and was the first to bring in both φ and F' and reveal their linkage. Einstein's signal f is on $[0, \infty]$, not $\mathbb{R}$. He defined F' essentially as had Schuster, but worked with the "periods" (or "wave lengths") $\theta = 1/\lambda$ instead of the "frequencies" λ. He linked the function $I(\theta) := F'(1/\theta)$ to θ by the equation

$$I(\theta) = \int_0^{2\pi} \varphi(t) \cos \pi \frac{t}{\theta} \, dt,$$

i.e. essentially by an inversion of (13). His derivation is not rigorous.

Taylor, unaware of Einstein's note, reintroduced the covariance φ in his 1920 paper {T3} but in a normalized form ψ, primarily because from ψ, ψ', ψ'', ..., the variances of the irregular signal f and of the derivatives could be found. He gave no spectral analysis.

The genesis of the different ideas went as follows:

Spectral density F' ("periodogram"): Schuster, 1889;
                                                                     Einstein, 1914.
Covariance φ: Einstein, 1914; Taylor, 1920.
Spectral distribution F: Wiener [28a].
Interconnections: inverse of (13) with F', Einstein, 1914;
                              (13) itself with F, Wiener [30a].

All the work was done independently. A detailed description of Einstein's paper and its nexus with G.H.A. can be found in the writer's 1986 article {M12}. An interesting stochastic process interpretation of Einstein's note, and its links to Khinchine's work of 1934, has been given by A.M. Yaglom {Y2} in 1987. He also gives the facts of the discovery of Einstein's paper by a school teacher in East Germany, and other interesting historical information.

Note 3: **Spectral synthesis** (cf. § 9E).

In his proof of the Tauberian theorem, Wiener had unconsciously developed and used the maximal ideal theory of the convolution algebra $L_1(\mathbb{R})$. Without knowing it, he had shown that subsets M_λ of functions f in $L_1(\mathbb{R})$ for which $\hat{f}(\lambda) = 0$ are, for different λ in $\mathbb{R}$, precisely the maximal ideals of the algebra $L_1(\mathbb{R})$. If we write $Z(\hat{f})$ for the set of zeros of $\hat{f}$, then the last result is restatable in the following form, which Wiener might have found even more esoteric: if $Z(\hat{f})$ is void, then *the smallest closed ideal containing f is equal to the intersection of maximal ideals containing f.* Does the italicized conclusion prevail when $Z(\hat{f})$ is non-void? Efforts begun in the late 1930s to answer this question culminated in a negative answer by P. Malliavin in France in 1959. These efforts have brought into being a whole new branch of mathematical analysis called *spectral synthesis*, with distinguished participants all over the globe. Its development is traced systematically in the treatises of Hewitt and Ross {H7, vol. II}, and J. Benedetto {B6}, and their accounts show how Wiener's ideas, in one form or another, keep lurking in the background.

Note 4: **Conditional Banach spaces** (cf. § 9E).

From the standpoint of functional analysis, the Wiener class S is a *conditionally* linear subspace of the Marcinkiewicz Banach space:

$$\mathfrak{M}_2(\mathbb{R}) = \{f \colon f \in L_1^{loc}(\mathbb{R}) \ \& \ \|f\|^2 = \varlimsup_{T \to \infty} (1/2T) \int_{-T}^{T} |f(t)|^2 dt < \infty\},$$

(cf. *Coll. Works*, II, pp. 333–379). K.S. Lau {L2} has shown that the f in S with $\|f\| = 1$ are in fact extreme points of the unit ball of $\mathfrak{M}_2(\mathbb{R})$. Recently, he has obtained an interesting extension of BMO spaces by considering the class in which the Marcinkiewicz $\varlimsup$ is replaced by sup, cf. {L3}.

Wiener's generalized Fourier transform s of the function f in S, cf. (15), gives rise to a helix in the Hilbert space $L_2(\mathbb{R})$. Every such helix is characterized by a single 'average vector' α in $L_2(\mathbb{R})$. This α turns out to be the Plancherel transform of $f \cdot \ell_0$, where $\ell_0(t) = 1/(t + i) \sqrt{\pi}$, and the Tauberian identity (17) holds with α replacing Wiener's s. Thus there is a simpler theory, accomplishing all of Wiener's objectives, but with the generalized Fourier transform in the Hilbert space $L_2(\mathbb{R})$.

At the more classical level, J. Benedetto has just obtained a full-fledged generalization of the Tauberian identity (17) for functions f in $\mathbb{R}^n$ in the Wiener class S, cf. J. Benedetto, G. Benke and W. Evans, "An n-dimensional Wiener-Plancherel formula", Adv. Appl. Math. (forthcoming).

10 Max Born and Wiener's Thoughts on Quantum Mechanics and Unified Field Theory

A The Uncertainty Principle in classical physics

From 1922 to 1927 Wiener traveled to Europe practically every summer. He found these trips intellectually refreshing and enjoyed the company of interesting people he met. During the 1920s his work was much better appreciated in Europe than in America, and he got considerably more encouragement from mathematicians abroad than he did from those at home, most of whom hardly knew him. During these visits, Wiener worked. The words: "The kind of work I do can be done anywhere. Why should I be less capable of reflecting about my problems on the Potsdam bridge than at home?" are Einstein's (cf. P. Frank {F4, p. 147}), but could have been Wiener's. His ideas on quantum mechanics were formed in the course of such travels and participation in seminar discussions in Europe.

Wiener was aware of an inequality in classical harmonic analysis, sometimes called "the time-frequency uncertainty principle". This asserts that

$$\int_{-\infty}^{\infty} |tf(t)|^2 dt \cdot \int_{-\infty}^{\infty} |\lambda \check{f}(\lambda)|^2 d\lambda \geq \frac{1}{4}, \quad \text{(cf. Weyl {W4, p. 393}).} \quad (1)$$

Here f is in $L_2(\mathbb{R})$, with norm 1, i.e. $\int_{-\infty}^{\infty} |f(t)|^2 dt = 1$, and the integrals $\int_{-\infty}^{\infty} |tf(t)|^2 dt$ and $\int_{-\infty}^{\infty} |\lambda \check{f}(\lambda)|^2 d\lambda$ are finite, $\check{f}$ being the (indirect) Fourier-Plancherel transform of f.

It follows that if a sound oscillation f of intensity (or loudness) 1 lasts only for a short time interval $[0, T]$, then the first factor on the left will be small; consequently the second factor will have to be large, i.e. the oscillation f will have to comprise a whole range of frequencies λ, and will not be a pure tone. Conversely, if the oscillation approximates a pure tone of frequency λ_0, i.e. its frequencies are all clustered around λ_0, then the second factor will be small; the first factor will now have to be large, i.e. the oscillation is spread over a long interval of time. In short, a pure tone of only momentary duration is impossible.

Wiener enjoyed reflecting on the practical consequences that this entailed in acoustics and music. In
musical notation for instance, the vertical position on the
staff indicates the pitch, i.e. frequency, and the horizontal interval the metronome duration of a full note○, say
1/2 seconds. The quarter note ♩ will only be 1/8th of a
second in duration. As this note is moved down the staff
(see Fig.), its frequency diminishes. Were this quarter
note (lasting 1/8th of a second) to be played on some
instrument, at the very low frequency of 5 oscillations per
second, not even one oscillation will be completed, and
the air will be pushed, not set into vibration. What we
will hear will sound not like a note "but rather like a blow
on the eardrum" [56g, p. 106].

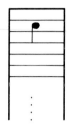

Thus for a composition to be musical, its notes must throughout
satisfy an inequality of the type:

$$(\text{metronome duration}) \cdot \text{pitch} \geq a,$$

where a is a positive threshold. Accordingly, low notes must be assigned
sufficiently long durations. In a fast jig, the notes are of short duration,
and as Wiener aptly put it: "A fast jig on the lowest register of an organ
is in fact not so much bad music but no music at all" [56g, p. 106].

Indeed, as Wiener emphasized, there is a more fundamental
paradox in music, for a pure tone (sinusoidal oscillation) "exists sub
specie aeternitatis" [56g, p. 106]. It is impossible to produce a strictly
pure tone, for to start or stop a note is inevitably to add new frequencies.
Indeed, it is impossible to produce physically any *purely periodic* oscillation, because of its eternity. The ideal concept of pure tone is, however,
essential to harmonic analysis as we saw (§ 9A) and to all its applications. What makes music possible is the fortunate cicumstance that our
ears are insensitive to the irregularities that accompany the commencement and termination of musical notes.

As Wiener observed, the same paradoxical circumstances prevail
in other fields: a change in the size of one quantity entails a disturbance
in that of another "complementary" quantity, and consequently certain
situations are unattainable. For instance, in optics, it is impossible to
produce a light ray passing through a definite point A in a definite
direction. For to ensure that the light passes through A, we have to
interpose in its path a screen with pinhole at A; but the latter causes the
emergent light to diffract, i.e. to spread out in a conical beam, and not
along a definite straight line. Another paradox occurs in the work of
Maxwell. According to the energy-equipartition principle of his statistical mechanics, a continuous substance (by virtue of its infinitely many

degrees of freedom) has infinite heat capacity, and the concept of temperature loses meaning. The combination of this principle and his theory of radiation by contiguous electro-magnetic propagation through the aether, entails the impossible "ultraviolet catastrophe". Wiener viewed this breakdown, and the drastic theoretical remedy proposed by Planck in 1900, as stemming from the complementarity of the notions: temperature and heat capacity, cf. [56g, pp. 102–104].

B The collaboration with Max Born

These unusual ideas formed the contents of Wiener's seminar talk at Göttingen in 1924 [56g, p. 106], and it seems that one of the listeners was the 27-year-old Werner Heisenberg.[1] At that very time Heisenberg and Max Born were grappling with the failure of the classical laws in atomic radiation, and were becoming increasingly aware of the limitations afflicting the simultaneous determination of complementary quantities such as the position and momentum of atomic particles. This cognate interest of the Göttingen circle led Wiener to learn quantum theory, and to collaborate with Born in the fall of 1925 when the latter came to MIT as Foreign Lecturer.

Wiener has described Max Born as a "calm, gentle musical soul, whose chief enthusiasm in life has been to play two-piano music with his wife" [56g, p. 107]. He was a "most modest" but profoundly creative scholar. Just before Born's arrival at MIT Heisenberg had enunciated his matrix mechanics to cover the motion of a closed system with discrete energy levels or frequencies, and Born raised the question as to its substitute for non-quantizable systems with continuous spectra, such as rectilinear motion, in which no periods are possible. They wrote a joint paper on this, [26d].

In the paper [26d] Born and Wiener introduced the operator Q, defined by

$$Q(f)(\,\cdot\,) = \lim_{T \to \infty} \frac{1}{2T} \int_{-T}^{T} q(s,\,\cdot\,)f(s)ds,$$

as a replacement for the Heisenberg matrix for configuration, and obtained commutation laws governing differential operators; they then proposed more general operators on function space. This was the first appearance in physics of the idea that *physical quantities are representable as operators on function space.* Unfortunately, the function space is left unspecified. It appears to be a subclass of S comprising at least the Besicovitch almost periodic functions on $(-\infty, \infty)$.

---

1   This bit of information appears in a paper {B1} by the late engineer-mathematician John L. Barnes.

Wiener with Max Born in 1925

The role of the Born-Wiener paper [26d] in the history of quantum mechanics is alluded to in Whittaker's history {W12, Vol. II, p. 267}, and discussed much more fully in J. Mehra and H. Rechenberg's recent comprehensive history of the subject {M15, Ch. 5}. The idea of a Hilbert space, still embryonic, was just not in Wiener's consciousness at that time. But the paper [26d], its limitations notwithstanding, had "an immediate impact on Heisenberg", as Mehra and Rechenberg point out {M15, p. 247}, and they conclude:

> At a time when just a few physicists struggled to develop a consistent theory of quantum mechanics, the Born-Wiener collaboration not only indicated the way for handling the problem of aperiodic motion but also contributed to the physical interpretation of the theory. {M15, p. 246}

The aftermath is well known. At about that very time, the question discussed by Born and Wiener, was also being addressed by Schrödinger, working on de Broglie's bold idea of the wave attached to the electron, and independently by Dirac. A little later John von Neumann provided a more rigorous and unified approach subsuming the work of Heisenberg as well, based on the operator theory of Hilbert spaces, cf. {V2}. This operational approach is of course superior to the Born-Wiener approach by operators on an (unspecified) function space. Wiener felt that he had no fresh insights to offer to such a galaxy of minds, and after Born's departure, he withdrew from steadfast participation in the field.

Wiener did, however, write two more papers [28d, 29e] along the lines of his work with Born. In these papers as well as in [30a] on G.H.A., Wiener pointed out that the linear-quadratic relationship prevailing in quantum theory also occurs in the branches of classical physics, e.g. white light optics or communications engineering, that demand generalized (rather than ordinary) harmonic analysis. In the theory of white light, for instance, the fundamental Maxwell equations are linear, but pertain to the intrinsically non-observable quantities $E(t)$ and $H(t)$, whereas the observable intensities are defined in terms of the squares of their amplitudes. Since all observation at this sub-Hertzian level is necessarily photometric, to take an observation amounts to reading $\varphi(0)$, where φ is the covariance function of the light signal f (cf. § 9F). But an apparatus which reads covariances destroys phase relationships. For instance, if

$$f(t) = \Sigma_1^n a_k e^{i\lambda_k t}, \quad \text{then} \quad \varphi(t) = \Sigma_1^n |a_k|^2 e^{i\lambda_k t}.$$

The phases of the complex numbers a_k are gone. Feeding this "observed light" into another optical instrument will not produce the same response as feeding in the unobserved light f. Thus observation affects the signal and thwarts prediction, much as in quantum mechanics. If we think of the light beam as a vector-signal, then in Wiener's words:

> ... if two optical instruments are arranged in series, the taking of a reading from the first will involve the interposition of a ground-glass screen or photographic plate between the two, and such a plate will destroy the phase relations of the coherency matrix of the emitted light, replacing it by the diagonal matrix with the same diagonal terms. Thus the observation of the output of the first instrument alters the output of the second. [30a, p. 194]

Thus, in vague analogy with the quantum situation, in Wiener's white light optics, observation has the effect of diagonalizing an operator and of enlightening the mind only by killing off a lot of information.

In [28d] Wiener also made a novel use of coherency matrices (§ 9F). With the time variable replaced by the fifth dimension of Kaluza, and with the space-time variables as parameters, he used it to solve the 5-dimensional Schrödinger equation.

C The work with Struik and Vallarta on unified field theory

During the late 1920s, when the last of his early quantum mechanical papers were being written, Wiener became interested in Einstein's general theory of relativity, primarily with an eye to subsuming Schrödinger's wave equation. As his early encyclopedia article [19a] on Non-Euclidean Geometry and his later essay [22a] on Kant's philosophy of geometry (§ 7B) testify, Wiener had an excellent general understanding of the nature of geometry. At the end of the article [19a], he discussed the question of the Euclidicity of physical space, but there was no allusion to Einstein. Presumably, with the Great War going, the news of the general theory had not reached Albany in 1918.

Now under the impetus of the geometer Dirk Jan Struik, his newly arrived colleague at MIT from Holland, Wiener began learning differential geometry, and soon began to write papers on relativity, specifically on the Klein-Gordon equation and on the unified theory of electricity and gravitation [27f–i, 28c, 29f, 29g]. These were in collaboration with Struik and also with M. S. Vallarta, a Mexican physicist who too had joined the MIT faculty. None of these papers has been particularly consequential.

The 1927 papers center on [27g], the aim of which is to subsume the Schrödinger wave mechanics within the framework of an extended general relativity. It goes beyond H. Weyl's unified theory of gravitation and electromagnetism in that the authors

> . . . carry the Weyl theory of natural gauge one step further, and thus provide a place for quantum phenomena. [27g, p. 3]

But, as Professor A. H. Taub has pointed out, certain assumptions made in the paper are not physically justified, and the subsequent evolution of relativistic quantum mechanics has not been along the lines envisioned by Struik and Wiener, cf. *Coll. Works*, III.

The paper [29f, g] with Vallarta owes its origin to Einstein's first paper on the unified theory of electricity and gravitation, which appeared in 1929, the year he became fifty, and which was eagerly awaited by the scientific community as well as by the public. In fact, Wiener and Vallarta were approached by Boston newspapers for help in understanding Einstein's equations. In [29f, g] they showed that the only spherically symmetric statical field satisfying the Einstein equations is zero. They communicated this information to Einstein, who thereupon reformulated his equations.

D The philosophical interlude. Leibniz and Haldane

> If I were to choose a patron saint for cybernetics, . . . I should have to choose Leibniz. [61c, p. 12]

In the 1930s Wiener, though disengaged from research in quantum mechanics, pursued its developments with much interest, and mused over their philosophical significance. The steady manifestation of a wave-particle dichotomy in the theory was the object of much philosophical concern.

Rejecting the naive view that when science "outlives a primitive stage, that stage is past for all time", Wiener recommended calling "history to our aid" in a paper entitled "Back to Leibniz" [32c]. The wave-particle dichotomy of quantum mechanics had its prototype in the 17th century controversy over the corpuscular and undulatory theories of light, and Leibniz's ideas could, Wiener felt, shed valuable light on the modern dichotomy. For Leibniz, while he "abjured the vacuum and all action at a distance", believed in the existence of discrete monads, which, however, were not dead but had the power of perception, albeit often a blurred one:

> It does not require much imagination to see an analogy between this mirroring activity of the monad which appears to our confused vision like a causal activity, emanating from one monad and impinging on the other, and the modern view in which the chief activity of the electrons consists in radiating to one another. [32c, p. 202]

Referring to the adaptations by O. Klein and V. Fock of Kaluza's five-dimensional theories to the needs of wave mechanics, Wiener wrote that the (insuperably difficult) many-body problem

> . . . can only be treated on the supposition that each electron carries with it its own three dimensions of spatiality, or what is more likely, its own complete space-time world. Thus, each electron possesses its own world of dimensions, which mirrors the many-dimensional universe of perfect cause and effect in an imperfect, four-dimensional, non-causal image. It is surely not fanciful to see in this a parallel to the Leibnizian monads, which live out their existences in a self-contained existence in pre-established harmony with the other monads, yet mirror the entire universe. [32c, p. 203]

Wiener gave an epistemological interpretation to Leibniz's principle of the identity of indiscernibles: what cannot be observed cannot be discerned and must be abandoned. Thus for him, Leibniz was the spiritual grandfather of the operational method brought into the theories of relativity and quanta by Einstein and Heisenberg and advocated as a general methodology by the physicist P. W. Bridgman {B20}, Wiener's co-student at Harvard. Wiener also referred to the central position of the principle of least action in modern physics, and tracing its origin back to Maupertius, suggested that it follows from Leibniz's principle of sufficient reason. If ours is the "best of all worlds", then surely it must be governed by a few very general and beautiful principles, such as Hamilton's.

Next, Leibniz exposes us to the idea that our world is just one of many possible worlds, viz. "the best" — a notion Voltaire found funny. If instead of assigning the probability 1 to the "best", and 0 to all other possible universes, a more equitable probability distribution is used, we arrive at the statistical conception of things, which J.W. Gibbs and L. Boltzmann propounded in the late 19th century, (cf. §§ 12A, C below). Thus, Leibnizian ideas have only to be modified appropriately to arrive at our modern position.

In 1934 a fresh impetus to Wiener to continue such reflections came from a speculative paper written by J.B.S. Haldane, the eminent British geneticist and versatile scholar. Wiener has vividly described his first encounter with Haldane during the memorable year, 1931–1932, which he spent at Cambridge University as Visiting Professor (see § 11C):

> One day I saw in *The Strand* a first-rate thriller called "The Gold-Makers". It was science fiction, with some very plausible science and economics in it, and it had an excellent plot, with conspiracy, pursuit, and escape. It was written by Professor J.B.S. Haldane, of Trinity College, Cambridge. There on the cover stood the photograph of a tall, powerfully built, beetle-browed man, whom I had repeatedly seen in the Philosophical Library. The next time I saw Haldane at the library, I summoned up my courage to speak to him, to introduce myself, and to express my appreciation of his story. There was one little point in his story, however, which I called to his attention. He had used a Danish name for a character supposed to be an Icelander.
> Haldane welcomed this impertinent suggestion of mine . . . [56g, p. 160]

The Haldanes and the Wieners become close friends.

In his paper Haldane adopts Lenin's rule: whenever materialism starts sinking, save it by redefining the word "matter" in a new way. But apart from this bit of easily eliminable semantic chicanery, (which Wiener easily spotted) witness his Leibnizian retort:

> I see no essential difference between a materialism, which includes soul as a complicated type of material particle, and a spiritualism that includes particles as a primitive type of soul. [34c, p. 480])

Haldane's paper is pithy. Entitled "Quantum mechanics as a basis for philosophy" {H2}, its central theme is that the wave-mechanics of de Broglie and Schrödinger can explain the phenomena of both life and mind, and admits as limiting extremes the billiard-ball atomism of Lucretius and Newton on the one hand, and on the other the ideal world of Plato, these limits being attained as the mass-energy of the system is allowed to tend to infinity or to zero, respectively. The fact that the universe is in-between these extremes is what makes life and mind possible.

More must be said about Haldane's paper because of its influence on Wiener and its place in the development of cybernetics. Living organisms self-regulate, self-repair and self-reproduce, i.e. act as organic units in seeming defiance of the laws of physics and chemistry. To incorporate the living organism into science holistic postulates of one sort or another had been proposed by vitalists. Haldane points out that the atoms of physics also exhibit an organic unity. For instance, they quickly "repair" the loss of an electron by picking up another—a behavior that the Rutherford, Bohr theories of the atom in the 1910s could not explain. Physicists healed this breach not by adding holistic postulates to these theories, however, but by discovering wave-mechanics (c. 1923). The de Broglie, Schrödinger theory was able not only to account for the observed atomic behavior, but to reveal the existence (then unknown) of two isomers of the hydrogen molecule. In his paper Haldane claims that this wave-mechanics serves to explain biological and psychological facts that were a mystery from the standpoint of classical physics and chemistry. The last forty years have borne out Haldane's claim remarkably well. His paper is a fine example of the potency of bold speculation when it comes from a learned and imaginative mind.

Haldane singles out the following deviations of the wave mechanical particle from the classical particle as especially germane:

A) The electron deviates from the classical particle not only in regard to the simultaneous indeterminability of its position and momentum, but in its loss of identity during interaction. Two electrons are extrinsically distinguishable, for instance by their different locations at a given moment. But their strange behavior when they approach one another precludes our keeping track of which is which in the future. In fact, the probability that the two exchange positions tends to 1 as their distance apart tends to 0. The expression "this electron" is therefore meaningless. Likewise for the helium nucleus, the so-called α particle.

B) For these particles the probability of spontaneously crossing ("leaking") over a potential barrier (i.e. attaining states of higher potential energy) is positive, as the phenomenon of radio activity reveals.[2] For the classical particle, such a cross-over is impossible without the supply of outside energy.

C) The stable quantum mechanical configurations—atoms, molecules, crystals—are states in which the probability of exchange is high, i.e. states of so-called high *internal resonance*, or *high degeneracy* in the sense of loss of degrees of freedom:

2 This quantum mechanical property is put to use in the tunneling transistors now in vogue.

In a degenerate system degrees of freedom are lost because certain periodic systems oscillate together instead of independently. This resonance gives rise to various observable phenomena. It is responsible for certain terms in the energy of a material system. As the resonators are removed from one another, the energy falls off very rapidly. {H2, p. 87}

Haldane points to some biological phenomena that exhibit the same sort of behavior. (a) During reproduction the gene loses its identity: it is hard to tell parent from daughter. (b) Quantum-mechanical resonance is behind the larger molecules, and larger building blocks of organic matter, e.g. the joining of amino acids that produces the protein molecule and chains thereof. Such resonance is what accounts for the self-regulatory and reproductive aspects of life.

Haldane draws an interesting analogy between crossing a potential barrier and purposive action on the part of brain-possessing organisms. Water in a cistern filled to one inch below the rim will not climb over the rim in order to spill out. But for a mouse placed in a cistern, who can see a piece of cheese lying outside, there is a positive probability that it will get out (or "leak") by climbing over the rim and jumping out. As with the quantum mechanical potential barrier, the probability of crossing will depend on the strength of the barrier, in our case, the height of the cistern.

Haldane gives a simple explanation for this analogous behavior of mice and electrons: "... the electron can penetrate its potential barrier because its wave system effectively extends beyond it" {H2, p. 89}. But mice and men and other objects too have wave-systems that extend beyond their bodies. In 4-dimensional spatio-temporal terms, the wave systems extend into the future as well. It is this that enables cerebrally organized creatures to act with reference to future and distant events such as the joys of munching cheese or winning a tennis match or observing a solar eclipse, i.e. to act purposively. Thus Haldane brings the mind into his quantum mechanical framework by interpeting it as a far-reaching wave system associated with the brain.

Haldane presses on with such hypostatizing. For instance, thought of the number 3 is a wave system of exceedingly low energy. The uncontemplated ideal 3 of Plato is its limit as this energy tends to zero. Nor does Haldane shirk harder concepts like the "divine mind" or those of Vedanta. As he well points out:

In truth mind is permeated with internal contradiction at every level. We can reason about eternity, but we are born and die. We measure in kiloparsecs and live in centimeters. We aspire and fail, we think and fall into error, we love and hate at once. A truly dialectical idealism would not meet these facts with remorse or denial, but with frank acceptance. {H2, p. 93}

Haldane's reflections are not limited to epistemology, they encompass the domain of human social action as well. He envisions social configurations under which the painful effort to cross a potential barrier becomes worthwhile, and treats duty, altruism and other ethical precepts in these terms.

By transgressing scientific frontiers and emphasizing purposive behavior, Haldane was treading on cybernetical soil. It is easy for us, knowing as we do the cybernetical direction in which Wiener was headed, to sense his elation with Haldane's paper. But Wiener's point of view was not quite that of Haldane's. His interesting response to the paper occurred in two steps:

1. In a long letter submitted to the *Philosophy of Science*, which got captioned "Quantum mechanics, Haldane and Leibniz" [34c], Wiener first points out the affinities between Haldane's position and that of Leibniz. He reminds us that resistence to the "dictatorship of mechanics", of which quantum mechanics is a beneficiary, began with Leibniz and Huygens. Once Leibnizian monadism is shorn off the ontological labels "pluralistic spiritualism" or "pluralistic materialism", its proximity to wave mechanics becomes clear:

> Only a small part of the monads are souls, the greater part being naked monads gifted with apperception rather than perception or consciousness. They mirror the universe, but do not integrate this activity of mirroring into a self-conscious consciousness. This mirroring is best to be understood as a parallelism, incomplete it is true, between the inner organization of the monad, and the organization of the world as a whole. The structure of the microcosm runs parallel to that of the macrocosm. [34c, p. 480]

To bring out the bearing of monadism on Haldane's theory of cognition, Wiener again refers to the ideas of O. Klein and V. Fock we mentioned earlier, and concludes:

> The more organized a system of elementary particles is, the less "naked" its electrons and other constituent elements, the better it mirrors the universe, and the better can we read back from our partial system to the universe at large. [34c, p. 480]

Moreover, from the global standpoint monadism was supportive of the variational formulation of natural laws, and it was the similarity between the variational formulations of the laws of optics and of mechanics (rather than their Newtonian formulations) that led de Broglie to his wave mechanics.

On the other hand, a monad, unlike the electron, preserves its identity, and monadic action, unlike the quantum mechanical, is deterministic. Wiener attributes these divergencies of monadism from

Gottfried Wilhelm Leibniz
1646–1716

quantum mechanics to the vicissitudes of history rather than to any intrinsic limitation of Leibnizian philosophy. Had modern data on electron interchange and on radio activity been accessible to Leibniz, he would have modified his monadic theory appropriately, like any good scientist.

Strangely, Wiener fails to mention an aspect of Leibnizian thought with which he was quite familiar, and which is quite relevant to Haldane's paper, viz. the de-alienation of matter and life in monadism. This is well emphasized in a 1941 paper by G.D. Birkhoff on "The Principle of Sufficient Reason" {B11}. The monads perceive and some

J.B.S. Haldane
(c. 1930)
1892–1964

are conscious. Birkhoff recounts the story of how Leibniz remarked to Hantschius over a cup of coffee, "There may be in this cup a monad which will one day be a rational soul!" {B11, p. 41}. There is, in fact, a remarkable concordance in the far-reaching significance that both Birkhoff and Wiener attribute to Leibnizian thought.

2. Wiener knew that nothing beyond imaginative speculation lay behind Haldane's conclusions. So of course did Haldane himself, who acknowledged that his opinions "may be rendered ridiculous by the progress of physics or biology" {H2, p. 97}. Consequently, Wiener sought to arrive at Haldane's generalizations without recourse to a "theory of the mind as a highly degenerate quantum mechanical system" [36g, p. 318]. This he did in the paper [36g] completed only in 1936 in China (§ 13C). In this paper Wiener sketches a scientific methodology wherein the coupling of the observer and the object observed is universal, i.e. one that can serve for all the sciences. The position he arrives at, based on four maxims ([36g, p. 318]), is close to that of complementarity

that Niels Bohr describes in his book {B14}. In the course of this paper Wiener describes science as a technique for wading through a sea of pointer readings: "Physics is merely a coherent way of describing the readings of physical instruments" [36g, p. 311]. This half-truth, clearly at variance with his earlier views, was soon to be abandoned. It is one of the solitary instances in which this very coherent thinker articulated badly.

E   Quantum theory and the Brownian motion. Hidden-parameters.

In the late 1940s a most interesting use of Wiener's Brownian motion in quantum theory was revealed by M. Kac's analysis of the "path integral" defined in R. P. Feynman's important thesis. This stems from the deep resemblance between the "path integration" employed in quantum mechanics and integration over the space $C[0, \infty)$ with respect to the "Wiener measure" induced over this space by the Brownian motion stochastic process. With "imaginary" time, i.e. with t replaced by $t\sqrt{-1}$, in the equations, it allows us to use the Brownian motion to prove theorems on Hilbert spaces germane to quantum field theory. This topic is too difficult to discuss here, and we would refer the reader to E. Nelson's commentary in the *Coll. Works*, III, pp. 565–579.

During the 20 years that intervened between Wiener's 1929 paper and this event, his methodological work on statistical mechanics had made him aware of the potentialities of the Brownian motion as a unifying concept. In the early 1950s it dawned on him that the Brownian motion might be used to explain in Gibbsian terms a question in quantum mechanics that had always intrigued him, viz. the strange appearance of probabilities as squares of amplitudes. Thus began a period of renewed activity in quantum mechanics that was to last over a decade.

Wiener undertook this research in collaboration with the younger colleagues A. Siegel and J. Della Riccia. This complicated research remains unfinished, and is hardly understood. However, Wiener's central idea is interesting, and well worth a brief presentation.

In the elegant formulation of classical quantum theory due to G. W. Mackey {M1} among others, the *states* are countably additive probability measures μ on the lattice $\mathscr{L}$ of projections P of a separable complex Hilbert space $\mathscr{H}$. By Gleason's fundamental theorem {G5}, to each μ corresponds a unique non-negative self adjoint operator A of trace 1 such that $\mu(P) = \text{trace } (AP)$, $P \in \mathscr{L}$. For the *pure states* μ, the extreme points of the state space, the corresponding A's turn out to be projections of rank 1. Accordingly, to any pure state μ corresponds a unit vector ψ in $\mathcal{H}$ such that

$$\mu(P) = |P\psi|^2_{\mathcal{H}}, \quad P \in \mathcal{L}. \tag{2}$$

This then is the probability of finding the quantum mechanical system in the state P: it is the square of the "amplitude" $|P\psi|_{\mathcal{H}}$.

Now to the Wiener-Siegel reformation. Given any quantum mechanical system, they exhibit a probability space $(\Omega, \mathcal{A}, \rho)$ and for any pure state μ, a function M_μ on the lattice $\mathcal{L}$ to the σ-algebra $\mathcal{A}$ such that

$$\mu(P) = \rho\{M_\mu(P)\}, \quad P \in \mathcal{L}. \tag{I}$$

In this representation, the choice of the space $(\Omega, \mathcal{A}, \rho)$ and the mapping M_μ are crucial. It will suffice to say that $(\Omega, \mathcal{A}, \rho)$ is the space of a multiparameter Brownian motion (Ch. 7, Note 5). In this way the Brownian motion enters the picture. For more on this and on M_μ, see the Note, p. 130.

By establishing (I) Wiener and Siegel fulfill their goal of showing that the probability $\mu(P)$ appearing in quantum mechanics as the square of the amplitude $|P\psi|_{\mathcal{H}}$ is the probability of a well-defined subset $M_\mu(P)$ of a well-understood probability space $(\Omega, \mathcal{A}, \rho)$, viz. a generalized Wiener space.

Unfortunately, it is not clear from the Wiener-Siegel work if an equality of the type (I) is available for impure states μ. Furthermore, Wiener and Siegel do not discuss the physical relevance of the Brownian motion as yielding "hidden parameters" in quantum mechanics, although this notion is central to their approach. This research is unfinished, and its significance is as yet far from clear.

The possibility that Brownian motion theory may have a role to play in quantum theory has occurred to several scientists, e.g. Bohm, de Broglie and Vigier, who follow Einstein and question the completeness of the existing quantum mechanics, and believe that the hidden-parameter problem is worth investigation, but perhaps in formulations differing from von Neumann's. As de Broglie has put it in his foreword to Bohm's book:

> To try to stop all attempts to pass beyond the present viewpoint of quantum physics could be very dangerous for the progress of science and would furthermore be contrary to the lessons we may learn from the history of science. This teaches us, in effect, that the actual state of our knowledge is always provisional and that there must be, beyond what is actually known, immense new regions to discover. Besides, quantum physics has found itself for several years tackling problems which it has not been able to solve and seems to have arrived at a dead end. This situation suggests strongly that an effort to modify the framework of ideas in which quantum physics has voluntarily wrapped itself would be valuable. {B13, px}.

Here, the words of Professor Nelson are particularly germane:

> What motivates Einstein, Schrödinger, de Broglie, Wiener, Bohm, and a host of others to reject the orthodox interpretation of quantum theory and to seek alternatives? I believe that this is not an isolated phenomenon but is the expression of a deep drive within science.
>
> When a science is young, its practitioners are struck by the differences which distinguish it from older sciences treating simpler phenomena. New forms of explanation are discovered which are quite different conceptually from the older forms. As these achieve successes in the new science, the strong temptation arises to elevate these new forms of explanation into philosophical principles which establish a fundamental cleavage between the new science and older theories. Later a countercurrent sets in, and attempts are made to reduce the new theory to older principles. When these attempts at reduction are successful—as, for example, in the synthesis of urea, the explanation of the physical nature of the chemical bond, and the discovery of the molecular structure of the gene—science is greatly enriched. This contrasts with the scientific aridity of antireductionist philosophical principles such as vitalism. {*Coll. Works*, III, p. 575}

The general idea in the attempts to resist the "temptation" and to build creatively on older forms has been to interpret the randomness in the behavior of atomic particles as the statistical manifestation of influence stemming from a causal order prevailing at a hitherto unpenetrated and unobservable sub-atomic level. The Brownian motion is of course the prime example of a situation in which chaos at a certain level is consistent with orderliness at another more microscopic level, and its appeal to investigators in the quantum field is not surprising. However, as Nelson points out, "such a reduction must be an explanation not an analogy. It must not lean on primitive concepts and must derive the value of Planck's constant from classical physics." This goal may, as he says, be unattainable. "That which is far off, and exceedingly deep, who can find it out?" (Ecclesiastes, 7).

Note: The mapping M_μ

Involved in the crucial mapping M_μ is the *homogeneous chaos* corresponding to the complex valued q-parametrized Brownian motion defined in Ch. 7, Note 5. (For the homogeneous chaos, see § 12E below.) Following Kakutani {K3} we may introduce this chaos, also called the *Brownian motion random measure* by the four conditions:

(a) Its domain is the δ-ring $\mathscr{D}_q$ of all Borel subsets D of $\mathbb{R}^q$ for which $m(D) < \infty$, m being the q-dimensional Lebesgue measure;

(b) for each D in $\mathscr{D}_q$, $\xi(D)$ is a real- or complex-valued random variable over a probability space $(\Omega, \mathscr{A}, \rho)$, which is normally distributed with mean 0 and variance $m(D)$; [3]

(c) ξ is countably additive under the norm of the space $L_2\Omega, \mathscr{A}, \rho)$;

(d) ξ is independently scattered, i.e. for disjoint $D_1, \ldots, D_n$ in $\mathscr{D}_q$, the random variables $\xi(D_1), \ldots, \xi(D_n)$ are independent.

An elegant definition can be given for the stochastic integral

$$\{\int_{\mathbb{R}^q} \varphi(x)\xi(dx)\}(\,\cdot\,), \quad \text{where } \varphi \in L_2(\mathbb{R}^q), \tag{3}$$

as a random variable on Ω. It subsumes the stochastic integration that Wiener introduced in the early 1930s, cf. § 7D(5).

The crucial mapping $M_\mu(\,\cdot\,)$ is defined as follows. Suppose for definiteness that the complex separable Hilbert space $\mathscr{H}$ of the quantum mechanical system is $L_2(\mathbb{R}^q)$, and that the pure state μ is represented as in (2) by the unit vector ψ in $L_2(\mathbb{R}^q)$. Let $(\Omega, \mathscr{A}, \rho)$ be the probability space appearing in (b), and define the complex-valued random variable $f_\mu(P)$, P in $\mathscr{L}$, on Ω by

$$f_\mu(P)(\omega) = [\int_{\mathbb{R}^q} \{P(\psi)\}(x)\xi(dx)]\,(\omega), \quad \omega \in \Omega,$$

the last being the stochastic integral mentioned in (3). Next, define

$$M_\mu(P) = \{\omega : \omega \in \Omega \ \& \ |f_\mu(P)(\omega)| \geq |f_\mu(P^\perp)(\omega)|\}$$

It then follows from [58i, pp. 78–84] and (2) that

$$\rho\{M_\mu(P)\} = |P\psi|^2_{L_2(\mathbb{R}^q)} = \mu(P).$$

Thus the equality (I) is established. Note that we can write (I) in the form

$$\mu(P) = \int_\Omega \upsilon_\mu(\omega)(P)\rho(d\omega) \quad \text{or} \quad \mu = \int_\Omega \upsilon_\mu(\omega)\rho(d\omega), \tag{II}$$

upon letting $\upsilon_\mu(\omega)(P) = 1_{M_\mu(P)}(\omega)$. Thus the establishment of (I) seems to be a new step in the search for so-called "hidden parameters", cf. von Neumann {V2, pp. 209–210, 323–}.

---

3 The normal distribution N over $\mathbb{C}$ with mean 0 and standard deviation σ (i.e. variance σ^2) is defined by

$$N(B) = \int_B \frac{1}{\sigma\sqrt{(2\pi)}} e^{-|z|^2/2\sigma^2} v(dz), \quad B \subseteq \mathbb{C},$$

where v is the Haar measure over $\mathbb{C}$, which assigns the value 1 to the unit square in $\mathbb{C}$. Kakutani, like Wiener, considered only real valued random variables $\xi(D)$. But the complex case is important, particularly for the Wiener-Siegel work on quantum theory.

11 The Collaboration with E. Hopf and R. E. A. C. Paley

A Radiative equilibrium in the stars

Wiener needed contacts outside mathematics for the inception of his own ideas. The electrical engineers at MIT had fulfilled this need during the 1920s, as had Max Born, cf. §§ 9A, 10B. But in the early 1930s it was G. D. Birkhoff's great discovery of the Individual Ergodic Theorem that brought Wiener into touch with one of his important collaborators. Soon after the appearance of Birkhoff's paper in 1931, a young German mathematical astrophysicist from the Potsdam Observatory, Eberhard Hopf, came to Harvard to study ergodic theory. As an ardent Gibbsian and Lebesguist, Wiener was quite excited by Birkhoff's discovery, and Hopf and he began discussing it in earnest. These conversations were instrumental in propelling Wiener to a proper appreciation of Birkhoff's theorem and toward a conception of statistical mechanics in which the Birkhoff theorem is central (cf. § 12A).

What first ensued from their discussions, however, were not theorems on ergodicity but on an equation that had appeared in the theory of radiative equilibrium in the stars. This equation governs the radiative transfer in a so-called plane parallel atmosphere of infinite optical depth with perfect isotropic scattering. The underlying assumptions introduce a "stationarity" into the system in the following way.

Our interest is in regions of the stellar atmosphere so far away from the surface of the radiating core that this surface is viewed as a plane, infinitely far away. The atmosphere is thus stratified into parallel planes, over each of which physical conditions are invariant. The physical conditions are thus determined by a single parameter that characterizes the plane. This parameter is not, however, the ordinary distance of the plane from the plane of observation, but the so-called *optical depth t* of the former.[1] At each t there is a "radiating source" of strength $f(t)$ equal to the ratio of the emission coefficient at t to the absorption coefficient at t. From the ideal scattering hypothesis, it follows that the

---

1   The optical depth of the plane at distance z from the origin is $\int_z^\infty \varkappa(z')\rho(z')dz'$, where $\varkappa(z')$ and $\rho(z')$ are the absorption coefficient and material density at depth z', cf. Hopf {H8}.

Eberhard Hopf
1902–1983

influence on the source at t due to that at τ does not depend on t and τ individually but on the difference $t-\tau$ alone, i.e. the dynamics is invariant under a translation of optical depth. In brief, the dynamics is depth-invariant. Consequently, the weighting factor to be assigned to $f(\tau)$ from the standpoint of t takes the form $W(t-\tau)$ rather than $W(t, \tau)$, and we arrive at the equation

$$f(t) = \int_0^\infty W (t - \tau) f(\tau) d\tau, \quad t \geq 0,$$

from which the unknown f can be found. This is the homogeneous case with $g = 0$ of the general equation

$$f(t) = g(t) + \int_0^\infty K(t - s) f(s) ds, \quad t \geq 0 \tag{1}$$

where $K(\cdot)$ and $g(\cdot)$ are given, and the unknown is $f(\cdot)$. The eminent astronomer E. F. Freundlich had apprised Hopf of the need for a mathematically more mature treatment than the one known to the astronomers. Wiener and Hopf proceeded to give such a treatment for the homogeneous case of (1).

Hopf's collaboration with Wiener was very short. They discussed the equation intensely one afternoon in Wiener's New Hampshire home. Wiener, who had come across similar equations in his study of electric circuits, saw the causal significance of the equation when t is interpreted as time. (In the radiation problem t is the optical depth, as we saw.) The next morning he came up with a solution. This turned out to be wrong, but the methods which he had employed proved extremely fruitful. Hopf was able to clear up the error and obtain a complete solution without much trouble.[2] The result was their famous joint paper [31a] *Über eine Klasse singulären Integralgleichungen.* Wiener wrote only one other paper on the subject [46a] in collaboration with A. E. Heins.

B Significance of the Hopf-Wiener equation; causality and analyticity.

It soon became clear that equations of the Hopf-Wiener type have wide applicability, as they cover many situations in which a barrier separates two different regimes, one of which can influence the other but not vice versa. In stellar radiation the barrier is the surface of the radiating core, and conditions inside influence those outside. Likewise with the atomic bomb, which is essentially the model of a star in which the surface of the bomb marks a change between an inner regime and an outer regime. But Wiener also perceived that in problems of Bergsonian temporal development, the "present" acts as a buffer between the influencing "past" and the indeterminate "future". The cogency of this perception emerged about ten years later when Wiener's war work on anti-aircraft fire control led him to his theories of prediction and filtering, and he readily came up with a Hopf-Wiener equation from his data. This equation, however, is of the form

$$g(t) = \int_0^\infty K(t-s)f(s)ds, \quad t \geq 0 \tag{2}$$

or its Stieltjes variant with $df(s)$ replacing $f(s)ds$. Here again $f(\cdot)$ is the unknown. This equation is called a Hopf-Wiener equation of the *first kind*, to distinguish it from (1), which is called a Hopf-Wiener equation of the *second kind*.

The Hopf-Wiener work reveals the deeper aspects of the area of analysis called *causality and analyticity*, which Wiener had uncovered in his papers [26c, 29c] by defining causal operators and studying the analyticity of their Fourier transforms, cf. § 9B, the underlying ideas of which have since been extended, cf. {F3}, {S1}. By a change of variables the equation for a causal filter, cf. § 9B(11), can be brought into the form

---

2   For this information I am grateful to Professor Hopf.

$$g(t) = \int_0^\infty f(t - s)W(s)ds, \quad t \in \mathbb{R} \tag{3}$$

which is akin to (2). But the complete lack of restriction on t in (3) makes it much easier to deal with than the equation (2).

C Paley and the Fourier transformation in the complex domain

The equations (1), (2), (3) share the important characteristic that though the variable t in them is in the real domain, the nature of their solutions depends on the analytical behavior of the Fourier transforms of f, g, K and W in the complex domain. But the solutions of (1) and (2) require the very important *factorization technique* that Hopf and Wiener introduced in [31a]. The ramifications of this technique have been important and widespread, as the commentaries of Professors Pincus and Kailath in the *Coll. Works*, III, testify. Strangely enough, it was this stellar radiation problem that led Wiener to undertake some deep, pure mathematical work on the border area between real and complex analysis in which his own mentors Hardy and Littlewood were leaders and whose work he had hitherto bypassed. Wiener has described this border area as follows:

> In the last hundred and fifty years, analysis, which represents the modern extension of the infinitesimal calculus, has split into two main parts. On the one hand we have what is known as the series of functions of a complex variable, which is a continuation of the eighteenth-century theory of series proceeding in the powers of a variable such as 1, x, x^2, etc. This theory is particularly applicable to quantities which change smoothly and gradually. At one time it was supposed that all important mathematical quantities changed smoothly and gradually; but toward the end of the eighteenth century, the study of harmonic analysis, which is analysis of vibrating systems, showed that curves pieced together out of parts which had nothing to do with one another were subject to an analysis of their own. This point of view first led to the theory of Fourier series and then to the general branch of study known as the theory of functions of a real variable.
>
> The theory of functions of a real variable and the theory of functions of a complex variable thus represent two separate but related subjects, which do not succeed one another like a freshman and a sophomore course but represent radically different insights into the nature of quantity and of the dependency of quantities upon one another. They have had a great deal of interplay in the course of the last hundred and fifty years. However, it is only recently that it has come to the attention of mathematicians and there are certain intermediate fields of work which share the methodology of both. [56g, pp. 90, 91]

Wiener brought many fine ideas to bear on this border area of analysis, but their fruition demanded detailed work for which he did not always have the patience to proceed single-handedly. Luckily, Wiener was able to find the right collaborator in a budding aspirant to the British mathematical establishment, during his visit to England.

Wiener spent the year 1931–1932 in Cambridge University where, acting as Hardy's "deputy", he gave lectures on the Fourier integral that were subsequently published by the Cambridge University Press [33i]. While his primary contacts in Cambridge were with Hardy and Littlewood, Wiener took the opportunity to mingle with several of their distinguished colleagues. Thus he got to know L.C. and R.C. Young, fellows of Peterhouse and Girton Colleges, son and daughter of the illustrious British mathematician W.H. Young. Indeed, R.C. Young joined Wiener in a paper [33d]. Wiener also had his fruitful encounter with J.B.S. Haldane during this period, as we saw in § 10D. But the contact that proved creatively most rewarding was with R.E.A.C. Paley, "a young Trinity don" with an "unlimited admiration for Littlewood" both as mathematician and Alpinist, whom he met thanks to Miss Mary Cartwright, mistress of Girton College and herself a fine mathematician [56g, pp. 152, 167].

Wiener started discussing the border area with Paley in England, and they talked again at Zürich in the summer during the International Congress of Mathematicians, but their real collaboration began only when Paley came to MIT on a British Commonwealth Fellowship in 1932–1933.

During the Christmas break 1931, Wiener lectured at Hamburg at the behest of Professor William Blaschke, the distinguished German mathematician. He then proceeded to Vienna to lecture at the invitation of Professor Karl Menger, the distinguished geometer and topologist. There Wiener met Menger's assistant, the 24-year old K. Gödel, who had just written his revolutionary paper {G8} of which we spoke in § 5B. Wiener then headed for the Germanic University of Prague on the invitation of the physicist and natural philosopher Philipp Frank, who was then holding the chair that Einstein had vacated. In Prague Wiener, apart from lecturing, visited President Thomas Masaryk, his father's old friend, and one of the last in the great liberal democratic tradition of Benedetto Croce. From Prague Wiener went to Leipzig to see Professor Leon Lichtenstein, a distant relative of his, and to give more lectures. Wiener then returned to Cambridge to lecture as "Hardy's deputy". In the summer he proceeded to the International Mathematical Congress in Zürich, and then returned to MIT.

At MIT Paley and Wiener worked on the analytical properties of the Fourier transform in the complex domain. Unfortunately, their collaboration lasted barely two terms, for Paley died in a skiing accident in April 1933. Their joint results are contained in several papers [33a,

R.E.A.C. Paley
1907–1933

33c, 33e] and in the American Mathematical Society Colloquium Lectures that Wiener delivered in 1934, and which he published under their joint names [34d]. This book, which covers a great many topics, is "unified by the central idea of the application of the Fourier transform in the complex domain". It demonstrated that a wide range of problems from complex function theory are amenable to a treatment by Fourier transformation. It was a very important contribution to the border area we spoke of, in which the methodologies of real and complex analysis have to be fused. This area continues to occupy a central position in research, and even advanced pedagogy, cf. Rudin {R3}.

Paley's untimely death came as a severe blow to Wiener. On him fell the sad duty of notifying Paley's mother and his British friends. "It took me some time to come back to a mental equilibrium sufficient to permit my further work . . ." [56g, p. 170]. Some of Wiener's words about his collaboration with Paley are worth quoting, since they reveal both his colossal indebtedness to Paley, as well as the difference between

Paley's establishmentarian attitude to difficult problems and his own
more lively approach to them:

> Paley and I used to work together on the great blackboard of a dusty,
> half-lighted, abandoned classroom . . . My role was primarily that of sug-
> gesting problems and the broad lines on which they might be attacked, and
> it was generally left to Paley to draw the strings tight.
>   He brought me a superb mastery of mathematics as a game and a
> vast number of tricks that added up to an armament by which almost any
> problem could be attacked, yet he had almost no sense of the orientation of
> mathematics among the other sciences. In many problems which we under-
> took, I saw, as was my habit, a physical and even an engineering application,
> and my sense of this often determined the images I formed and the tools by
> which I sought to solve my problems. Paley was eager to learn my ways, as
> I was eager to learn his, but my applied point of view did not come easily to
> him, nor I think, did he regard it as fully sportsmanlike. I must have shocked
> him and my other English friends by my willingness to shoot a mathematical
> fox if I could not follow it with the hunt. [56g, pp. 167–168]

A very important engineering application that Wiener saw but Paley
perhaps did not, is the *Paley-Wiener criterion* for the physical realizability, i.e. causality, of a time-invariant linear filter for which only the
absolute value of the frequency response $\hat{W}$ is prescribed, viz.

$$\int_{-\infty}^{\infty} \left|\log|\,\hat{W}(\lambda)\,|\,\right|/(1 + \lambda^2)d\lambda < \infty.$$

12 Birkhoff's Theorem and the Consolidation of Wiener's Ideas on Statistical Physics

A The need for ergodicity in Gibbsian statistical mechanics

Birkhoff's Individual Ergodic Theorem, the discovery of which in 1931 had fortuituously resulted in Wiener's important collaboration with Hopf, became a permanent article in Wiener's scientific thought.

Concerning this theorem Wiener has written:

> The fundamental assumption of Gibbs, for example, that an ensemble of dynamic systems in some way traces in the course of time a distribution of parameters which is identical with the distribution of parameters of all systems at a given time, was first stated in a form not merely inadequate, but impossible. This is the famous ergodic hypothesis. It is one of the greatest triumphs of recent mathematics in America, or elsewhere, that the correct formulation of the ergodic hypothesis and the proof of the theorem on which it depends have both been found by the elder Birkhoff of Harvard. [38e, p. 63]

Wiener's high estimation of this theorem stemmed from his appreciation of the crucial part played by the ergodic hypothesis in Gibbs's statistical mechanics, and from the fact that it beautifully confirmed his long-standing conviction that the mathematical drawbacks in Gibbs's creative work would find their circumvention in the ideas of Lebesgue. In his *Cybernetics* Wiener alludes to the differences between Gibbs and Lebesgue as scientists, but adds:

> Nevertheless, the work of these two men forms a single whole in which the questions asked by Gibbs find their answers, not in his own work but in the work of Lebesgue. [61c, p. 45]

Wiener also valued the Individual Ergodic Theorem for the new light it shed on his generalized harmonic analysis. The result that any strictly stationary stochastic process is equivalent to one governed by a measure-preserving flow, cf. {D5, pp. 509–511}, allows us to re-enunciate the theorem in Wienerian terms:

Ergodic Theorem (G. D. Birkhoff, 1931). *Let $\{f(t, \cdot): -\infty < t < \infty\}$ be a complex-valued strictly stationary stochastic process over the probability space $(\Omega, \mathscr{B}, P)$, and let $f(0, \cdot) \in L_2(\Omega, \mathscr{B}, P)$. Then for P almost all ω in Ω, the signal $f(\cdot, \omega)$ belongs to the Wiener class S, and its covariance function $\varphi(\cdot, \omega)$ satisfies the equality*

$$\varphi(\tau, \omega) = E_{\mathscr{I}}\{f(\tau, \cdot)\overline{f(0, \cdot)}\}(\omega), \quad \tau \in \mathbb{R} \tag{1}$$

where $E_{\mathscr{I}}\{\,\cdot\,\}$ is the conditional expectation with respect to the σ-algebra $\mathscr{I}$ of sets in $\mathscr{B}$, invariant under the measure-preserving flow associated with the process. In case the process is ergodic ("metrically transitive"), i.e. $\mathscr{I}$ is trivial,

$$\varphi(\tau, \omega) = E\{f(\tau, \cdot)\overline{f(0, \cdot)}\}(\omega), \quad \tau \in \mathbb{R}, \tag{2}$$

where $E\{\,\cdot\,\}$ is the (unconditional) expectation.

Wiener valued the ergodic theorem not just for its justification of the deep equalities (1) and (2), but also for its validation of his long-cherished belief that

$$\lim_{T \to \infty} \frac{1}{T} \int_{-T}^{0} f(t + \tau)\overline{f(t)}dt = \lim_{T \to \infty} \frac{1}{2T} \int_{-T}^{T} f(t + \tau)\overline{f(t)}dt. \tag{3}$$

The average on the left involves only the past values of the signal f and is thus in principle estimable from observation, whereas the average on the right is not, since it involves the unknown future.

The Birkhoff theorem thus gave Wiener just what the Gibbsian in him wanted to know: *past observations of the signal suffice for the estimation of its covariance function, and this function gives information intrinsic to the stochastic process from which the signal hails.* The fact, that ergodicity has to be postulated in order that the covariance may have an unconditional probabilistic significance, did not deter Wiener at all. This assumption was just one of many idealizations that his scientific philosophy permitted. Moreover, von Neumann's celebrated 1932 theorem on the disintegration of regular measure-preserving flows over complete metric spaces into ergodic sections {V4} gave Wiener a good excuse to deal almost exclusively with the ergodic case, cf. for instance, [61c, pp. 55–56].

B Wiener's work in Ergodic Theory

The first evidence of Wiener's active involvement in this area appears in the last chapter, devoted to ergodic theory, of his book with Paley [34d]. There Paley and Wiener demonstrate the existence of a flow T_t, $t \in \mathbb{R}$, which preserves Lebesgue measure over [0, 1] and is ergodic, and such that for almost all α in [0, 1],

$$x(b + t, \alpha) - x(a + t, \alpha) = x(b, T_t\alpha) - y(a, T_t\alpha), \quad t \in \mathbb{R}, \tag{4}$$

where $x(\,\cdot\,,\,\cdot\,)$ is the Brownian motion stochastic process defined in § 7D, cf. [34d, § 40]. The use of this result, along with Birkhoff's, appreciably simplifies certain proofs in the memoir [30a] on G.H.A. but it is a fundamental result in its own right.

In the late 1930s Wiener began a deep study of ergodic theory. In the paper [39a], he deduced the Birkhoff theorem from von Neumann's mean ergodic theorem, by ingenious use of a neglected lemma of Birkhoff himself. He also extended the ergodic theorems to measure-preserving flows with *several* parameters, i.e. flows T_λ, where $\lambda \in \mathbb{R}^n$, $n > 1$, thereby making these theorems available in the study of spatial or spatio-temporal *homogeneous random fields*. For the latter fields, $\lambda = (x, y, z, t)$ represents the space-time coordinates of an evolving random process, in a certain quantity $f(\lambda, \omega)$ of which we are interested. In collaboration with A. Wintner, Wiener also proved that almost all signals f, which emanate from an ergodic stationary stochastic process in L_2, are cross-correlated with the characters e_λ, $e_\lambda(t) = e^{it\lambda}$, i.e. possess generalized Fourier coefficients, [41a, b]. More fully, they showed that if in the Birkhoff theorem the invariant σ-algebra $\mathscr{I}$ is trivial, then for P almost all ω in Ω and *every* λ in $\mathbb{R}$

$$\lim_{T \to \infty} \frac{1}{2T} \int_{-T}^{T} f(t,\omega) e^{-i\lambda t} dt \quad \text{exists.}$$

They also gave conditions under which such a signal $f(\,\cdot\,, \omega)$ will be a Besicovitch almost periodic function. These results are significant in view of the presence of lots of functions in the unrestricted Wiener class S that are not cross-correlated with the characters e_λ.

In an earlier paper [38a] Wiener embarked on an even greater generalization of the ergodic theorem. With an eye on the intractable problem of turbulence, he brought ergodic concepts to bear on the stochastic integrals of the Brownian motion which he had introduced much earlier. The inception of these ideas depended, however, on the much clearer perception he had by then acquired on the nature of cosmic contingency. To this we must now turn.

C The contingent cosmos, noise, and Gibbsian statistical mechanics

The understanding that Wiener acquired from his ergodic work was instrumental in deepening his earlier appreciation of Gibbsian mechanics.

Gibbsian mechanics is an offshoot of classical mechanics. In the latter we assume that the material system placed in a field of force is *deterministic*, i.e. given the laws governing the field of force and the "state" of the system at an instant t, its "state" at any subsequent instant τ $(\tau > t)$ is determined. As the system moves under the action of the force-field, its state s traces out a trajectory in the *state-space* $\mathcal{S}$.

In the formulation of mechanics given by W. R. Hamilton in the 1830s the *state* of a material system of N degrees of freedom is prescribed by its *phase*:

$$(q, p) = (q_1, \ldots, q_N; p_1, \ldots, p_N) \in \mathbb{R}^{2N},$$

where $q_1, \ldots, q_N$ are the N generalized coordinates of configuration and $p_1, \ldots, p_N$ are the corresponding generalized components of momenta. The state-space $\mathcal{S}$ (now called *phase space*) is a subregion of $\mathbb{R}^{2N}$. A fundamental theorem of Hamilton tells us that when the force-field is the negative gradient of a potential Φ, the differential equations of the trajectory in $\mathcal{S}$ are the *canonical equations*:

$$\frac{dq_i}{dt} = \frac{\partial H}{\partial p_i}, \quad \frac{dp_i}{dt} = -\frac{\partial H}{\partial q_i}, \quad i = 1 \ldots, N,$$

where H is the function on $\mathcal{S} \times \mathbb{R}$ such that $H(q, p, t)$ is the total energy of the system in the phase (q, p) at the instant t. H is called the *Hamiltonian* of the system. (For its definition see Note 1, p. 157).

It follows that for time-independent force-fields, H cannot depend on t. It also follows that the field is *conservative*, i.e. the total energy of the system does not change as it moves. Consequently, the phase trajectory of any system with energy E_o is confined to the E_o-*energy hypersurface*

$$\Sigma: \quad H(q, p) = E_0,$$

in the phase space $\mathcal{S}$.

For macroscopic systems with a small number N of degrees of freedom, whose phase at some initial moment can be determined with reasonable accuracy, we can obtain the equations of the phase trajectory by solving Hamilton's canonical equations. But this procedure is clearly impossible (even today with the best of computers) for a system such as a cubic centimeter of gas at room temperature, which has approximately 3×10^9 molecules, and consequently about 18×10^9 degrees of freedom.[1]

---

1   This is because at best the molecules are rigid, and a rigid body has 6 degrees of freedom. Its configuration is determined by fixing three non-collinear points A, B, C, which we may suppose are at unit distance apart. This gives 9 coordinates. But these are not independent since $AB = BC = CA = 1$. The number of independent coordinates is 6.

It was Gibbs's key idea to define the "state" of such systems for which N is enormous not by the phase (q, p), but by a *probability distribution* P over the phase space. Now *the state-space is the class of probability distributions P over the phase space*. The problem of statistical mechanics is to deduce the evolution of the probability P, given the laws governing the force-field and the probability P_0 at some initial moment t_0.

Even with a deterministic cosmos the Gibbsian attitude of bringing in probabilities has considerable scientific merit. But it assumes a compelling inevitibility when judged from the perspective of the molecular kinetic conception of the world that emerged from the efforts of Maxwell and Boltzmann in statistical mechanics. As we saw in § 7D, this perspective was accepted without skepticism after the confirmation of Einstein's theory of the Brownian motion in the 1910s.

This molecular kinetic theory of matter has a pre-history that goes back to Democritus. It tells us that all atoms perform at all times a disorderly motion, which is, so-to-speak, superposed on their more orderly movement. This chaos does not allow the happenings between a small number of atoms to involve themselves according to any recognizable laws. Only in the cooperation of an enormously large number of atoms do statistical laws begin to operate, their accuracy increasing with the number of atoms. All the supposedly *causal laws* of physics and chemistry that operate at the *macroscopic* level are really of this statistical kind. At the *microscopic level*, however, we have patently *probabilistic laws*. These govern the kind of random phenomena of which the Brownian motion and the shot effect are good examples.

Random phenomena, like the Brownian motion, which occur at the microscopic level, tend to disturb and dampen the systematic movement of matter and energy due to a definite stimulus. We therefore describe them as *noise* when we wish to refer to their disturbing role. For instance, the messages (electromagnetic pulses) transmitted over the radio are disturbed by atmospheric noise. The messages (chemical reactions) flowing across a nerve in the human body are disturbed by noise caused by sporadic movements of particles within the body, and so on.

The systematic movement of mass and energy in an instrument, (e.g. a chemical balance) due to a definite stimulus (e.g. a specimen in one pan and weights in the other) is likewise disturbed by noise, and this affects the equilibrium-position of the pointer. This explains why when we perform repeated trials of a controlled experiment, our results keep fluctuating. To reduce the effects of noise we will have to make the instrument insensitive, thereby diminishing its accuracy and utility. To make it accurate, we will have to increase its sensitivity, but then it will begin to register noise along with the message, and again lose its useful-

Four great thinkers who influenced Wiener's views on statistical physics.

J.-B. Fourier
1768–1830

H. Lebesgue
1875–1941

J.W. Gibbs
1839–1903

G.D. Birkhoff
1884–1944
(c. 1930)

ness. Clearly, a *perfectly accurate and universally useful measuring instrument is impossible, even in principle.*

Such examples, and the intrinsically stochastic aspects of subatomic phenomena, suggested by quantum theory (Ch. 10) (unknown to Maxwell, Boltzmann and Gibbs) force us to recognize that there is a chance or random element in the very texture of Nature, and that *the orderliness of the world is incomplete.* For scientific purposes we can no longer regard the universe as a strictly deterministic system, the phase of which at any instant t is determined exclusively by its phases at all previous instances $t' < t$. It was Wiener's position that the universe is still a cosmos, and that the Principle of the Uniformity of Nature still reigns. But it reigns at a stochastic level: it is the probability measures, engendered by the statistical aspect of Nature, that remain invariant under time-translations. (See [50j, Introd.] and [55a, pp. 251–252].)

Wiener held that Gibbs's idea of defining the "state" of a system as a probability distribution is indispensable in view of the contingency of the cosmos. This belief came from his realization that our observations of such systems are confined to a few macroscopic parameters, such as pressure or temperature, and that even their measurements are inevitably inaccurate because of the random impacts of Brownian motion on our instruments, especially the more sensitive ones. But a study commencing with a probability distribution can only yield stochastic quantities: new distributions, expectations, variances and the like. On the other hand, our observations on the system can only tell us how it behaves over different time-intervals. To fulfill Gibbs's program, i.e. to be able to judge the hypothetical probability distributions P empirically, some assurance is needed that the time-averages are equal to the corresponding phase-averages. This assurance the ergodic hypothesis was supposed to provide.

This line of thought began with Maxwell and Boltzmann, who viewed the gas enclosed in an insulating vessel (with perfectly adiabatic, perfectly rigid and infinitely massive walls) as a conservative system with a very large number N of degrees of freedom. The phase space is now $\mathbb{R}^{2N}$, and the orbit of the gas is confined to the energy hypersurface

$$\Sigma: \quad H(q, p) = \text{constant},$$

where H is the Hamiltonian. Their ergodic hypothesis asserted that each orbit over Σ is Σ-filling, and that the *mean-time of stay* in a region R of Σ, viz.

$$\bar{t}_R(x) = \lim_{T \to \infty} \frac{1}{T} \int_0^T 1_R(T_t x)\,dt, \quad x \in \Sigma,$$

is equal to $v(R)/v(\Sigma)$, where v is the hypervolume induced over Σ by the invariant Liouville measure over $\mathbb{R}^{2N}$ given by Hamilton's canonical equations.

The orbits here considered, unlike Wiener's Brownian curves and Peano's space-fillers, have continuous derivatives satisfying the canonical equations, and this invalidates the ergodic hypothesis, as Plancherel and Rosenthal showed in 1913. The dubiousness of this hypothesis had already been sensed by P. and T. Ehrenfest in 1911, and they proposed instead a *quasi-ergodic hypothesis* to the effect that all "non-exceptional" orbits in Σ are everywhere dense in Σ, and satisfy the equality $\bar{t}_R(x) = v(R)/v(\Sigma)$. In this, the qualification "non-exceptional" was, of course, very important. It is the possibility of this amended hypothesis that the Birkhoff theorem ensures, but only by employing the measure-theoretic ideas of Lebesgue.

D The Second Law of Thermodynamics. Entropy

The step from statistical mechanics to thermodynamics, classical and statistical, came naturally for Wiener. The concept of entropy began to permeate many facets of his work. This concept is bound up with the Second Law of Thermodynamics, which emerged from the pioneering studies of steam engines by the French engineer Sadi Carnot during the Industrial Revolution.

In classical thermodynamics one starts with the thermodynamic state space $\mathscr{S}$ of a thermodynamic system α, such as a mass of gas enclosed in a cylindrical vessel with a piston. For this simple system the state s is just given by (p, v) where p is its pressure and v its volume. The entropy of a system α in the state s is given by $S_\alpha(s)$; the definition of the functions S_α, which requires a preliminary discussion of temperature, is indicated in Note 2 (p. 157). With the understanding that the temperature is measured by a standard hydrogen thermometer, it can be demonstrated that *the entropy $S_\alpha(s)$ of the system α cannot decrease in any adiabatic transformation of s, and must increase in all non-quasi-static adiabatic ones.*[2]

This law of increasing entropy answers the question why so many thermodynamical transformations, unlike those in Newtonian mechanics, are irreversible. It is because many of them are adiabatic, and the entropies in their final states exceed the entropies in their initial states; thus their reversability would violate the law. The law denies the possibility of drawing useful energy from the oceans by cooling them. It

---

2 An *adiabatic transformation* is one during which the system α is thermally insulated from its surroundings. A *quasi-static transformation* is an ideally slow one in which α has a well-defined state s at each intermediate instant t.

justifies Carnot's principle that the efficiency of all steam engines is less than one. It enables us to grade energy into high, medium and low grades. It makes available the concept of the (Helmholtz) *free energy*, viz. $U - TS$, i.e. energy ideally convertible into useful work, a concept of considerable ecological and technological importance. Here S is the entropy, T the temperature we spoke of, and U the internal energy, cf. Note 2, p. 157.

Classical thermodynamics gives way to *statistical thermodynamics* when its concepts are interpreted in terms of the molecular kinetic theory. From the standpoint of the latter, a thermodynamic system α is an assembly of a very large number of atoms, which perform at all times a completely disorderly motion, in addition to any orderly movement they may acquire from the action of a definite external field of force. The mechanical energy of this chaotic movement is the *quantity of heat* in the system. The mean kinetic energy of the random motion of the atoms is its *absolute temperature*. Many familiar thermodynamic facts receive a very natural explanation in these molecular statistical terms. One instance is the flow of heat from a body at high temperature to one at low temperature; another is Avogadro's Law to the effect that all gases having the same volume, pressure and temperature have the same number of molecules. But many other facts, incomprehensible from the classical thermodynamic angle, such as the Brownian motion, are also explainable.

Especially relevant to Wiener's work is the statistical interpretation of the thermodynamic entropy S_α or briefly S, that Boltzmann proposed in the 1870s. Consider a mass of dilute gas consisting of n identical molecules, each of r degrees of freedom, enclosed in a cylinder with piston. The molecular phase space Ω is contained in $\mathbb{R}^{2r}$, and the gas phase space Ω^n in $\mathbb{R}^{2rn}$. An observable macrostate s of the gas, (i.e. a thermodynamic state such as the one in which its volume is v and its pressure is p) will comprise an enormous number of phases of the gas, and these will form a subregion s^* of Ω^n having a large $2rn$-dimensional volume $V(s^*)$.

By the Ergodic Theorem it turns out that the *mean-time of stay* of the gas in the macrostate s depends on this volume $V(s^*)$, cf. § 12A. Boltzmann defined the *statistical entropy* of the gas in the macrostate s in essence (apart from unimportant constants) by the equation

$$\text{Ent}(s) = k \cdot \log V(s^*), \tag{5}$$

k being a constant, subsequently called the Boltzmann constant. See Note 3, p. 158.

Boltzmann interpreted $\text{Ent}(s)$ as a measure of the *internal disorder* of the thermodynamic system in state s. For many systems it can be shown that $\mathrm{Ent}(s_2) > \mathrm{Ent}(s_1)$ when $S(s_2) > S(s_1)$, S being the thermodynamic entropy. This authenticates Ent as the statistical correlate of S. Boltzmann attempted to derive the law of increasing entropy by showing that

$$d\mathrm{Ent}(s_t)/dt \geq 0,$$

s_t being the state of the system at instant t ("H-theorem"). Criticism forced him to attribute only a probabilistic validity to the inequality and to its consequences. Even so, the controversies surrounding the H-theorem persist to this day. The fruitfulness of this definition of Ent, however, is beyond dispute.

E The homogeneous chaos and the Wiener program in statistical mechanics

By the late 1930s, Wiener had assumed a position that might be termed "ergodic stochasticism", i.e. he almost invariably treated signals in the class S of G.H.A. as trajectories of a stationary stochastic process governed by a measure-preserving flow over a probability space, and used the Birkhoff ergodic theorem wherever possible. This position also reflected the influence of A.N. Kolmogorov and A.L. Khinchine, who had made stochastic processes a central theme in probability theory. This new position amounted to a consolidation of earlier attitudes rather than a sudden shift, for Wiener's early work on Brownian motion had involved a stochastic process. The Gibbsian standpoint, to which he had always subscribed, he now reinterpreted as one in which the basic observable quantities are random fields, possibly random measures, over probability spaces, exhibiting different degrees of stationarity and ergodicity.

Wiener now felt that he had the right viewpoint and equipment to tackle the problems of statistical mechanics, in particular the problem of turbulence that he had found intractable in 1920. He had one piece of equipment, all his own: a profound grasp of the Brownian motion, and considerable proficiency in ergodic theory. He felt that he could launch a successive effort by fusing together these two important strands in his thought.

The paper [38a] is devoted to this end. In it Wiener introduced the idea of homogeneous chaos and extended the Birkhoff theorem to such chaoses. A *homogeneous chaos*, in current terminology a *stationary random measure*, is a finitely additive measure μ on a ring $\mathcal{R}$ of subsets of a group X, such that all its values $\mu(A)$, A in $\mathcal{R}$, are real- or complex-valued random variables over a probability space $(\Omega, \mathcal{B}, P)$, and furthermore the random variables $\mu(A)$ and $\mu(A + x)$, have the same probability distribution over $\mathbb{R}$ or $\mathbb{C}$, for different x in X. A simple instance is the measure μ defined by

$$\mu(a, b](\alpha) = x(b, \alpha) - x(a, \alpha), \quad \alpha \in [0, 1] \tag{6}$$

where $x(\cdot, \cdot)$ is Wiener's Brownian motion (§ 7D), and (a, b] stands for the interval $\{x : a < x \leq b\}$. This follows from § 12B (4).

Wiener made a far-reaching extension of the notion of ergodicity as well as of Birkhoff's Theorem to such chaoses, [38a, Theorem 1]:

Theorem. *Let μ be a real-valued homogeneous chaos on a ring $\mathscr{R}$ over $X = \mathbb{R}^n$, and let f be a measurable function on $\mathbb{R}$ such that*

$$\int_\Omega f\{\mu(A)(\omega)\} \cdot \log^+ |f\{\mu(A)(\omega)\}| \, P(d\omega) < \infty,$$

then for all A in $\mathscr{R}$ and P almost all ω in Ω,

$$\lim_{r \to \infty} \frac{1}{v(r)} \int_{V(r)} f\{\mu(A + t)(\omega)\} dt \quad exists,$$

where $v(r)$ is the volume of the ball $V(r)$, centre 0, radius r, in $\mathbb{R}^n$, $t = (t_1, \ldots, t_n)$ and $dt = dt_1 \ldots dt_n$. Furthermore, this limit is equal to the (absolute) expectation $E[f\{\mu(A)\}]$ in case the chaos μ is "metrically transitive" (or "ergodic"), i.e. in case

$$\lim_{|t| \to \infty} P\{\omega: \mu(A)(\omega) \in G \quad \& \quad \mu(A + t)(\omega) \in H\} = P\{\omega: \mu(A)(\omega) \in H\},$$

for any Borel subsets G and H of $\mathbb{R}$.

The measure μ given by (6) is not merely stationary: it is *independently scattered*, i.e. $\mu(a, b]$ and $\mu(c, d]$ are stochastically independent for disjoint intervals (a, b] and (c, d]. Wiener called this μ the *1-dimensional pure chaos*. Repeated stochastic integration (cf. § 7D) with respect to this μ yield *derived* or *polynomial chaoses*. For instance, over $\mathbb{R}^3$ we get a polynomial chaos v given by

$$v(B)(\alpha) = \iiint_B K(t_1, t_2, t_3) dx(t_1, \alpha) dx(t_2, \alpha) dx(t_3, \alpha)$$

for any K in $L_2(\mathbb{R}^3)$; here $B \subseteq \mathbb{R}^3$ and $\alpha \in [0,1]$. Wiener's theorem [38a, § 12] is that *every ergodic homogeneous chaos μ over $\mathbb{R}^n$ can be "approximated" (in a reasonable sense) by a sequence $\mu_1, \mu_2, \ldots, \mu_k, \ldots$ of polynomial chaoses over $\mathbb{R}^n$.*

Wiener's hope was to bring these considerations to bear on statistical mechanics. What he cherished may be summed up very roughly as follows. In continuum statistical mechanics (e.g. in the statistical theory of fluids or plasmas) the initial distubance f_0 is a random phase field, and the dynamical transformation U_t such that $U_t(f_0)$ is the disturbance at instant t, is ergodic. The ergodic action is well understood when it is applied to the 1-dimensional pure chaos μ of (6) or to one of the chaoses v derived from it; from § 12B(4) we get the equality, $U_t\{v(B)\}$ = $v(B + t)$. Consequently, if f_0 can be expanded in a series of multiple stochastic integrals with respect to such chaoses, we get a handle on finding $U_t(f_0)$.

In subsequent work which appeared in [58i], nearly 20 years later, Wiener showed that a large class of functions f admit such series representation. In this work he was guided at least indirectly by the important contributions of Cameron and Martin {C1} and Kakutani {K3}. While the purely mathematical value of this work initiated in [38a] and in a cognate paper with A. Wintner [43a] on the discrete chaos, is beyond question, its import in statistical mechanics is not yet clear. This issue is discussed fully in the survey of Drs. McMillan and Deem in the *Coll. Works*, I, pp. 654–671. Wiener was well aware of the difficulties in his approach, and clearly pointed them out in [38a]:

> In many cases, such as that of turbulence, the demands of chaos theory go considerably beyond the best knowledge of the present day. The difficulty is often both mathematical and physical. The mathematical theory may lead inevitably to a catastrophe beyond which there is no continuation, either because it is not the adequate presentation of the physical facts; or because after the catastrophe the physical system continues to develop in a manner not adequately provided for in a mathematical formulation which is adequate up to the occurrence of the catastrophe; or lastly, because the catastrophe does really occur physically, and the system really has no subsequent history. [38a, p. 935]

In the light of this, it is fair to say that Wiener's work in this difficult area of statistical physics is less a solid piece of research than it is of a deeply thought-out program for the investigation of these immensely complicated phenomena. Although much headway has been reported since Wiener's death, his program is far from fulfilled.

Interestingly, a present school of thought, led by Professor J. Bass in Paris, holds that it is more pertinent to regard a turbulent velocity simply as a function in the Wiener class S rather than as a trajectory of some hypothetical random field. The work of this school has revived an interest in the paper [30a], and has also brought to light some hitherto unnoticed connections of Wiener's G.H.A. to H. Weyl's earlier work on *equidistributions*, and to the so-called Monte Carlo method. We would refer the interested reader to *Coll. Works*, II, pp. 359–372, to the references cited therein and to the publications {B2} and {B3}.

F Anisotropic time and Bergsonian time

Most of Wiener's work in statistical physics during and after World War II concerned time-series and communication. Wiener's contributions in this area were shaped by his perception of how our conception of time has been affected by the advent of thermodynamics and of the molecular kinetic theory during the 19th and 20th centuries.

The law of increasing entropy (§ 12D) imposes on events an ordering, past → future, determined by increase in entropy. It shows that the distinction of past from future, of which we are all mentally aware, is not illusory, and that our subjective temporal ordering accords with the objective ordering. The law thus provides the foundation for the objectivity of the *anisotropy of time*, or "the arrow" in time as Eddington put it.

Many familiar phenomena, of much interest to Wiener, hinge on the anisotropy of time. For instance, our *memory* consists in retaining the changes in nervous structure brought on by stimuli received in the *past*, not in the *future*. The concept makes sense only with anisotropic time. Likewise for *communication*. We communicate primarily in order to bring our *past* experiences and knowledge to bear on the future course of events. Wiener pointed out the impossibility of communication between two individuals having opposite time-senses, cf. [61c, pp. 34, 35]. The same is the case with the concept of a *controlled experiment*. In such an experiment (e. g. the spectral analysis of a candle flame) we collimate the input (candle light), then pass it through an apparatus which alters it in a controlled way (spectrometer), and pass the output into a recording instrument (camera). Time-reversal of the process would produce an "Alice in Wonderland" situation.

Wiener also realized that our acknowledgment of the contingency of the cosmos demands an even more radical change in our conception of time. The contingency of the universe, and its indeterminism opens up the possibilities of noise and orderliness, freedom, innovation, growth and decay, error and learning. Stochastic prediction and filtering rest on the possibilities of contingency, innovation and noise, as does modern control theory, military science, meteorology and a host of other fields. The writings of the French philosopher H. Bergson on time and evolution, although somewhat imprecise, were poignant in stressing these non-extensive (non-spatial) novelty-creating aspects of time. Wiener therefore spoke of *Bergsonian time* in contrast to *Newtonian time*, when he wanted to emphasize this last aspect of time, cf. [61c, Ch. I].

An important concept simply inconceivable in Newtonian time is that of *purpose*. However, in Bergsonian time a clear definition of a purposive or teleological mechanism can be given (§ 15B). Likewise with the concept of *learning*. In learning, a system (living or non-living) changes its behavior pattern on the basis of a record it keeps of the effects on the environment of its own past performance (§ 18E). In a fully determined (non-contingent) cosmos, the relationship between the system and its environment would be pre-established, and the system would be unable to change its behavior pattern on the basis of its observations of its past performance.

G Information, negentropy and Maxwell's demon

Wiener repeatedly pointed out that the transmission of messages via a medium (or channel) is an intrinsically statistical phenomenon in Bergsonian time. We have to deal with a collection of messages, not prescribed by definite laws but only by a few statistical rules. (Think of the ensemble of messages that cross a telephone exchange in the course of a day.)

The messages convey information, and their transmission via any channel requires energy. Generally speaking, the more informative a message, the longer it will be, and the more the energy needed for its transmission. Methods have to be devised to transmit a given information-content in the most economical way. A proper numerical measure of the *informative-value of a transmitted message* is clearly imperative for any systematic attack on the problem of energy-efficient transmission. Wiener and Shannon provided such measures suitable for telephony and telegraphy, respectively.

Imagine that the recipient of the message knows the probability distribution p of the different outcomes of a repeatable experiment being performed far away, and the message he expects to receive tells him which of these outcomes has occurred in a particular trial of the experiment. Then the *informative value* of the message, "the outcome x has occurred", is deemed to be

$$I(x) = -\log p\{x\} = \log[1/p\{x\}].$$

The more unlikely the outcome x, the smaller will be the value $p\{x\}$, and the greater the value $\log[1/p\{x\}]$, i.e. $I(x)$. Thus the definition meets the reasonable criterion that *the more a message removes uncertainty, the greater its informative value*.[3]

For the simplest p, viz. one with two outcomes of equal probability $\frac{1}{2}$, the message that the first (or second) of these outcomes has occurred has the value log 2. It is convenient to define this to be *unit information*, and accordingly to take 2 as the base for the logarithm. To transmit this information, it suffices to flash just one of the digits 0 or 1 according as the first or second alternative has occurred, i.e. flash exactly one binary digit. The unit is therefore called the *bit* (for "binary digit"). If p happens to have exactly 2^n atomic outcomes, then the

---

3 Cognate ideas appeared earlier in the maximum likelihood method of the statistician R.A. Fisher, c. 1920.

message of the occurrence of any one of these can be transmitted by flashing an n-tuple drawn from $\{0, 1\}$, i.e. by flashing n binary digits. In general, for an experiment with r atomic outcomes, it will suffice to flash n binary digits, where n is the integer equal to or just exceeding $\log_2 r$. Let X be the set of all atomic outcomes of the experiment, then what concerns the transmission engineer are not the individual values $I(x)$, but rather their average:

$$\text{Inf}(p) = E[-\log p\{ \cdot \}] = - \sum_{x \in X} p\{x\} \cdot \log p\{x\}. \tag{7}$$

We may call this *the average informative value of the probability distribution p*. It gives the average energy and (average cost) of transmission.

This definition, in which X is finite or at most countable, is due to Dr. C.E. Shannon in 1947 or 1948 {S7}, who was then at the Bell Telephone Laboratories and was concerned with the energy-efficient coding of telegraphic messages over noisy channels. It presaged his deep work on channel capacities, encoding and decoding.

In the summer of 1947 Wiener was led to the same problem for an absolutely continuous probability distribution p over the real line $\mathbb{R}$ by the needs of filter theory. Since an infinite sequence of binary digits is required to transmit a real number, a limiting approach, starting with the Shannon concept, will assign an infinite average information to such a p. Wiener's starting point was the observation that we may forget all digits after a fixed number because of noise. By an argument, very obscurely presented, he arrived at the following definition for the *average informative value of p*:

$$\text{Inf}(p) = E\{-\log p'(\cdot)\} = -\int_{\mathbb{R}} p'(x) \cdot \log p'(x) dx, \tag{8}$$

where $p'(\cdot)$ is the probability density, cf. [61c, p. 62, (3.05)].[4] He too took logarithms to the base 2. This definition is widely used in the theory of information processing for continuous time.

Wiener's Inf. unlike Shannon's, can become negative. Nevertheless both Inf's have essential common features. For instance, both Inf's are additive for independent information-sources. We may therefore speak of a single *Shannon-Wiener conception of average informative value*. It provides the foundation of a large and important branch of contemporary engineering.

The nexus between communications engineering and statistical mechanics, which Wiener had dimly discerned in the mid-1920s,

---

4   Wiener omitted the minus sign, but this is of little consequence since $\text{Inf}(p)$ can take either positive or negative values for different p.

C.E. Shannon

(cf. § 9C), is deep indeed, for the Shannon-Wiener concept of information has turned out to be a disguised version of the very concept of statistical entropy to which Boltzmann was driven 70 years earlier. For the proof, see Note 4, p. 158.) Let it also be noted that to prove his H-theorem, $d\mathrm{Ent}(s_t)/dt \geq 0$, (§ 12D), Boltzmann had to replace his summation by an integral. This integral is precisely the Wiener average information of a multidimensional probability distribution, cf. [61c, p. 63] and Born {B17, pp. 57 (6.25), 165}.

Whereas the Boltzmann entropy of a gas is best interpreted as a measure of "internal disorder", Shannon's average information is most naturally interpreted as a measure of "uncertainty removed". Their equality has suggested the term *negentropy*, or measure of *internal order*, as a substitute for the term *information* in certain contexts.

The cogency of this viewpoint has become clear from the work of Szilard, Brillouin and also Wiener [50g, 52a] on the Maxwell demon. Dr. J.R. Pierce {P5, pp. 198–200, 290} defines this demon as a hypothetical and impossible creature, who without expenditure of energy, can see a molecule coming in a gas which is all at one temperature, and act on the basis of this information.

Seated in a cup of water with an insulating partition having a tiny door, it can, by intelligently opening and shutting the door, warm the water on one side of the partition and cool it on the other. To perform its miracles, the demon must receive information about impending molecular movements, and for this, electromagnetic radiation must be available in the gas. But each time the demon draws information from a photon of light, it degrades its energy and (by Planck's law) also its frequency, and creates an equal amount of entropy. The entropy of the matter-radiation mixture is not reduced. For the entire mixture nothing miraculous occurs.

But Wiener's imaginative mind was not satisfied with this rather easy disposal of the demon. Granted that the demon fails in its overall mission, it still scores a local success: it enhances the negentropy in its immediate neighborhood by degrading the photons of light. Does not such enhancement occur when a piece of green leaf uses sunlight to produce portions of a molecule of a carbohydrate from the carbon dioxide and water in its midst, and to release a molecule of oxygen, following the chemical formula:

$$\text{Light} + n\text{CO}_2 + n\text{H}_2\text{O} \xrightarrow{\text{chloro} \atop \text{phyll}} (\text{CH}_2\text{O})_n + n\text{O}_2?$$

If so, a piece of green leaf in an environment of carbon dioxide and water, irradiated by the sun, is a thermodynamic machine studded with "Maxwell demons" (particles of chlorophyll), all of which enhance negentropy *locally* by degrading the sunlight impinging on them.

Briefly, no demons, no life. But Wiener also noticed the temporariness of the demon's local successes. It could perform locally only as long as it had usable light, i.e. light from a source at a temperature higher than that of the gas. In a gas-radiation mixture in equilibrium, it will be as impotent as in a gas devoid of light. Ultimately, it too will fall into equilibrium and its intelligent activities cease. In short, it will die. Wiener felt these reflections on the demon impinged on the biological issues of life, decay and death.

These reflections led Wiener to the idea of a comprehensive quantum theory of entropy, embracing both matter and radiation, in which photons would carry information, and the Second Law of Thermodynamics would become "rigidly true", [52a]. In this thought-provoking paper Wiener mentioned his intention to pursue seriously this line of inquiry. But he did not keep this promise.

Note 1: Definition of The Hamiltonian (cf. § 12C)

If Φ is the potential, then by definition

$$H(q, p, t) = \Phi(q, t) + T(q, p),$$

where

$$T(q, p) = \sum_{i=1}^{N} \sum_{j=1}^{N} a_{ij}(q) q_i p_j.$$

is the kinetic energy of the system in the phase (q, p). Here $a_{ij}(q)$ are the *coefficients of inertia*; they form an $N \times N$ non-negative symmetric matrix $[a_{ij}(q)]$.

Note 2: Thermodynamic entropy (cf. § 12D)

In classical thermodynamics one starts with the thermodynamic state space $\mathscr{S}$ of a thermodynamic system α, such as a mass of gas enclosed in a cylindrical vessel with a piston. In the elegant Carathéodory treatment (cf. Born {B17, pp. 38–39, 143}, and Buchdahl {B21}), the Zeroth Law on thermal equilibrium makes possible the introduction of the concept of an *empirical temperature* $\theta_\alpha(s)$ for the system α in the state s in $\mathscr{S}$. The First Law asserts that the work done on α in any quasi-static adiabatic transformation depends only on its terminal states. It allows us to define the notions of its *internal energy* $U_\alpha(\cdot)$ on $\mathscr{S}$ as well as its *heat-gain* ΔQ for a large class of transformations. (ΔQ is a function, not on $\mathscr{S}$, but on a class of paths in $\mathscr{S}$.) The Second Law is best formulated as an *Inaccessibility Principle*: given a point in $\mathscr{S}$, every neighborhood of s contains a point s' inaccessible to s by adiabatic transformation. From this principle and a pure mathematical theorem of Carathéodory it follows that the Pfaffian equation $dQ = 0$ has an integrating divisor of the form $T\{\theta_\alpha(\cdot)\}$, i.e. there exists a function $T(\cdot)$ on $\mathbb{R}$ such that

$$dQ/T\{\theta_\alpha(\cdot)\} = dS_\alpha(\cdot) \quad \text{on } \mathscr{S},$$

where $S_\alpha(\cdot)$ is a function on $\mathscr{S}$. The function $T\{\theta_\alpha(\cdot)\}$ on $\mathscr{S}$ is called the *absolute temperature*, and $S_\alpha(\cdot)$ the corresponding *entropy* of the system.

The function $T(\cdot)$ on $\mathbb{R}$ is not entirely unique. The simple determination $T(t) = c \cdot t$, $c = $ const., suffices, provided that the empirical temperature $\theta_\alpha(s)$ of α in the state s is measured by the "perfect gas thermometer". (This is defined as the thermometer which assigns to a perfect gas in the state (p, v) the temperature pv/R. In the laboratory, hydrogen is used as the "perfect gas"). With this standardization of $T(\cdot)$, it can be demonstrated that *the entropy* $S_\alpha(\cdot)$ *of the system α cannot decrease in any adiabatic transformation, and must increase in all non-quasi-static adiabatic ones.*

Note 3: **Boltzmann statistical entropy** (cf. § 12D(5))

The reason for the logarithm is to make the entropy additive for non-interacting systems. If $s = (s_1, s_2)$ is the macrostate of the mixture of two such systems (e.g. gases with $s_1 = (v_1, p_1)$ and $s_2 = (v_2, p_2)$, then $s^* = s_1^* \times s_2^*$, whence, on taking the appropriate hypervolumes, we get

$$V(s^*) = V_1(s_1^*) \cdot V_2(s_2^*),$$

and

$$Ent(s) = Ent(s_1) + Ent(s_2).$$

Note 4: **Proof of equality of entropy and information** (cf. § 12G)

To see this, recall from § 12D(5), Boltzmann's expression for the entropy of a thermodynamic system in a macrostate s, viz.

$$Ent(s) = k \cdot \log V(s^*),$$

where $V(s^*)$ is the $2rn$-dimensional volume of the region s^* in the gas phase space Ω^n spanned by s. Notice if the volume of a subregion Δ of Ω^n is below a certain threshold, there will be no observable way of knowing that the phase of the gas is in Δ. Accordingly, with Boltzmann, let us partition the gas phase space Ω^n into a large number of cells Δ of equal volume, say V_0. Each of these cells is called a *microstate* of the gas. Their collection is a finite partition Π of Ω^n. For each macrostate s of the gas, s^* will comprise a large number of the microstates, i.e.

$$s^* = \bigcup_{\Delta \in \Pi_s} \Delta, \quad \text{where } \Pi_s \subseteq \Pi. \tag{9}$$

It follows that

$$V(s^*) = \sum_{\Delta \in \Pi_s} V(\Delta) = V_o \cdot \text{card } \Pi_s,$$

and that

$$Ent(s) = k \cdot \log(V_o \cdot \text{card } \Pi_s) \tag{10}$$

where "card A" denotes the number of elements in the set A.

Now let the partition Π of Ω^n be the product π^n of a finite partition π of the molecular phase space Ω into cells σ of equal $2r$-dimensional volume v_0, say. Then each cell Δ in Π is a "rectangle" in $\mathbb{R}^{2nr}$ of the form

$$\sigma_{k_1} x \ldots x \sigma_{k_n}, \quad \sigma_{k_i} \in \pi,$$

and clearly $V_0 = v_0^n$. Let the numbers of molecules with phases in the cell σ be $\lambda(\sigma)$. Then the function λ on π to $\{0, 1, \ldots, n\}$ is called a *complexion* of the gas. Attached to each complexion λ is a probability measure $p_\lambda(\cdot)$ over the set π given by

$$p_\lambda(\sigma) = \lambda(\sigma)/n, \quad \sigma \in \pi.^5$$

Boltzmann assumed that each complexion λ is a macrostate of the gas. Hence by (9)

$$\lambda^* = \bigcup_{\Delta \in \Pi_\lambda} \Delta, \quad \text{where } \Pi_\lambda \subseteq \Pi.$$

A straightforward calculation yields

$$\text{card } \Pi_\lambda = n!/ \prod_{\sigma \in \pi} \lambda(\sigma)!.$$

Assuming that each $\lambda(\sigma)$ is large enough to allow Stirling's approximation to $\lambda(\sigma)!$, we find that

$$\log\{\text{card } \Pi_\lambda\} = -n \sum_{\sigma \in \pi} \frac{\lambda(\sigma)}{n} \cdot \log \frac{\lambda(\sigma)}{n} = n \cdot \text{Inf}(p_\lambda) \cdot \log_e 2.$$

Thus by (10),

$$\text{Ent}(\lambda) = kn \cdot \log_e 2 \cdot \text{Inf}(p_\lambda) + k \cdot \log V_0.$$

In words, *the Boltzmann entropy of the gas in the complexion λ is equal, apart from an additive constant, to $kn \cdot \log_e 2$ times the Shannon average information of the associated probability measure p_λ, on the partition π of the molecular phase space.* Here k is Boltzmann's constant and n is the number of molecules.

---

5   It is assumed that each σ, though very small from our macroscopic standpoint, is very large from the atomic, i.e. an enormous number of molecules will have phases lying in σ.

13 Lee, Bush, and Wiener's Thoughts on Networks and Computers

A The analogue computer program at MIT. The Wiener integraph.

Wiener's work on the Heaviside operational calculus (§ 9B), undertaken at the behest of his engineering friends, was mathematical and did not represent his more tangible engineering predilections and flair for things electrical. The latter showed themselves, however, in his excitement over the computer building program at MIT from the mid-1920s to the 1940s, executed under the leadership of Dr. Vannevar Bush. This had three basic targets: (1) a *network analyzer* for solving systems of algebraic equations that appear in power engineering, (2) an *intergraph* for the evaluation of integrals depending on a parameter t, e.g.

$$\int_a^t f(x)g(x)dx \quad \text{or} \quad \int_o^a f(t-x)g(x)dx,$$

and (3) a *differential analyser* for solving ordinary differential equations.

All three were to be *analog machines*, i.e. numbers are to be represented on them by measured physical quantities rather than by digits as on a desk or pocket computer. As Wiener has stated, "These physical quantities can be currents or voltages or the angles of rotation of shafts or quantities of still different sorts" [56g, p. 136]. The principle underlying such computation is to replace the quantities that we wish to compute "by other physical quantities of a different nature but with the same quantitative interrelations" [56g, p. 136]. For instance,

> In an electrodynamometer, two coils are attracted to one another in proportion to the product of the current carried by these two coils, and this attraction can be measured by an appropriate sort of scale instrument. If, then, one coil is carrying seven units of current and the other five units, the scale will read something proportional to thirty-five. This sort of instrument for multiplying is known as an analogy instrument, because we are replacing our original situation, in which certain quantities are to be multiplied, by a new situation, in which we set up two currents in analogy to the original quantities and read off the product by a physical situation analogous to the original one. [56g, p. 236]

In a similar vein, the *potentiometer*, an instrument designed to measure the voltage of a battery, can be viewed as an analog machine for computing a fraction of a given number, e.g. the 33/57 part of 293. The simplest example of an analog machine for multiplication is the *slide rule*, consisting of two sliding rulers graduated according to the logarithms of numbers. (To multiply the numbers a and b, we need only read off the graduation for the sum log a + log b.) Many of the fine astronomical instruments designed by Islamic scientists of the 14th century are analogue machines for specific computations. Lord Kelvin's *harmonic analyzer*, to compute the first few Fourier coefficients of the periodic rise and fall of tides at a fixed spot, is another example of an analog computer.

Wiener was fascinated by Bush's program, and his close touch with it significantly molded his thought. His interest in the fabrication of thinking machines had begun in college when he studied Leibniz's ideas of this question (cf. Ch. 4), and he had used computers during his army service at the Aberdeen Proving Grounds (Ch. 6). From the engineers, Wiener now learned of problems involved in the designing of workable computers. This was to contribute to his understanding of the problems of automatization. For their part, the engineers, aware of the service rendered to computer science by philosophical minds such as Pascal and Leibniz, found in Wiener a useful person to have around. His flair for electrical and mechanical devices, his practical computing experience, hs mathematical abilities and philosophical training were a rare combination, and he was known to throw out interesting ideas on all sorts of subjects, many of them worth a try.

Wiener's first concrete contribution came in his advice on the *continuous integraph*. In Bush's 1927 model {B24}, a sheet of paper bearing the graphs $y = f(x)$, $y = g(x)$ was attached to a moving table. Presiding over the table were two operators, each holding over one of the graphs a pointer attached to a vertically movable slider of a potentiometer carrying a constant current. The ordinates of the graphs were thus turned into voltage drops, and these were fed into the potential coil and the current coil of a modified Watthour meter. By a relay mechanism, the reading of the meter was then continuously recorded on another sheet of paper attached to the moving table. Thus as the table moved, the stylus traced out the desired curve $y = \int_o^t f(x)g(x)dx$. (The ordinary household Watthour meter performs the integration $\int_o^T v(t)i(t)dt$, where $v(t)$ and $i(t)$ are the voltage and current, and thus gives the energy consumed during the time-interval $[0, T]$.) This ingenious device fails to deliver a single graph, however, when the parameter t occurs in the integrand, as for instance in the integral $I(t) = \int_o^a f(t - x)g(x)dx$.

In early 1926 Wiener found a solution. This is how it happened:

One time when I was visiting the show at the Old Copley Theatre, an idea came into my mind which simply distracted all my attention from the performance. It was the notion of an optical computing machine for harmonic analysis. I had already learned not to disregard these stray ideas, no

> matter when they came to my attention, and I promptly left the theater to
> work out some of the details of my new plan. The next day I consulted Bush.
> [56g, p. 112]

Scientists get their new ideas at odd moments, much as people in other
walks of life. In this respect Wiener was no different. The Englishman
Charles Babbage, one of the great minds in the history of computing, hit
upon the idea of automating computation also in curious circumstances,
as the following anecdote (c. 1825) reveals:

> One evening I was sitting in the rooms of the Analytical Society at Cam-
> bridge, my head leaning forward on the table in a kind of dreamy mood, with
> a Table of logarithms lying open before me. Another member, coming into
> the room, and seeing me half asleep, called out, "Well, Babbage, what are
> you dreaming about?" to which I replied, "I am thinking that all these Tables
> (pointing to the logarithms) might be calculated by machinery." {G10, p. 11}

Wiener, like Leibniz, liked to see his ideas worked out in practice. But myopic and manually clumsy, he had to rely on colleagues for
this. Dr. Bush put a student to try out Wiener's idea "in-the-metal", and
in 1929 came a paper, "A new machine for integrating functional
products", by K.E. Gould, which opens with the words: "This paper is
intended to describe a new machine conceived by Dr. Wiener, which
performs certain integrations which are important in engineering, ..."
{G12, p. 305}. Nearly a dozen papers and theses were written on his
machine by Bush's younger colleagues, starting in 1929 and ending with
Hazen and Brown {H5} in 1940, by which time the apparatus was
equipped with a moving film and called the *cinema integraph.*

The Wiener integraph is optical in nature. Radiation from a
plane source S of uniform intensity is made to pass through the aperture
in a plane opaque screen A, parallel to S, and very close to S and of the
same width a and the same height h, the aperture having the shape of
the region bounded by the (horizontal) x-axis, (the vertical) ordinates
$x = 0$ and $x = a$, and the curve $y = f(x)$. This filtered radiation then
passes through the aperture in another plane opaque screen B, parallel
to A, and distance ℓ away, of width $a/2$ and height h, this aperture
having the shape of the region bounded by the (horizontal) x-axis, the
(vertical) ordinates $x = 0$ and $x = a/2$, and the curve $y = g(2x)$. But
this screen B is swung through $180°$ about the y-axis, and then slid
parallel to the x-axis a distance $t/2$. The radiation from the second
aperture, so displaced, falls on a narrow vertical receiver R distance ℓ
away from B, of height $3h$, and very small width. The intensity of
illumination of R is nearly $\int_o^a f(x)g(t - x)dx$, the accuracy improving as
h/ℓ and a/ℓ decrease. For a diagram and proof, see Note 1, p. 175.

The Gould apparatus of 1929 {G12} was superceded in 1931 by
another of T.S. Gray {G14}, embodying several interesting technical

features which enabled the final reading to be made accurately by a "null" or "balance" procedure. The H. L. Hazen and G. S. Brown apparatus of 1940 {H5} in essence replaced the opaque screens A and B in the Gray apparatus by transparent ones into which could be fed motion-picture films with transparent regions shaped as in the earlier apertures and opaque remainders. This considerably increased the versatility of the apparatus. Not only could f and g having negative values be handled, but even integrals of the form

$$F(y) = \int_a^b f(x, y)g(x)dx$$

could be conveniently evaluated, by preparing a large number of cinema-frames for the functions $f(\cdot, y_1), \dots, f(\cdot, y_n)$, attaching them on film and stepping-in this film to the screen A. The whole operation and recording was automated and fairly rapid.

These ingenious analog devices appear pathetically crude when set beside our present-day analog machines. But they represented the pioneering effort of the 1920s and 1930s to which the computer age at least partially owes its existence. Actually, the integraph was put to all sorts of uses during its lifetime. These ranged from the evaluation of Fourier, Legendre, and other orthogonal series representations of functions to the solution of integral equations (Fredholm and Volterra) appearing in different branches of engineering.

B The Lee-Wiener network

Around 1929 Wiener became interested in the synthesis of linear electric networks, the performance characteristics of which are prescribed in advance. He surmised that for a large class of such characteristics, synthesis would be possible by a proper cascade of copies of a few "standard" circuits.

To appreciate Wiener's thought, we must recall the concept of causal operator and the result that many a time-invariant linear causal filter, which turns the input $x(t)$ into the output $y(t)$, is given by

$$y(t) = \int_{-\infty}^t W(t - s)x(s)ds, \quad t \text{ real}, \tag{1}$$

where W is a suitable "weighting function" vanishing on $(-\infty, 0]$, cf. § 9B (11). Taking the direct Fourier transformation, we get

$$\hat{y}(\lambda) = \hat{W}(\lambda)\hat{x}(\lambda), \quad \lambda \text{ complex}, \tag{2}$$

and $\hat{W}(\,\cdot\,)$, called the *frequency response function* of the filter, has a holomorphic extension[1] to the lower half plane. When the filter in question is an electrical network with a finite number of lumped passive elements, i.e., resistors, capacitances (storage tanks) and inductances (coils), the input $x(t)$ is an impressed voltage-difference, and the output $y(t)$ is a current, it turns out that

$$\hat{y}(\lambda) = R(i\lambda)\hat{x}(\lambda), \quad \text{i.e. } \hat{W}(\lambda) = R(i\lambda), \tag{3}$$

where R is a *rational function*, i.e. a function of the form

$$R(\lambda) = \frac{a_0 + a_1\lambda + a_2\lambda^2 + \ldots + a_m\lambda^m}{b_0 + b_1\lambda + b_2\lambda^2 + \ldots + b_n\lambda^n}$$

where m, n are positive integers, and the coefficients a_k, b_k are real numbers. In (3), however, we have not $R(\lambda)$ but $R(i\lambda)$, i being $\sqrt{(-1)}$. How (3) comes about is indicated in Note 2, p. 176.

A filter with a weighting W for which $\hat{W}$ is not a rational function, is only approximable by means of networks of the type mentioned, with $\hat{W}(\lambda) = R(i\lambda)$. To get better and better approximations we may have to take larger and larger values of m and n, i.e. in physical terms, networks with more and more components.

All this was well understood by the late 1920s. Wiener's contribution lay in his bringing to bear on this design problem the fundamental ideas of "harmonic" analysis in a broad sense, viz. "tones" and their orthogonal organization. Briefly, he conceived the filter as a musical chord—an array or "cascade" of well-chosen "tones" (i.e. standard circuits). These "tones", viz. the weightings

$$W_0(t), W_1(t), \ldots, W_n(t), \ldots$$

have to yield realizable electrical circuits, i.e. they must satisfy the requirement that the $\hat{W}_0(\lambda)$, $\hat{W}_1(\lambda)$, etc. be of the form $R_0(i\lambda)$, $R_1(i\lambda)$, etc., where R_0, R_1 etc., are rational functions. This rules out the sinusoids $\cos nt$, $\sin nt$, $n = 0, 1, 2, \ldots$ as "tones" W_n. Wiener chose his W_n to be the rather complicated functions introduced by the French mathematician E. Laguerre:

$$\begin{cases} L_n(t) = e^{-t} \sum_{k=o}^{n} (-1)^{n-k} 2^{k+\frac{1}{2}} \dfrac{n!}{(k!)^2(n-k)!} t^k, & \text{for } t \geq 0 \\ L_n(t) = 0, & \text{for } t < 0. \end{cases} \tag{4}$$

---

1   I.e. roughly speaking, a smooth extension not taking infinite values.

On Fourier transformation, however, their complexity gives way to a beautiful simplicity of relationship, viz.

$$\hat{L}_n(\lambda) = \left(\frac{1 - i\lambda}{1 + i\lambda}\right)^n \hat{L}_0(\lambda), \quad \hat{L}_0(\lambda) = \frac{1}{\sqrt{\pi}} \cdot \frac{1}{1 + i\lambda}. \tag{5}$$

This, Wiener felt, would be reflected in a corresponding simplicity and elegance in engineering design.

These ideas were very tentative. Wiener, his flair for the electrical notwithstanding, needed assistance to transcribe them into metallic form. He approached Bush to put him in touch with a good student who could try them out. Bush brought Wiener together with Y.W. Lee, an engineering student from China. "This was one of the finest things that Bush has done for me . . . " Wiener was to write later [56g, p. 142], for his fruitful association with Lee lasted till the end.

Lee was not only able to assimilate Wiener's suggestions, but succeeded in implementing an efficient arrangement of the cascade, which secured the best utilization of hardware. This came to be known as the *Lee-Wiener network*. For a description and diagram, see Note 3, p. 177. Lee completed a doctoral dissertation on the topic, which was published in 1932, under the title "Synthesis of electric networks by means of Fourier transforms of Laguerre's functions" {L4}. Lee also helped in completing their designs and getting them patented. One patent was obtained in 1935 and two more in 1938, cf. Lee {L5, p. 459}.

The Lee-Wiener network was to play a crucial part in Wiener's war work in the 1940s on anti-aircraft fire control (§ 14A), and also in his later work on the analysis and synthesis of arbitrary time-invariant black boxes—work which in turn was to mould his ideas on self-reproducing machines and self-organizing systems (§§ 18E, F).

C Further work with Lee. The refugee problem. Visit to the Orient.

Lee's thesis contained another important idea that he has attributed to Wiener, cf. {L4, p. 85}. The design of any electrical filter has to reckon with certain intrinsic mathematical constraints. We noted one such constraint on the frequency-response function $\hat{W}$ at the end of Ch. 11. Let us write Y, instead of $\hat{W}$, for the *frequency-response* function. The functions $|Y(\cdot)|$ and ph $Y(\cdot)$, related by the equality $Y(\lambda) = |Y(\lambda)| \cdot e^{iphY(\lambda)}$, λ real, are called the *amplitude response* and *phase response* of the filter. Wiener realized that these two functions were directly related by the Hilbert transformations

$$\log|Y(\lambda)| = \frac{1}{\pi} \int_{-\infty}^{\infty} \frac{\text{ph } Y(u)}{\lambda - u} \, du,$$

Vannevar Bush
1890–1974

$$\text{ph } Y(\lambda) = \frac{1}{\pi} \int_{-\infty}^{\infty} \frac{\log|Y(u)|}{\lambda - u} \, du.$$

In this the integral $\int$ is "improper", and has to be carefully defined.
Under auspicious conditions we have the beautiful "Parceval equality":

$$\int_{-\infty}^{\infty} |\log|Y(\lambda)||^2 d\lambda = \int_{-\infty}^{\infty} |\text{ph } Y(\lambda)|^2 d\lambda.$$

In the late 1920s these relations began showing up in the mathematical
literature. But engineers were unaware of them and of their engineering
significance.

After finishing his doctorate at MIT, Lee returned to China and
joined the engineering faculty of Tsing Hua University in Peking, and
in the fall of 1931 Wiener embarked on his one-year visit to Cambridge,
England, and to the International Mathematical Congress in Zürich in
the summer of 1932, and then onto his collaboration with Paley at MIT,
as we recounted in Ch. 11.

Y.W. Lee

The political storm clouds that were gathering began to affect academic life. Scientists were being evicted for racial or political reasons. Others, finding fascism intolerable, were leaving voluntarily. An effort was underway in England and the United States to find placement for the refugee scientists from central Europe. Wiener joined in this effort, which in America was led by Professors Oswald Veblen and J. R. Kline. He became a member of the Emergency Committee in the Aid of Displaced German Scholars and of the China Aid Society.

Wiener's efforts to find a place for Professor L. Lichtenstein of Leipzig were frustrated by the latter's death from a sudden heart attack. But he was able, thanks to the cooperation of his old friend from the 1920s, Professor I. A. Barnett, to give Professor Otto Szasz, an acquaintance from his Göttingen days, a haven in the University of Cincinnati. Another beneficiary of Wiener's efforts was Professor Karl Menger from Vienna, whom he had met there in 1932. Wiener also joined in the support of the Loyalist side in the Spanish Civil War. In this he was guided by Professor Walter B. Cannon of the Harvard Medical School, his family friend, who had taken a great interest in the scientific revival in Spain, especially in neurophysiology.

In 1935, amid the worsening political climate, Wiener brought a mature perspective to bear on the mounting problem of the placement of refugee scientists in two articles [34e, 35f] appearing in the *Jewish Advocate*. In these articles Wiener makes an important distinction between "short-time" and "long-time" problems, which was to become a part of his later thought (§ 21E). Quite possibly, this awareness came from his reflecting on Haldane's philosophical article {H2}, (cf. § 10D).

In 1934 Wiener received an invitation for a visiting professorship at Tsing Hua University for 1935–1936. He accepted, with high hopes of resuming his collaboration with Lee. Wiener looked forward with much anticipation to this, his first visit to the Orient, which was to include an initial stop-over in Japan:

> My prospective trip to China filled me with great enthusiasm. Not only have I always loved travel for its own sake, but my father had brought me up to consider the intellectual world a whole and each country, however exalted its position might be, as a mere province in that world. I had actually been a witness and a participant in the rise of American science from a provincial reflection of that of Europe into a relatively important and autonomous position, and I was sure that what had happened here could happen in any country, or at least in any country which had already shown in action an aptitude for intellectual and cultural innovation. I have never felt the advantage of European culture over any of the great cultures of the Orient as anything more than a temporary episode in history, and I was eager to see these extra-European countries with my own eyes and to observe their modes of life and thought by direct inspection. In this I was thoroughly seconded by my wife, to whom national and racial prejudice have always been as foreign as they have been to me. [56g, pp. 181–182]

On his way to China with his family, Wiener spent two weeks in Japan, and renewed contacts with S. Ikehara, his former student and collaborator on the Prime Number Theorem, who had become a professor at Osaka University. Wiener also delivered lectures at Osaka and Tokyo. The Wieners then headed for China. The first days were spent in settling down, getting their young children enrolled in an English-speaking elementary school, and getting reacquainted with Lee and his Canadian wife. Wiener and Lee then began work. In Wiener's words:

> What Lee and I had really tried to do was to follow in the footsteps of Bush in making an analogy-computing machine, but to gear it to the high speed of electrical circuits instead of to the much lower one of mechanical shafts and integrators. The principle was sound enough, and in fact has been followed out by other people later. What was lacking in our work was a thorough understanding of the problems of designing an apparatus in which

Wiener (center) with colleagues at Tsing Hua University in Peking, 1935

> part of the output motion is fed back again to the beginning of the process
> as a new input.
>
> . . . .
>
> What I should have done was to attack the problem from the beginning and
> develop on my own initiative a fairly comprehensive theory of feedback
> mechanisms. I did not do this at the time, and failure was the consequence.
> [56g, p. 190]

This is the first indication we have of Wiener's active interest in the
concept of *feedback*, which was to play such a vital part in his later
thought and work.

Wiener's other duty at Tsing Hua was to lecture on G.H.A., the
Paley-Wiener material and their applications. Wiener's attention turned
to the *quasi-analytic functions*, he had studied with Paley. These are
functions or curves "which are smooth enough so that the whole course
of these curves is known from any one part, but which are not smooth
enough to be treated by the classical theory of the functions of a complex
variable" [56g, p. 191]. It was Wiener's good luck that during his tenure
at Tsing Hua, Professor Jacques Hadamard of the Sorbonne, who he
had met at Strasbourg in 1920, paid a visit to the university. His great
eminence as a French analyst was well reported in the Chinese press, as

Wiener learned when his rickshaw man asked him "if Ha Ta Ma Hsien-Sheng was as great a mathematician as the newspapers said he was" [56g, p. 193]. Through Hadamard, Wiener arranged to meet his Polish pupil and distinguished colleague S. Mandelbrojt, a great authority on quasi-analytic functions. This materialized, and resulted in joint work by Wiener and Mandelbrojt on quasi-analytic functions in 1936 and later, [36b, 36c, 47a].

During his stay in China Wiener picked up a smattering conversational knowledge of Chinese, which he once in a while enjoyed showing off. As in his other travels, he was a shrewd but empathic observer, and has left us with interesting reflections of life in these lands in his autobiography [56g].

Wiener returned home from China via Europe, where he participated in the International Mathematical Congress in Oslo in the summer of 1936. On his trip Wiener and his wife made a brief trip to Cairo. In Clermont-Ferrand, Wiener met Mandelbrojt and put in five days of hard work with him on a joint paper on quasi-analytic functions. Enroute to Oslo, Wiener looked up the Haldanes in London and friends in Copenhagen. At the Congress, Wiener renewed his friendly contacts with Canon G. Lemaître, the great cosmologist whom he knew during the latter's visit to MIT for advanced study. Later, Lemaître was to become the prime mover of the view of the universe with a finite past, a modified version of which is currently in vogue under the innane nomenclature "big bang".

D Wiener's 1940 memorandum on an electronic computer for partial differential equations

In 1940, with war raging in Europe, Dr. Vannevar Bush sought from the MIT faculty research proposals that might contribute to national defense, as well as suggestions for its organization. Wiener responded promptly. On the organizational question, he favored scientific collaboration which would cross scientific frontiers and would at the same time be voluntary, thereby preserving a large measure of the scientists' initiative and individual responsability. He accordingly suggested

> ... the organization of small mobile teams of scientists from different fields, which would make joint attacks on their problems. When they had accomplished something, I planned that they should pass their work over to a development group and go on in a body to the next problem on the basis of the scientific experience and the experience in collaboration which they had already acquired. [56g, pp. 231, 232]

But bureaucracies, prone to big size, big budgets and static locations tend to shirk off compactness, mobility and efficiency, and nothing came of Wiener's suggestion. Moreover, as Wiener has noted [56g, p. 232], many an inventor falls in love with his gadget, viewing it as his private possession rather than as a small step in an ongoing evolution that began with the caveman.

On the specific topic of war research, Wiener proposed the design of an electronic computer to solve partial differential equations. This, like the integraph, was to be a *special purpose computer*, designed to carry out a specific mathematical task, viz. solve a PDE. But unlike the integraph, it was to be a *digital computer*, which counts digits as we do in our mental arithmetic, and does not measure quantities as in an analog machine.

The earliest example of a digital computer is the abacus, which until recently was used extensively in the Orient. The first mechanical computers built by Pascal (c. 1642), Leibniz (c. 1672) and Babbage (c. 1830), for addition, multiplication, etc. were digital, as were the desk computers in vogue some decades ago. Digital computers are much more accurate than the analog, and can yield as accurate an answer as we need, if we are patient. But before World War II, they were notably slower.

What prompted Wiener to make this defense proposal was the question that Bush had raised with him in the early 1930s of the possibility of mechanizing the solution of PDE's which, as we saw in § 9A, are fundamental to field physics, and thus to a vast body of engineering, and which are accurately solvable by pure mathematics only in the rarest cases. Their mechanical solution to high degrees of approximation was thus a vital practical need, and one of defense potential in view of their appearance in naval and aeronautical engineering.

In the PDE, the numerical data is spread over a planer or higher dimensional region. In the Dirichlet problem, for instance, the data consists of the values $T(x, y, z)$ of the temperature at each point (x, y, z) inside a room, cf. § 7E. Wiener saw at once the futility of analog devices in such a situation, and realized that it was crucial to replace the function of many variables by one with just one variable. His 1923 work with Phillips on the Dirichlet problem [23b] and his fine technological intuition gave him a clue as to how this could be done in two steps. For simplicity suppose that we have just two variables, i.e. the data are spread over a planer region, say a square. First the continuous data must be "discretized", i.e. replaced by the array of their values at a large number of points that form a grid. Second, these discrete data must be read off by a line-to-line scanning from left to right as in television (or indeed as we do at a much slower pace when reading a page with our eyes).

If there are ℓ points on each line of an n-dimensional cubical grid, there will be ℓ^n points in the grid, each of which will have to be scanned. Consequently, the time required to scan such a grid of "length" ℓ is the same as that required to scan a single line of "length" ℓ^n; e.g. with $\ell = 40$ and $n = 3$, we have 64,000 points to scan. Correspondingly large will be the number of intermediate operations that must be carried out to get the solution of the PDE. For the fruition of such a procedure, extreme high speeds are an imperative necessity, as Wiener was quick to see.

In mechanical or electromechanical computing, devices such as wheels or punch cards or switches have to be moved, and this is slow by virtue of their inertia. On the other hand, in a vacuum tube what has to move is the electron, a subatomic particle whose inertia is roughly $1/10^{27}$ of that of relay contacts. Moreover, the attenuation of atomic movement due to resistance is considerably smaller than that due to friction in macroscopic movement, and can be more readily compensated for by electrical amplification. Thus as early as 1940 it was clear to Wiener, and to a few kindred spirits, such as V.J. Atanasoff, that the time was ripe for the electronic digital computer. Wiener's proposal was for an electronic computer with magnetic taping for storing data.

Another important feature of Wiener's projected machine was the use of the binary scale instead of the decimal for the representation of numbers. In this scale, a number is represented by the array of its coefficients in a series of powers of 2 instead of powers of 10; these coefficients being 0, 1, instead of 0, 1, . . . , 9. Thus the number ninety-four has to be written not 94 but 1011110, since $94 = 1 \cdot 2^6 + 0 \cdot 2^5 + 1 \cdot 2^4 + 1 \cdot 2^3 + 1 \cdot 2^2 + 1 \cdot 2^1 + 0 \cdot 2^0$. Because of its extremely simple multiplication tables, viz. $0 \times 0 = 0 \times 1 = 1 \times 0 = 0$ and $1 \times 1 = 1$, the binary scale lends itself to mechanization much more readily than any other. Indeed, Wiener saw a practical exhibition of its merits in the digital computer to multiply complex numbers built by Dr. G.R. Stibitz of the Bell Laboratories and his associates, which was on display at the summer meeting of the American Mathematical Society in Dartmouth in 1940. This was unique not only in its use of the binary scale but also for its electromechanical relays. On the way back home from Dartmouth Wiener told Dr. Levinson, his colleague and former student, of his growing conviction that when equipped with proper scanning devices, "electronic binary machines would be precisely the devices required for the high speed computation needed in partial-differential-equation problems" [56g, p. 231].

Thus, when Bush's circular came in 1940, Wiener had merely to articulate in a concrete proposal the ideas that had been incubating in his mind over a 10-year period. This proposal, entitled *Memorandum on the mechanical solution of partial differential equations*, has 12 pages and was accompanied by a 4-page covering letter to Dr. Bush, dated September 21, 1940, and is reprinted in the *Coll. Works*, IV, as [85a] and [85b] respectively.[2] Its abstract reads:

> The projected machine will solve boundary value problems in the field of partial differential equations. In particular, it will determine the equipotential lines and lines of flow about an airfoil section given by determining about 200 points on its profile, to an accuracy of one part in a thousand, in from three to four hours. It will also solve three-dimensional potential problems, problems from the theory of elasticity, etc. It is not confined to linear problems, and may be used in direct attacks on hydrodynamics. It will also solve the problem of determining the natural modes of vibration of a linear system. [85b, p. 125]

As D.K. Ferry and R.E. Saeks point out in the *Coll. Works*, IV, what Wiener proposed was a high-speed electronic special purpose computer, modern in all respects except for a stored program. As they say, Wiener's proposed machine employs (1) a discrete quantized numerical algorithm for solution of the PDE, (2) a classical Turing machine architecture, (3) binary arithmetic and data storage, (4) an electronic arithmetic logic unit, (5) a multitrack magnetic tape.

Wiener's appreciation of the novelty of the computational solvability of *partial* as opposed to *ordinary* differential equations appears in the last section:

> The fundamental difficulty in the solution of partial differential equations by mechanical methods lies in the fact that they presuppose a method of representing functions of two or more variables. Here television technique has shown the proper way: *scanning, or the approximate mapping of such functions as functions of a single variable*, the time. This technique depends on *very rapid* methods of recording, operating on, and reading quantities or numbers. [85b, p. 133] (emphasis added)

On how this rapidity was to be brought about, and on the economics of the venture, Wiener wrote:

> By giving up our present dependence on mechanical parts of high inertia and friction, and resorting to electrical devices of low impedance, it is easy to perform arithmetical operations at *several thousand times the present speed*, with but a *slight increase over the present cost*. Where operations are so multiplied in number as is the case with partial differential equations as distinguished from ordinary differential equations, the economic advantage of high-speed electronic arithmetical machines, combined with scanning processes, over the multiplications of mechanical parts, becomes so great that it is imperative. [85b, p. 133] (emphasis added)

More specifically,

> With magnetic scanning and printing, it does not seem too much to hope that an entire set of reading, adding, and printing operations may be completed in 10^{-4} seconds. [85b, p. 130].

2 The documents together with comments by Drs. B. Randell, R. Saeks, D.K. Ferry and the writer are reprinted in the Annals of the History of Computing 9 (1987), 183–197.

As Ferry and Saeks remark, such speeds were attained only by the first generation of transistorized computers (IBM 7090, 7094, etc.) in the late 1950s.

The report then alludes to the adaptability of the type of computer envisioned in domains in which the solution of important equations are so complex that they become almost impossible to solve rigorously, and which "suffer greatly from the lack of computational tools"—domains such as the theories of turbulence, shock waves, supersonic propulsion, explosions and ballistics [85b, p. 134]. In Section 2 Wiener illustrated in detail the scanning procedure for the parabolic equation

$$4 \frac{\partial u}{\partial t} = \frac{\partial^2 u}{\partial x^2} + \frac{\partial^2 u}{\partial y^2} \tag{6}$$

subject to initial and boundary conditions. (For details, see Note 4, p. 179.)

In Section 3, Wiener listed the component elements, five in all, required to compute the solution mechanically. The first two are: "quick" mechanisms for imprinting, and reading "numerical values on a running tape". On the third component "a rapid adding mechanism", Wiener wrote "I am told that vacuum-tube capacitance mechanisms of the type exist with an overall speed of 1/50 000 seconds per operation" [85b, p. 130]. The fourth and fifth were designed for specific steps in the algorithm. The last was for the erasure of unneeded data.

But Wiener was not an engineer, and the vividness of his perception was no assurance that the projected machine could be built with the resources that were then available. More than anything else what Wiener now needed was the dissemination of his proposal among his engineering friends, and lively discussions with them on all its facets. Unfortunately, this was not to be the case. For Bush, believing that computer technology was too weak to accomplish what Wiener had in mind and therefore to bear fruit in the war effort, recommended deferment of the project to the post-war years. In this evaluation Bush was cautious, but not unduly. For even though Harvard, the University of Pennsylvania, (the ENIAC), the Aberdeen Proving Grounds and the Institute for Advanced Study began building electronic computers during the war, embodying one or more of the very features that Wiener had advocated, those used in the war effort were smaller relay machines of older vintage. What Wiener wanted materialized only in the transistorized computers (IBM 7090, etc.) of the late 1950s as Ferry and Saeks point out. Thus, in this as in several other situations, Wiener's thinking had been far ahead of its time, and his over-all general conceptions were to be borne out brilliantly, but only a decade and a half later.

What was extremely unfortunate was that Wiener's memorandum was not brought to the attention of the computer builders at Harvard, Pennsylvania and elsewhere. (Dr. H.H. Goldstine, at least, was unaware of the memorandum until the writer brought it to his notice in early 1982.) In his autobiography Wiener has suggested that his computer proposal was rejected because Bush's "estimation of any work which did not reach the level of actual construction was extremely low" [56g, p. 239]. But this explanation is not convincing, for, as we have seen, Dr. Bush had treated Wiener's earlier tentative suggestions with considerable respect.

A more convincing explanation is that Bush felt that the line of approach that Wiener had advocated, while drawing away manpower from other important projects, would only marginally assist the war effort. One has also to take into account the recent investigations of Dr. Brian Randell {R1} of pre-war memoranda in the Bush archives in Washington, which reveal that Bush was himself involved in an all-purpose digital computer project at that time—a fact barely known to his colleagues. This relevation casts doubts on Dr. H.H. Goldstine's suggestion that it was perhaps an over-commitment to the analog point of view and a failure of vision that stopped Bush and his MIT colleagues from pursuing electronic digital devices, until J.W. Forrester broke the trend, cf. {G10, p. 91}. The memorandum shows that in engineering vision Wiener had hardly a peer, and could glimpse far into the future.

Note 1: **The Wiener integraph** (cf. § 13A)

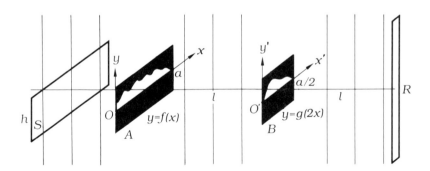

Fig. 1. The Wiener integraph

Imagine a wedge made up of two vertical planes passing through the mid-section of R and making there a small angle $\Delta\theta$. The wedge will cut the aperture at A in a narrow vertical strip, centered at x, say, of short width Δx and height $f(x)$. Its illumination is proportional to $f(x)\Delta x$, since the source S is uniform. All this energy $f(x)\Delta x$ will fall on the narrow vertical strip, with center $x/2$ and height h, in which the wedge cuts the screen B. But the part of this strip above height $g(2 \cdot x/2)$, i.e. $g(x)$, is opaque; hence only $g(x)f(x)\Delta x$ of the radiation in the wedge gets through the aperture in B and falls on the receiver R. Now rotate the wedge about the midsection of R from the position $x = 0$ at A to $x = a$ at A. Then both apertures are fully swept in the rotation, and we see that the total intensity at R is $\int_o^a f(x)g(x)dx$.

Next, imagine a contraption whereby the screen B can be made to swing through $180°$ about the y-axis, and then slid parallel to the x-axis a distance $t/2$ thus affecting the transformation $x/2 \rightarrow -x/2 + t/2$. If the illumination of the receiver R is measured with the screen B in this position, the reading will clearly be the desired $\int_o^a f(x)g(t - x)dx$.

Note **2: Proof of the relation § 13B(3)**

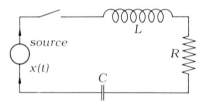

Fig. 2.   LRC circuit

Each of the passive elements causes the impressed voltage $x(t)$ to drop. For simplicity, suppose that we have only one resistor, inductor and capacitor of strengths R, L, C. (See Fig. 2). The voltage drops, according to the Laws of Ohm, Faraday and others, are

$$Ry(t), \quad Ly'(t), \quad \frac{1}{C}\int_o^t y(s)ds,$$

where $y(t)$ is the electric current at instant t. By Kirchoff's Laws, there is no accumulation of voltage, and the current $y(t)$ is the same at all points. Consequently, we have

$$Ly'(t) + Ry(t) + \frac{1}{C}\int_o^t y(s)ds = x(t),$$

or equivalently

$$Ly''(t) + Ry'(t) + \frac{1}{C}y(t) = x'(t).$$

Taking the direct Fourier transformation, we get

$$L \cdot (i\lambda)^2 \hat{y}(\lambda) + R \cdot i\lambda y(\lambda) + \frac{1}{C} \hat{y}(\lambda) = i\lambda \hat{x}(\lambda),$$

whence

$$\hat{y}(\lambda) = \frac{i\lambda}{L \cdot (i\lambda)^2 + R \cdot (i\lambda) + \dfrac{1}{C}} \hat{x}(\lambda),$$

We thus have § 13B(3), with $R(\lambda) = \lambda/(L\lambda^2 + R\lambda + 1/C)$.

Note 3: **The Lee-Wiener network** (cf. § 13B)

Let W in $L_1 \cap L_2$ be the time-domain transfer or weighting function of a physically realizable, time-invariant linear filter. Then, cf. § 9B(10), (11), W has to vanish on $(-\infty, 0)$. As Paley and Wiener discovered [34d, p. 8] for a function W in $L_1 \cap L_2$, this requirement is equivalent to the one that the Fourier transform $\hat{W}$ of W belongs to the Hardy class H_2 on the lower half-plane $\varDelta_-$. If this $\hat{W}$ is a rational function, we can in principle construct an electrical network with lumped passive elements (resistors, inductors, condensors) and amplifiers-attenuators, which realizes $W*$, i.e. which converts the input f in L_∞ into the output $W*f$, the convolution of W and f. But if $\hat{W}$ is not a rational function, we cannot so construct the convolution filter $W*$. All we can do is intelligently approximate to W by functions W_n in $L_1 \cap L_2$ vanishing on $(-\infty, 0)$ for which $\hat{W}_n$ are rational functions, and construct networks for the filters W_n*.

The merits of the Lee-Wiener network lay in the choice of the approximating functions W_n. They noticed that the Laguerre functions $L_n(\cdot), n = 0, 1, 2, \ldots$, cf. § 13B(4), had two very convenient features. (i) The set $\{L_n\}_{n=1}^\infty$ is an ortho-normal basis for $L_2[0, \infty)$, and consequently our W can be represented by an infinite series

$$W = \sum_{k=0}^\infty c_k L_k, \quad \text{with } c_k = \int_0^\infty W(t) L_k(t) dt. \tag{7}$$

(ii) The Fourier transforms $\hat{L}_n$ are rational functions of the simple form, cf. § 13B, (5),

$$\hat{L}_n(\omega) = \frac{1}{\sqrt{\pi}} \left(\frac{1 - i\omega}{1 + i\omega}\right)^n \cdot \frac{1}{1 + i\omega}, \quad \omega \text{ real.}$$

The right-most factor is realizable by a simple RL circuit of the type shown in Fig. 3.

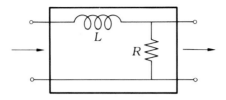

Fig. 3.   An *LR* network

The voltage ratio for this, viz. $R/(R + iL\omega)$, can be converted to $1/(i + i\omega)$ by suitable choice of the parameters R and L. The other factor (of absolute value 1, since ω is real) is realizable by cascading n identical "lattice" circuits of the type shown in Fig. 4.

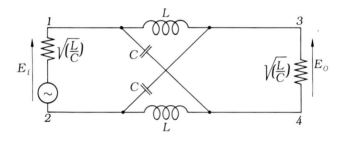

Fig. 4.   A form of phase-shift network

The voltage ratio for this circuit is $(1 - i\omega\sqrt{LC})/(1 + i\omega\sqrt{LC})$, which again with suitable choice of the parameters L and C yields $(1 - i\omega)/(1 + i\omega)$.

By making good use of the simple form of $\hat{L}_n$, Lee arrived at an economical design for a single network which realizes any weighting W, given by (7), to any desired degree of approximation. This network, the so-called *Lee-Wiener network*, is as shown in Fig. 5. In it, the number n is chosen large enough to secure the desired approximation. N_0 is the simple circuit shown in Fig. 3, and $N_1 \ldots, N_n$ are replicas of the circuit shown in Fig. 4. The A_k's are instruments (amplifier/attentuators) which convert an input f into the output $c_k \cdot f$. They have very high impedances, so that the current they draw away (from the top horizontal branch of the network) is negligible. The lower horizontal branch is a summing

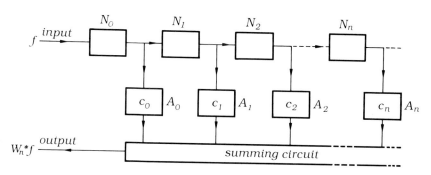

Fig. 5.   Block diagram of Lee-Wiener network

circuit. If a varying voltage f is plugged in at the input end of the network, we will receive at the output end the voltage W_n*f, where

$$W_n = \sum_{k=0}^{n} c_k L_k.$$

The network thus realizes the convolution filter $W*$ to any degree of approximation. To realize a different filter $\mathring{W}*$, no rebuilding has to be done; only the knobs on the instruments $A\mathring{\,}_1, \ldots, A_n$ have to be turned in order to secure the coefficients $\mathring{c}_1, \ldots, c_n$ of the new weighting $\mathring{W}$.

Note 4: Scanning procedure to solve the PDE (cf. § 13D(6))

As Wiener stated in his covering letter to Bush, in his scheme the proposed computing machine

> solves a partial difference equation involving the time, and asymptotically equvalent to a partial differential equation involving the time, ... This partial difference equation is solved by an apparatus which repeatedly scans a collection of data recorded on some very inexpensive device and replaces these data by new data. [85a, p. 122]

Thus, the equation (6) is replaced by the corresponding difference equation obtained by taking a fixed rectangular lattice of points in the txy – space. Without loss of generality, the lattice may be assumed to be that of the integers, and thereby the difference equation given the form:

$$4\{u(x, y, t+1) - u(x, y, t)\} =$$
$$\{u(x+1, y, t) - 2u(x, y, t) + u(x-1, y, t\}$$
$$+ \{u(x, y+1, t) - 2u(x, y, t) + u(x, y-1, t)\}$$

i.e.

$$4u(x, y, t+1) = u(x+1, y, t) + u(x-1, y, t)$$
$$+ u(x, y+1, t) - u(x, y-1, t)\}. \qquad (8)$$

The scanning procedure that Wiener had in mind to solve (8) does not lend itself to easy description. For simplicity suppose that the curve over which the boundary conditions are prescribed is the periphery of a rectangle in the xy – plane with sides parallel to the coordinate axes, and that the lattice has v points in each row. Assume that these points are counted from left to right, row after row, from top to bottom, and put into a single sequence P_n, $n = 1, 2, \ldots, v^2$. Then (6) yields for points P_n, not on the periphery,

$$u(P_n, t + 1) = \frac{1}{4}[u(P_{n+1}, t) + u(P_{n-1}, t) + u(P_{n-v}, t)$$
$$+ u(P_{n+v}, t)].$$

This is because the points in the sequence just above and just below P_n are clearly P_{n-v} and P_{n+v}.

> Thus if we have a record of the values of $u(P_n, t)$ on a linear tape, if we can scan these values so as n to obtain $u(P_{n\pm1}, t)$, $u(P_{n\pm v}, t)$, if we have a rapid mechanism for adding these values together and dividing by four, and if we can imprint this on the tape for the next run, leaving all boundary values unaltered, say by some switch-off mechanisms, we can solve the system of equations (4) (i.e. our (8)) mechanically. [85b, p. 129]

It is apparent from this that what Wiener meant by "scanning procedure" is a partial difference reduction adapted to transcription on a tape. In [23b] (with H. B. Phillips) he had employed this very net to attack the Dirichlet problem by a corresponding difference equation. The idea goes back to Runge (1911) and reappears in the late 1930s in the relaxation technique of D. G. Christopherson and R. V. Southwell, cf. B. Randell Coll. Works IV, pp. 135–136.

14 Bigelow and Anti-Aircraft Fire Control, 1940–1945

A Project D.I.C. 5980: its military and scientific significance

Soon after the shelving of his computer project, Wiener was drawn into the study of the control of servo-mechanisms that was being carried out by engineers at MIT. His work during World War II grew from a suggestion he made to the servo-engineers in early November 1940, that networks with frequency responses of a certain kind, into which the positional data of an airplane's flight trajectory is fed, might provide a means for evaluating its future locations, and so assist in the improvement of anti-aircraft fire. Anti-aircraft fire control was a problem of tremendous military importance at that time because of German air superiority over England.

The problem of anti-aircraft fire control deviates from similar problems in land and naval ballistics in that the speed of the target (airplane) is a substantial fraction of that of the missile fired at it. Consequently, to score a hit the gun must be aimed at a future position of the plane. The future path of the plane has therefore to be extrapolated from its observed path in the past before the firing can begin. The extrapolation must be done by a high-speed computer, and the aiming and triggering done automatically. Briefly, the gunner and his range computation tables have to be supplanted by a speedier and more accurate automaton that does its own radar tracking, anticipating, aiming and firing.

The servo-engineers under the leadership of Professor S. H. Caldwell went to work on Wiener's idea using the Bush differential analyser at MIT. These efforts were voluntary, for the group had no financial support from the National Defense Research Committee (NDRC). But as Dr. Caldwell wrote in his proposal to Section D-2 of the NDRC on November 22, 1940:

> The preliminary results obtained . . . indicate a much higher probability of success, and it is the conclusion of all the members of that group . . . that a more extensive exploratory program should be initiated. [Ia, pp. 1, 2]

The Caldwell proposal was accompanied by a four-page memorandum from Wiener entitled "Principles governing the construction of prediction and compensating apparatus". It was summed up by Wiener in these words:

> The proposed project is the design of a lead or prediction apparatus in which, when one member follows the actual track of an airplane, another member anticipates where the airplane is to be after a fixed lapse of time. This is done by a linear network into which information is put by the entire past motion of the airplane and which generates a correction term indicating the amount that the airplane is going to be away from its present position when a shell arrives in its neighborhood . . . The proposal is, first, to explore the purely mathematical possibilities of prediction by any apparatus whatever; second, to obtain particular characteristics which are suitable for apparatus of this type and are physically realizable; third, to approximate to these characteristics by rational impedance functions of the frequency; then to develop a physical structure whose impedance characteristic is that of this rational function, and finally, to construct the apparatus. [Ia, pp. 2, 3]

By "rational impedance function of the frequency" Wiener was referring to the frequency-response function, which, as we saw, cf. § 13B(3) et seq, is a rational function for lumped passive networks. The project led to the appearance of a classified monograph in February 1942, entitled "The interpolation, extrapolation of linear time series and communication engineering", which was affectionately nicknamed "the yellow peril" by the engineers, because of its yellow cover and its abstruse mathematics. As we flip over the pages of Wiener's monograph (declassified at the end of the war) and his later reports, we see the remarkable clarity of Wiener's initial perception of what his goals had to be, and how thoroughly he accomplished them by 13 months of hard work.

The modest appropriation of $2,325 that was requested, was granted by the NDRC in December 1940, and so was launched the Project D.I.C. 5980 on *Anti-aircraft (A.A.) Directors*. Wiener's solution of this purely military problem in collaboration with Julian Bigelow, was extremely creative, and has had wider ramifications than any of his other work. Its overall significance only became clear after its declassification at the end of the war, and its appearance in book form [49g]. Briefly, and quite apart from its pure mathematical value, *the Wiener-Bigelow work transformed the ad hoc art of filter-designing into a branch of statistical physics; it contributed substantially, along with the work of Dr. C.E. Shannon, to the creation of modern communication theory; and it initiated the new field of medical cybernetics.* On the other hand, its military uses were limited to improvements in minimizing the noise in radar tracking, for the Wiener-Bigelow anti-aircraft filter itself proved to be too elaborate for use in the field, and offered no practical improvement over the non-statistical "memory-point" method of prediction worked out by Drs. H.W. Bode and C.A. Lovell of the Bell Telephone Laboratories. In Wiener's candid words, at the end of his

Julian Bigelow

Photograph taken
in 1942

Final Report to Dr. Warren Weaver of December 1, 1942, on Section
D-2, Project No. 6:

> Present field results are much closer to the best obtainable by any imaginable
> apparatus than I had expected. . . . Accordingly, there is less scope for furth-
> er work in this field than we had believed to be the case. [IX, p. 7]

Very candidly, he advised that "all efforts in this field should be concentrated on features leading to more rapid production and simplicity
of use in the field" [IX, p. 8].

To backtrack a little, in January 1941, an engineer from IBM,
Mr. Julian Bigelow, was assigned to assist Wiener in his work. As
Wiener wrote in his Final Report to Dr. Weaver, Bigelow rendered
considerable help in the theoretical work as well as in numerical computation, "but his assistance was greatest and indeed indispensable in
the design of apparatus, and in particular in the development of the
technique of electric circuits of long time constant". (For more on
Bigelow's contribution and on the Lee-Wiener network, see Note 1,

p. 195.) In 1946 Bigelow became the chief engineer for the computer at the Institute for Advanced Study, Princeton, that von Neumann was building.

B Idealization of the flight trajectory and resulting tasks

As stated in Wiener's accompaniment to the Caldwell proposal of November 22, 1940, the anti-aircraft predictor had two components, the first a membrane to track the airplane, the other a filter that gave out its future position. The value $f(t)$ of the input $f(\cdot)$ of this filter is the position of the airplane at instant t as indicated by radar; and for a given lead $h > 0$, the value $g_h(t)$ of its output $g_h(\cdot)$ is the forecast position of the plane's location at the instant $t + h$. Wiener found that a filter which works well when the input signal $f(\cdot)$ is smooth, responds erratically when $f(\cdot)$ has sharp bends, as it would in the case of a zig-zagging airplane, and vice versa. And he soon realized that this difficulty is intrinsic. Not surprisingly (cf. § 10A) he found this antagonism "to have something in common with the contrasting problems of the measure of position and momentum to be found in the Heisenberg quantum mechanics" [61c, p. 9]. As there is no way of knowing whether an enemy pilot is going to follow a smooth course or start to zig-zag, a good single predictor must give a reasonably good forecast $g_h(\cdot)$ for both types of input $f(\cdot)$. Sometime in early 1941, Wiener and Bigelow realized that it was hopeless to try for perfect prediction (for which $g_h(t) = f(t + h)$) in all cases, that flight prediction, like meteorological prediction, is statistical, and that the merits of a particular prediction filter (predictor) must be judged in the same way as the merits of a particular system of weather-forecasting—by probabilistic criteria of overall performance.

What sort of statistical hypotheses could possibly fit the flight trajectories of military aircraft on bombing runs? The trajectory depends of course on when and how the pilot decides to exploit his plane's maneuverability. The plane and the pilot are both involved. The project would have been stymied had not Wiener brought to bear on it his full understanding of harmonic analysis. He made the ideal assumption that the flight signals f belong to the class S of covariance-possessing functions, which he had demarcated in his 1930 work on Generalized Harmonic Analysis, cf. § 9D. Indeed, this is a considerable idealization. For one thing, signals in S (like the sinusoids), are eternal, i.e. defined on $(-\infty, \infty)$, whereas actual flight trajectories are defined only on short bounded time intervals $[a, b]$. The idealization was based on the supposition that "the tracking runs are of sufficiently long duration so that the disturbances of their initiation has no great effect by the time the predictions are to be used" [VII, p. 4].

Stated differently and in more musical terms, Wiener was hoping that his apparatus would be as good as the human ear, i.e. possess just the right amount of insensitivity to do its job. As we saw (§ 10A), while the ear is insensitive enough to make the $C^{\#}$ struck up by the violinist sound like a purely periodic (and therefore eternal) oscillation, it is not so bad as to make us tone-deaf. Wiener was hoping that his apparatus too would be insensitive enough to ignore the fact that its readings began at $t = -10$ seconds, instead of $t = -\infty$, but not so intensitive as to totally misjudge the plane's whereabouts at $t = +20$ seconds. The Air Force wanted Wiener's apparatus to do its wonders within a 30 second interval, of which 10 were for tracking, and 20 for aiming and firing. These constraints were dictated by the limitations of the tracking gear and the anti-aircraft guns available in the early 1940s.

Only towards the end of the project did it transpire that these hopes were too good to be true. The pioneer does not know what lies ahead. By faith he conjectures, and with faith he presses on. Hence, with their idealization in place, Bigelow and Wiener went on to carry out the following tasks:

(a) They worked out a mathematical theory for the prediction of functions f in the class S. The prediction g_h of f, for a lead h, was expressed in the form

$$g_h(t) = \int_o^\infty f(t - \tau)dW_h(\tau),$$

and the (unknown) weighting W_h was expressed by a simple expression involving the covariance function φ of the signal f. (c. January, February 1941)

(b) They designed an electric filter that could realize this weighting W_h in-the-metal, i.e. which, when fed with an input voltage f, would give out the voltage g_h, to a high degree of approximation. (c. March–June 1941)

(c) They built a contrivance which would generate "an entire family of curves having the same sort of randomness as a maneuvering airplane". (c. July–December 1941)

(d) They studied the statistics of actual airplane tracks, and for this visited several army and navy centers in eastern United States where such work was being done, receiving full cooperation from the authorities. (The so-called "runs 303 and 304" made at the Anti-aircraft Board at Camp Davis, in particular, assumed a conspicuous role in the later phases of their work.)

C The mathematical theory

Let $f(t)$ be the incoming signal from the airplane at the instant t. By Wiener's ideal assumption f is in the class S, and with the ergodicity assumption, the covariance φ of f can be found from our knowledge of $f(t)$ for $t < 0$, cf. § 12A(3). Actually f is a mixed signal:

$$f(t) = f_1(t) + f_2(t),$$

where $f_1(t)$ is the plane's true position at instant t (the "message"), and $f_2(t)$ is the tracking error (the "noise"). Idealizing further, Wiener and Bigelow assumed that both f_1, f_2 are in S, and are cross-correlated, and that the noise-noise autocovariance φ_{22} and the message-noise cross-covariance φ_{12} are known from practical or theoretical considerations, cf. § 9F(22). It follows that the message-signal cross-covariance φ_1 of f_1 and f, obtainable from the formula

$$\varphi_1(t) = \varphi(t) + \varphi_{22}(t) - \overline{\varphi_{12}(t)},$$

is also known.

Given $h > 0$, the problem is now to determine from the known covariances φ, φ_1, the weighting W_h such that the average

$$f_h(t) = \int_0^\infty f(t - \tau)dW_h(\tau), \quad \text{cf. } (a) \text{ above}, \tag{1}$$

which involves only the observed data prior to instant t, approximates to $f_1(t + h)$, the true position at instant $t + h$ in the future. Wiener adopted the root-mean-square error criterion

$$\lim_{T \to \infty} \frac{1}{2T} \int_{-T}^T |f_1(t + h) - \int_0^\infty f(t - \tau)dW_h(\tau)|^2 dt = \min. \tag{2}$$

By using the calculus of variations he showed that $W_h(\cdot)$ must satisfy the integral equation

$$\varphi_1(t + h) = \int_0^\infty \varphi(t - \tau)dW_h(\tau), \quad t \geq 0. \tag{3}$$

This is a Stieltjes variant of the Hopf-Wiener equation of the first kind with unknown W_h (cf. § 11B(2)), and Wiener was able to solve it by extending the methods he had discovered with Hopf (cf. § 11A).

Wiener and Bigelow realized that their formulation and solution of the equation (3) led to a general recipe for the *design of filters,* or equivalently, the design of *instruments.* Ingenious guesswork, though useful, was no longer necessary. An *instrument* (such as a speedometer) is a filter, the output of which is a reading that informs the operator of the condition of one or more factors affecting his machine, and so helps in its control. The input f of the filter is causally related to the condition of the machine. Suppose that we want to procure from each signal (or input) f of a certain class $\mathscr{S}$, a desired signal $\mathring{f}$ bearing a definite and causal relation to f, i.e. such that for each real t

$$f_1 = f_2 \text{ on } (-\infty, t) \quad \text{implies} \quad \mathring{f}_1 = \mathring{f}_2 \text{ on } (-\infty, t). \tag{4}$$

For the speedometer, $f(t)$ is the distance traveled at instant t, and $\mathring{f}(t)$ is its speed at instant t. Wiener and Bigelow gave a recipe to build a filter to accomplish this task as best as the laws of mathematics and physics will allow, i.e. to build an instrument which, when subjected to input f, will give a reading g for which $\| g - \mathring{f} \| = $ minimum, $\| \cdot \|$ being a suitable measure of the *error of performance* $g - \mathring{f}$.

The significance of this Wiener-Bigelow contribution to engineering science is best stated in Professor Lee's words:

> Wiener's theory of optimum linear systems is a milestone in the development of communication theory . . . The problems of filtering, prediction, and other similar operations were given a unity in formulation by the introduction of the idea that they all have in common an input and a desired output. Then the minimization of a measure of error, which is absent in classical theory, was carried out . . . The entire theory from its inception to the final expressions for the system function and the minimum mean-square error is invaluable to the communication engineer in the understanding of many communication problems in a new light. {L6, p. 23}.

See Note 2, p. 195, for more on the equation (3) and its role in the theory of instrumentation.

D Operation of the anti-aircraft predictor

In this book we can say nothing on the very complex problem of designing the anti-aircraft predictor, and can only touch on its operation.

The input signals comprised the so-called "intrinsic coordinates" of the airplane, i.e. the acceleration (positive or negative), the rate of change of horizontal curvature, and the rate of climb. From these signals the apparatus first filtered out the noise emanating from random disturbances in the plane's geometric path (due to mechanical deficiencies, pilot errors, uneven air currents, etc.), from tracking errors (caused by instrumental and personnel deficiencies) and from so-called "cranking errors", i.e. transients initiated when the apparatus is started. Only the filtered values were used in computing the plane's future position.

The problem of transient cranking errors led Dr. G. R. Stibitz of the Bell Telephone Laboratories, and Chairman of Section D2, to propose in February 1942 a modification of the RMS error criterion by the introduction of "a weighting factor for the merit of the prediction, depending on the time from the beginning of the run", cf. Wiener's Report of June 10, 1942 (NDCac-83) p. 8. Wiener was struck by this idea, and it was to permeate some of his later work, but it is too technical to discuss here.

E Resolution of the man-machine concatenation

The Wiener-Bigelow predictor network had of course to be tested before any outlay could be made on producing anti-aircraft directors containing it. But where was the accurate flight data needed for such testing to come from? The seriousness of this issue is evident from Wiener's military reports:

> Little information on the nature of the paths of combatant aircraft was accessible either in terms of actual space trajectories, or observational data from tracking apparatus. Furthermore, it then appeared that radical new developments in both aircraft and tracking equipment were soon to come, so that it would be unwise to found any new theory or practice upon brief samples of transitional data. [VII, p. 2, para. 2]

Outstanding was the thorny question as to what "sort of input function" to use to test the predictor:

> This was not easy to answer, since the theory clearly indicated that the successful predictor of any single curve of simple geometrical properties will have little significance as an index of the performance of a predictor on a more general input. The obvious answer was that the test input must consist of what amounts to a class of curves rather than of any single curve. *We needed some sort of contrivance which should generate an entire family of curves having the same sort of randomness as a manoeuvering airplane.*
> * * * * * * * *
> Accordingly we developed a very simple random curve generator, consisting (1) of a light-spot projector executing a smooth but non-uniform reciprocating motion across the wall of the laboratory at a rate of about fifteen seconds per traverse; and (2) of a sluggish contrivance, with which an operator (simulating the pilot) attempts to follow the "perfect manoeuver" of the independently moving spot. [VII, pp. 5, 6] (emphasis added)

Wiener saw to it that "the spot had to be moved by a control which was complicated to begin with, and furthermore felt completely wrong" [56g, p. 250].

From operating with the random curve generator, Wiener and Bigelow found to their surprise that it was "highly sensitive to individual nervous reactions of the operator". "The recorded performance of each operator showed an autocorrelation curve definitely similar to that of his other performances, and differing from that of other operators" [VII, p. 7, para. 2]. Wiener and Bigelow were thus led to believe that although the pilot is ostensibly a free agent, the environment in which he is placed during a bombing run effectively constrains his behavior. His individual movements are auto- and cross-correlated, as a result of which the course his airplane executes in the sky is itself autocorrelated. To sum up, they realized that

> ... the "randomness" or irregularity of an airplane's path is introduced by the pilot; that in attempting to force his dynamic craft to execute a useful

manoeuver, such as a straight-line flight, or a 180 degree turn, *the pilot behaves like a servo-mechanism,* attempting to overcome the intrinsic lag due to the dynamics of his plane as a physical system, in response to a stimulus which increases in intensity with the degree to which he has failed to accomplish his task. A further factor of importance was that the pilot's kinaesthetic reaction to the motion of the plane is quite different from that which his other senses would normally lead him to expect, so that for precision flying, he must disassociate his kinaesthetic from his visual sense. [VII, p. 6] (emphasis added)

All this reinforced their initial supposition that the observed trajectory of the airplane belongs to the class S that Wiener had demarcated in the late 1920s.

F The problem of transients

Unfortunately, Wiener's hopes that his apparatus would be blind to the discrepancy involved in his ideal assumption that the flight signals are in the class S were not fulfilled. The acoustic analogy we suggested is invalid: the response of Wiener's apparatus to the incoming signals was a far cry from that of the human ear in response to sound waves. To quote Wiener:

> Our original theoretical development was based on infinite runs for the determination of autocorrelation coefficients as well as on infinite runs for prediction. Of course, we have known from the beginning that these assumptions are not strictly true and we have found out that the practical consequences of their falsity are not always negligible. This is particularly true of the error in determining the autocorrelation coefficient. The transients due to a sudden starting and a sudden stopping of our catalogued data may completely mask the true behavior of our statistical system for high frequencies. [VII, p. 8, para. 1]

It was this intrangiency of cranking errors that made Wiener's stochastic approach to flight prediction less effective militarily than the non-stochastic approach of Bode and Lovell, cf. § 14A. This was a major factor in ending the Wiener-Bigelow project.

We may explain the failure of the Wiener predictor rather crudely as follows. A motorist knows that he must not change gear and start driving after starting the engine, but must first let it settle down to a steady hum. If his is an old jalopy that quakes and rattles on starting, then he must be patient. The Wiener apparatus could not settle down within the 30 seconds alloted to it. Had present-day technology been available, this difficulty might perhaps have been averted. If so, then Wiener's problem, as with his computer project, was that he wanted to do in the early 1940s what only the technology of the 1950s and 1960s could deliver.

Did the termination of the project lead to resentment on Wiener's part and to his estrangement with his engineering and military colleagues? The documentary evidence is crystal clear.

As our quotations from his reports show, Wiener openly analyzed the limitations in his approach. Obviously he was the one best qualified to know whether his project merited continuation. And on this point he was quite clear, as the last page, marked "Recommendations", from his Final Report of December 1942 shows:

> The author finds that an optimum mean square prediction method based on a 10 second past and with a lead of 20 seconds *does not give substantial improvement over a memory-point method, nor over existing practice.* He proposes to check this result with a similar method in which the more unpredictable parts of the course are discarded[1], and to discover the maximum lead for which his method yields an effective improvement in prediction. He also wishes to examine the theoretical effect of better tracking. He considers that if these investigations do not yield a much more favorable result than those already carried out, and *he does not anticipate that they will,* he will have established that *new developments in the design of long-time predictors have already reached the point of diminishing returns,* and that *all efforts in this field should be concentrated on features leading to more rapid production and simplicity of use in the field.* [IX, p. 8] (emphasis added)

In his follow-up letter of January 15, 1943, after describing the further work that would be needed to improve flight prediction, and the merits of perusing it, Wiener concluded:

> I hesitate to make any such recommendation because I frankly do not know to what extent work of this sort may tie up present working sources of the country, and because *the present expectation of great improvement is too distant to be significant in the present war.* I most definitely do recommend that such study be made *within some part of the long time program of our armed services.* [X, p. 5] (emphasis added)

Thus Wiener advocated ending attempts to design better flight predictors as a part of the war effort, but maintained that further studies of the many interesting issues that had come up during the project would be in the long-range interests of the armed forces.

Did the Bell engineers and defense officials who watched this project feel that the effort was wasted? Far from it. As late as July 1, 1942, Dr. Stibitz, the Chairman of Section D2, noted in his diary: "I feel more strongly than ever that any attempt to predict curved flight, which is anything more than a stop-gap, must make use of Wiener's theory" [VIII]. That this whole undertaking was favorably evaluated is clear from the official Preamble, entitled *Statistical method of prediction in fire control,* that accompanied the limited circulation of Wiener's Final

---

1   This, by the way, was Stibitz's suggestion of February 12, 1942.

Report. Pointing out the limited value of straight line prediction (were it to succeed, enemy aircraft would not fly straight courses), the Preamble continues:

> Thus it is important to investigate how one can, by various methods, predict other than a pure linear signal. For analytical curves of varying amounts of complexity (horizontal circles, helices with dive or climb, etc.) the prediction procedures can be worked out by simple geometrical methods. But to investigate prediction problems for signals which contain noise due to tracking or to flight errors, or for signals corresponding to various general types of curved flight, that is a *much deeper and more difficult matter.*
>
> This is, in very general terms, the problem with which Professor Wiener has been concerned. In a previous memoir (Report to Services No. 19) he applied powerful analytical tools to develop a statistical method of predicting. In the present report he indicates, in two papers dated December 1, 1942 and January 15, 1943, the result of applying his method to certain definite cases. *That these particular applications did not turn out to be of practical importance does not, in our judgment, mean that the study was not well worthwhile. The general theory will doubtless have other applications; and it was a matter of importance to know just how successful this statistical method would be for the antiaircraft problem.* [Preamble to IX & X] (emphasis added)

This last suggestion, about "other applications", was indeed borne out, and the Department of Defense was a beneficiary. For as Professor O. K. Moore, the sociologist, has written:

> ... in the early 1950s Naval Research Laboratory (NRL) set up a Systems Analysis Branch that had as its main mission the simulation of complex man-machine systems in terms of the cybernetical agenda. Researchers there used Wiener's "Yellow Peril"—the classified lithographed book on time series that appeared in 1949 [49g]. Ideas implicit there and in earlier classified documents of Wiener on the latency (i.e., delays) in human response were put to practical use in the "quickening" of control systems in ships, submarines, airplanes, and elsewhere, such as marking dials ahead on instrument panels to compensate for latency. {*Coll. Works,* IV, p. 821}

Professor Moore adds "I was a member of the System Analysis Branch. Much of the work was classified, so I cannot cite appropriate references". But this was not all. It soon transpired that the nervous systems of animals, big and small, are replete with these quickenings (Wiener's "lead mechanisms" of 1940 or "anticipatory feedbacks" [61c, p. 112]) as well as "lag mechanisms" to slow down, of which Wiener was also quite aware.

G Secrecy, overwork and tension

Wiener was very appreciative of his exchanges with the engineers of the Bell Telephone Laboratories from whom he benefited considerably, as we just saw. He was much irked by security regulations which forbade such discussions. He must have complained about this to President Karl T. Compton of MIT, for there is a letter from Dr. Compton to Dr. Warren Weaver dated May 13, 1941, in which we read:

> Norbert Wiener is much disturbed over the instructions given him not to discuss his work on the gun-laying project with the others working in this field who were at the Fortress Monroe conference. He feels that discussion with several of the men, notably Lovell, and also with Stewart[2], would be decidedly valuable to all concerned. He says that it is evident that Lovell has instrumental methods of accomplishing some of the operations which would be desirable in the best embodiment of his mathematical analysis and which he had not known were available. He also believes that some aspects of his analysis can be decidedly helpful in steering future development of the practical equipment. [II]

Actually, in this instance, as we gather from the same letter, the clamp was put on Wiener not from the fear that he would spill some valuable beans, but from the fear that his "highly mathematical" discussions "might well have prevented the close get-together of the military and civilian personnel which was a primary objective of this conference"! But Wiener had a strong and lasting aversion to secrecy. Reflecting, later on, on the overall meaning of secrecy in public affairs, he was to write:

> This demand for secrecy is scarcely more than the wish of a sick civilization not to learn of the progress of its own disease. [50j, p. 127]

Fortunately, later conferences allowed for free exchanges between Wiener and the Bell engineers, and his 1942 "yellow peril" received a wide circulation among those engaged in related war work. Without a doubt, this dissemination of ideas enhanced both military and scientific research.

Some other human problems arose during the project. Wiener began to overwork. He stayed awake with benzedrine, computing numerical data all night—a task for which he was physiologically and tempermentally unsuited, cf. [56g, p. 249]. He became nervous and irritable, and could not meet his obligations. This, and some squabbling with official patent attorneys, strained his relations with the Radiation Laboratory. There is a letter from Dr. Warren Weaver to Professor J.C. Boyce of MIT of March 24, 1942, [VI] referring to Wiener's nervous condition and his failure to submit promised reports. Professor Boyce replied that he had advised Wiener to take a few days rest, and he was hopeful that "this part of the zoo will be quiet again". Such tempests in the "human zoo" pale into insignificance, however, when set beside the path-breaking research and new ideas that came from Wiener's intense activity.

2   Dr. C.A. Lovell of the Bell Laboratories and Mr. Duncan Stewart from Barber-Colman Company of Rockford, Illinois.

H Kolmogorov's paper and prediction theory

As Wiener stated in his Final Report on Section D-2, project #6 of December 1, 1942, p. 6, his objective in undertaking all this work was "to find the design of the predictor which will produce the maximum number of effective hits" in an anti-aircraft barrage. It is remarkable that the mathematical theory emerging from this very practical engineering motivation turned out to be equivalent to another developed by A.N. Kolmogorov, titled "On stationary sequences in Hilbert space" {K8}. This paper appeared in 1941 but the results had already been announced in 1939, {K6}. Kolmogorov states that "all the new problems that are studied and solved in this paper have originated in the theory of probability and mathematical statistics", but the paper, unlike Wiener's, deals with the geometry of Hilbert spaces and has nothing on filters. Kolmogorov refers the reader to another paper of his, in the Izvestiya Akad. Nauk {K9}, for the applications.

Kolmogorov's treatment places heavy emphasis on a fundamental 1938 theorem of H. Wold of Sweden, of which Wiener did not know. Wold announced it as a basic theorem on time series, but as is now amply clear, it is a basic theorem on operators on a Hilbert space. Kolmogorov's paper is written with a meticulous exactitude and thoroughness, in sharp contrast to Wiener's [49g], which leaves many crucial mathematical questions unanswered and contains several lacunae. In his war work Wiener was in a hurry, and operated at what Professor A.M. Yaglom {Y1} has aptly termed "a heuristic level of rigour".

What has emerged from the efforts of Wold, Kolmogorov, and Wiener is nowadays called *prediction theory*. It has had very interesting mathematical ramifications, which have been commented upon extensively in the review {M9, pp. 102–105} and in the *Coll. Works*, III. In a nutshell, the Kolmogorov-Wiener theory in its systematized multivariate version, which was worked out during Wiener's visit to the Indian Statistical Institute in 1955–1956, and independently in the Soviet Union and the United States at about the same time, occupies a central position in the general theory of stationary and causal systems. As the authors B. Sz. Nagy and C. Foias point out in their book {N1, p. v}, prediction theory has offered new insights into the extension of the spectral theory of operators on a Hilbert space to the non-normal operators, a topic on which the Hungarian and Odessa schools of mathematicians had been engaged. Wiener also made some significant inroads in the much more difficult realm of *non-linear prediction,* [55a, 58i, 59b]. Interesting reviews of these papers also appear in the *Coll. Works*, III.

I The Kolmogorov-Wiener concatenation

Wiener became aware of Kolmogorov's work on prediction only in the last days of 1941. Long before that he had been questioned by United States military authorities as to who abroad was likely to duplicate his ideas on flight prediction. Wiener replied that he doubted if anyone in Germany could, that his probabilist friends H. Cramer in Sweden and P. Lévy in France were possibilities, but that "if anyone in the world were working on these ideas it would most likely be Kolmogorov" [56g, p. 262]. Earlier in his autobiography, Wiener had described his symbiotic relationship with Kolmogorov, with whom he had never communicated and whom he was to meet only in 1960, as follows:

> Khintchine and Kolmogoroff, the two chief Russian exponents of the theory of probability, have long been involved in the same field in which I was working. For more than twenty years, we have been on one another's heels; either they had proved a theorem which I was about to prove, or I had been ahead of them by the narrowest of margins. This contact between our work

Academician
Andre Kolmogorov
1903–1987

came not from any definite program on my part nor, I believe, from any on theirs but was due to the fact that we had come into greatest activity at about the same time with about the same intellectual equipment. [56g, p. 145]

In § 7D we spoke of the importance of Wiener measure in the evolution of Kolmogorov's thought. Kolmogorov's 1940 paper on "Wiener's spiral and some other interesting curves in Hilbert space" {K7} shows that parts of this early work on the Brownian motion lend themselves to Hilbert spatial formulation. Wiener's work on Generalized Harmonic Analysis can be phrased in terms of these very "spirals" or "helices" in Hilbert space, (cf. § 9E, end). Likewise, parts of his work on the Homogeneous Chaos can be explicated in terms of stationary measures with values in a Hilbert space. Finally, after Kolmogorov, Wiener's theory of multivariate prediction can be viewed as a theory of stationary curves or sequences in a Hilbert module $\mathcal{H}^q$ over the linear algebra of $q \times q$ complex matrices. Thus, in the light of Kolmogorov's contributions, it appears that a good deal of Wiener's deepest mathematical work really concerns the geometry of stationary varieties or stationary measures in Hilbert spaces and Hilbert modules.

Note 1: Bigelow's design (cf. § 14A)

Specifically, Wiener described Bigelow's contributions as follows:

> By the use of an amplifier of extremely great input impedance and a very accurate voltage-ratio of unity, he was able to show that circuits with any impedance characteristic realizable at all were realizable with the use only of amplifiers, capacities and resistances, completely ignoring the use of inductances. The inductances are well known to be the least satisfactory element of ordinary circuit design . . . It was therefore a step of enormous importance to free the technique of circuit design from the use of such objectionable elements. [IX, p. 3]

In terms of our circuit diagram of the Lee-Wiener filter shown in Ch. 13, Note 3, Wiener was referring to the improved designing of the instruments $A_1, \ldots, A_n$.

Note 2: The equation § 14C(3); theory of instrumentation (cf. § 14C)

By the theorems of Wiener and Bochner, φ and φ_1 are Fourier-Stieltjes transforms of a non-negative distribution F on $(-\infty, \infty)$ and of a complex-valued distribution F_1 on $(-\infty, \infty)$, respectively. Assuming that the signals f and f_1 are non-deterministic, it follows that F and F_1 are absolutely continuous, and moreover that $F' = |\Psi|^2$, where Ψ is holomorphic on the upper half-plane. Exploiting fully the properties of Ψ and the holomorphism inherent in the situation, Wiener found the

Fourier-Stieltjes transform $\hat{W}_h$ of W_h by the factorization method he discovered in his work with Hopf (§ 11B). W_h was recovered by the Fourier-Lévy inversion from $\hat{W}_h$.

To turn to the suggested theory of instrumentation, suppose that we want to build a time-invariant linear filter, which, when subjected to input f, will give a response g for which $\|g - \mathring{f}\| = $ minimum, $\mathring{f}$ being the desired reading and $\| \cdot \|$ a suitable measure of the error of performance $g - \mathring{f}$. Granting the equation § 14C(4), the Wiener-Bigelow method provides a recipe for finding the weighting W of this optimal filter.

If the inputs f are energetic, i.e. in $L_2(-\infty, \infty)$, define the covariance functions φ of f by $\varphi(h) = (f(\cdot + h), f(\cdot))_{L_2}$. If the inputs f are powerful, i.e. in the Wiener class S, define φ as in § 9D(12). Similarly define the cross-covariance function $\varphi_1(\cdot)$ for the pair of signals $f, \mathring{f}$. Then our goal translates once again into the Hopf-Wiener integral equation § 14C(3) we met above, but for the weighting W instead of W_h. By solving this equation we will again get W. Once W is found, we can synthesize the filter or instrument by adapting the Lee-Wiener network technique (§ 13B).

15   Arturo Rosenblueth and Wiener's Work in Physiology

A   Rosenblueth, friend and philosopher

The year 1933, that witnessed the untimely death of R.E.A.C. Paley, also saw Wiener's introduction to Dr. Arturo Rosenblueth, a Mexican physiologist, of Hungarian descent, trained at the Sorbonne and working with Dr. Walter Cannon of the Harvard Medical School. Wiener became a regular participant in an interdisciplinary seminar on scientific method which Rosenblueth was conducting at the school, the "Philosophy of Science Club". The role of this seminar in the genesis of cybernetics and in Wiener's life can hardly be exaggerated. The seminar ended in 1944 when Rosenblueth left for Mexico to become the Head of the Department of Physiology of the new National Institute of Cardiology. But Rosenblueth's contacts with Wiener lasted much longer, and he was to become one of Wiener's closest collaborators, and also one with whom Wiener had the warmest relationship. Wiener's description of Rosenblueth reads:

> He had not started as a scientist but as a musician, and had earned his living for some time by playing classical piano music in a Mexico City restaurant. Arturo is also a first-rate chess player and a superb bridge player, so good that in neither of these games does he allow me to play with him. He is a great enthusiast for the climate and arts of his native land. . . He is a hard worker, and he makes the greatest demands on the sincerity and industry of those about him, demands which are only exceeded by those he makes upon himself. [56g, p. 278]

Rosenblueth's uplifting influence was felt by many. The book *Libro homenaje, Arturo Rosenblueth,* by J. Garcia Ramos {G3}, that appeared after his death, contains tributes from many distinguished medical scientists. To cite just one, from Dr. George Acheson, former head of the Department of Pharmacology of the University of Cincinnati:

> I first came under Dr. Rosenblueth's influence in 1934 as a first-year medical student in Dr. Cannon's Department of Physiology. I was fortunate to be assigned to a special seminar Dr. Rosenblueth conducted in the physiology course. Dr. Rosenblueth had a most important influence over my development. The clarity of his thinking, writing, and speaking and the vigor and imaginativeness of his attack on physiological problems had a strong effect

on my career. I like to think that, as Dr. Cannon was my scientific grandfather, so Dr. Rosenblueth is my scientific father. {G3, p. 24}

Indeed, there was much disappointment when Rosenblueth was not
appointed to the Chair that Dr. Cannon vacated in late 1942.

Wiener was much concerned that overwork would irreparably
damage Rosenblueth's health, and expressed this worry in letters to
Rosenblueth's physician, Dr. S.A. Simeone. In reply, the doctor, a
mutual friend, pointed out that an intensive pace of work was necessary
to Rosenblueth's well-being and that there was no cause for alarm [MC,
125]. Reciprocally, Rosenblueth was an uplifting force on Wiener's life
in both its intellectual and emotional phases. A letter he wrote to Wiener
in August 1944, is indicative of his buoyant spirit:

> You sounded rather pessimistic in your last letter. That is wrong. The war
> could not possibly be going better . . . Your work seems to be going along
> beautifully from what you tell me. Your family is doing handsomely. Our
> projects, although still in the realm of the "we shall see" are alive and kicking
> (or maybe I should say, and wagging their caudal appendage). Your novel
> is still trying to crack its shell. You have friends and they don't forget you
> —witness thereof, the pleasant time I'm having writing to you. What the
> Avernus can you crab about, anyhow? It's really a great world and a great
> life, my dear Norbert, notwithstanding their occasional infirmities. [MC, 66]

Rosenblueth was also a moral force, one that counteracted Wiener's
querulous tendencies in moments of noisy childishness. On touchy issues
he urged Wiener to be generous and forgiving, cf. [MC, 77].

What drew Wiener and Rosenblueth together, notwithstanding
their different vocations and training, was commonality on basic methodological issues of science:

(1) a faith in the unity of science, and the belief that interdisciplinary
     work is vital to its survival;
(2) a positive attitude towards philosophical analysis that has the
     potential to accelerate the forward movement of science;
(3) acceptance of the epistemological principle that knowledge can
     only be knowledge of relation-structure.

Dr. Rosenblueth's interdisciplinay seminar at the Harvard
Medical School was concerned with philosophical problems that actually affected the scientific research being pursued by its participants.
Issues such as the *brain-mind dualism* were approached from a standpoint that was cognizant of fresh neurological discovery, and could thus
aid neurological investigations in progress. The seminar's orientation
was thus rather different from that of the run-of-the-mill "philosophy of
science" circle, in which philosophical cud is chewed and rechewed with
imperceptible effect on scientific advancement. This difference becomes

Arturo Rosenblueth
1900–1970

Photograph taken
in 1945

Courtesy of Mrs.
Virginia Rosenblueth

obvious when Dr. Rosenblueth's book *Mind and Brain* {R2} is compared
with more stereotyped tracts on the philosophy of science.

Dr. Rosenblueth's book clearly reveals that Wiener's philosophical mentors were also his own, and that a central concept in his scientific
philosophy was what Whitehead and Russell have called *relation-structure*. Affirming the position taken by Russell in his book *Human Knowledge, Its Scope and Limits* {R7} that all we can know of the objects and
events in the universe is their structure, Rosenblueth wrote:

> I fully concur with this statement, and believe that this limitation is basically
> due to the fact that as soon as the messages we receive from the outside reach
> the afferent sensory fibers, they are phrased in a code which has nothing in
> common with the original objects or events except for a common structure.
> When a person hears a symphony, the messages sent by the orchestra reach
> the listener as air vibrations. These vibrations stimulate mechanically the
> receptors of the organ of Corti, and these receptors set up nerve impulses
> along the fibers of the VIIIth nerve. It is clear that at this stage the physical
> events that are taking place are of an entirely different kind from those

occurring in the instruments of the orchestra. Yet the message is preserved because there are similarities in certain features of the two series of events —sounds emitted by the orchestra and nerve impulses traveling over the auditory nerves. The existence of these similarities of relations is precisely what is called a common structure. The mental decoding, which is the perception of the symphony, again preserves the corresponding relations. A common structure thus implies the quantitative preservation of the relations that exist between the independent constituents of an event or message through a set of transformations. {R2, p. 55}

Rosenblueth then demarcates the concept of *structure* in terms of *isomorphism,* precisely as done by the great mathematician H. Weyl {W5, p. 25}. Briefly, a system comprising a set X and n relations $R_1, \ldots, R_n$, thereon (each R_k being an r-tuple of elements of X, $r \geq 1$ depending on k)[1] has the same structure as another system comprising a set Y and n relations $S_1, \ldots, S_n$ thereon, if and only there is a one-one function T on X onto Y such that

$$(x_1, \ldots, x_r) \; \varepsilon \; R_k \quad \text{if \& only if} \quad (Tx_1, \ldots, Tx_r) \; \varepsilon \; S_k.$$

Thus the relational-predicates "has the same structure as" and "is isomorphic to" are synonymous.

By way of illustration, Rosenblueth cites the large number of such isomorphic (or similarly structured) systems involved in the transition from the set of events that constituted the composition of the piano sonata, Opus 111, by Beethoven in 1823 to the set of events in the mind of a listener today of a Schnabel recording of the sonata made in 1932. Here the relational structures that remain invariant are the ratios among the fundamental frequencies of the simultaneous or successive sounds, their intensity, duration, timing and the like. And an important transformation T is between these and the musical notation on sheets of paper (coding), another important T being the final mental decoding that enables us to appreciate the sonata. But for optical perception, among the relations that must be preserved are also some that are topological in nature. In all such transitions from sets of events in the outside world to those in our minds "... the only features that we can perceive are those which remain invariant under the physiological transformations which our organisms impose on them" {R2, p. 57}.

This strong endorsement of some of his own early philosophical ideas by an active and independent neurologist had a tonic effect on Wiener. From Rosenblueth's seminar Wiener became conscious of the importance of brain wave encephalography as a tool for understanding the mind. Also, Rosenblueth's thought on the preservation of structure gave a foundation to the theory of pattern-recognition that McCulloch,

---

1   We may look upon properties as unary relations.

Pitts and Wiener advanced around 1945 (§ 16C). But the first input from Rosenblueth came in the course of Wiener's war work in the stormy year 1942 (next section).

B Teleology in the animal and the machine

Affected as he was by the pressures of war, Wiener did not lose sight of the broader aspects of his military research. His defense reports refer, for instance, to the medical implications of his work. Thus an unabridged version of the quote we made on p. 188, from his report on the random curve generator, reads:

> An interesting point which came apparent early in the shakedown trial and correction stage was that the random signal generator was highly sensitive to the individual nervous reactions of the operator. The recorded performance of each operator showed an autocorrelation curve definitely similar to that of his other performances, and differing from that of other operators. This suggests the use of such apparatus in the diagnosis of individual differences in reflex behavior, and of pathological conditions affecting the reflex arc. Many other extensions of these ideas will suggest themselves to the physiologist, the neuropathologist, and the expert in aptitude tests. [VII, p. 7, para. 2]

This experimental evidence led Wiener and Bigelow to suppose that feedback occurs in the human sensory and motor abilities that constitute purposeful activity. To test this supposition they proceeded on the heuristic principle that "the pathology of an organ throws very great light on its normal behavior" [56g, p. 252]. Now in servomechanisms based on feedback there is an aberration called *hunting,* in which excessive feedback causes the mechanism to overshoot its mark and to fluctuate wildly. We witness "hunting" when an elevator, instead of coming smoothly to a stop, goes up a little too much, and overcompensates by going down too much, and so keeps oscillating until stopped by external input. Wiener and Bigelow, wondering if a similar pathology afflicted human behavior, posed the following question to Rosenblueth:

> Is there any pathological condition in which the patient, in trying to perform some voluntary act like picking up a pencil, overshoots the mark, and goes into an uncontrollable oscillation? Dr. Rosenblueth immediately answered us that there is such a well-known condition, that it is called purpose tremor, and that it is often associated with injury to the cerebellum. [61c, p. 8]

As in important discoveries in the past, this small but startling bit of evidence was enough to convince Wiener and Bigelow that they were on the right trail, that the neurophysiological system does function by negative feedback. This in turn necessitated a new view of the human nervous system:

... the central nervous system no longer appears as a self-contained organ, receiving inputs from the senses and discharging into the muscles. On the contrary, some of its most characteristic activities are explicable only as circular processes, emerging from the nervous system into the muscles, and re-entering the nervous system through the sense organs, whether they be proprioceptors or organs of the special senses. This seemed to us to mark a new step in the study of that part of neurophysiology which concerns not solely the elementary processes of nerves and synapses but the performance of the nervous system as an integrated whole. [61c, p. 8]

But these developments also gave new insights into communication engineering. For if the engineering concept of "feedback" has a place in neurophysiology, might not neurological concepts such as "purpose", "message", "memory" be definable usefully within engineering? Wiener and Bigelow were thus led to the tentative formulation of these concepts within engineering, and by 1948, this included the concept of "information" as well, as we saw in § 12G.

The collaboration with Rosenblueth led to an important joint paper by them on "Behavior, purpose and teleology" [43b]. In it is suggested, perhaps for the first time, that *voluntary activity involves negative feedback,* and that certain activities of servomechanisms as well as of animals can be studied in a unified way by considering the goals of the activities. This led to the recognition that in the consideration of natural phenomena for which the Entropy Principle is significant and time is directed and "Bergsonian" rather than "Newtonian" in the sense discussed in § 12F, the concept of *purpose* is fruitfully definable and its use leads to significant unification.

In [43b] a mechanism M is called *purposive,* if there is a coupling of M with an element E in its environment, which makes M move toward a stance S_0 bearing a certain definite relation to E. A mechanism M is called *teleological,* if it is purposive and if its movement towards the stance S_0 is controlled at each moment by the deviation of its actual stance S from S_0, i.e. if its movement is governed by continuous *negative feedback.* A simple example of a teleological mechanism is the compass needle. When displaced from the SN direction it overshoots, then over-compensates, and eventually the oscillations die down and it assumes the SN "stance". Another example is a target-seeking missile. A lizard poised for striking a fly is a purposive, non-teleological mechanism: its action is too quick to permit intermediate negative feedback. Such mechanisms are called *passive,* or *active,* according as all, or only a part, of the output energy comes from the input. The compass needle is passive. Servomechanisms are active. In the steersman of a ship, for instance, the energy expended by the man at the wheel merely activates steam valves, and it is the mechanical energy, derived from steam pressure in the boiler, that turns the rudder.

This new point of view—the restoration in science of Aristotle's "final cause" suitably explicated—has far-reaching philosophical and religious consequences, of which Wiener was increasingly to become aware during the late 1950s and the 1960s.

C Mexico, and the return to physiology. The heart.

> Almost from the moment I crossed the frontier, I was charmed by the pink-and-blue adobe houses, the bright keen air of the desert, by new plants and flowers, by the indications of a new way of living with more gusto in it than belongs to us inhibited North Americans. The high, cool climate of Mexico City, the vivid colors of the jacaranda blossoms and the bougainvillaeas, the Mediterranean architecture, all prepared me for something new and exciting. The many times I have returned to Mexico have not belied these first impressions. It will be a sad day for me when I come to feel that I have no further chance to renew my contacts with that country and to participate in its life.
>
> . . . .
>
> The Indian element in Mexico and the Spanish differ in many ways. It is the Spaniard rather than the Indian who has the romantic *élan* which we associate with the South. On the other hand, the Indian is unsurpassable where steadiness, loyalty, and conscientiousness come into consideration. Thus, each race furnishes qualities which the country needs. It is a splendid thing that the Indian has come to his own and is participating in the development of a new middle class, which stands on the tripod of the Spaniard, the Indian, and the foreigner.
>
> My Mexican friends came from all of these three elements of society, and it was an exciting thing to see how this new middle class of diverse origins constituted a cordial, friendly, and well-organized body of people. [56g, pp. 276–280]

Wiener was speaking of his first visit to Mexico in 1945[2] at the invitations from the Mexican Mathematical Society to speak at their meeting in June of that year, from Dr. Rosenblueth of the National Institute of Cardiology to continue their collaboration, and from his former colleague Dr. Vallarta, then in the Science Commission (Comisión Instigadora y Coordinadora de la Investigación Científica). Wiener spent ten weeks in Mexico, during which he collaborated with Rosenblueth in an area in physiology proper. Wiener's biological curiosities, held dormant since 1912 when he shifted from the study of zoology to that of philosophy, now found expression. He could now show what his "teachers did not recognize", that despite all his "grievous faults", he might still have a contribution to make to biology. (See the full quotation on p. 40–.) With encouragement from Dr. Walter B. Cannon, they sought to find out *how muscular contractions during epileptic convulsions are related to movements of the heart* [46b]. (To appreciate the physiolog-

---

2 The "Mexico, 1944" in the title on p. 276 of [56g] should read "Mexico, 1945".

ical work of Wiener, a knowledge of the nervous system, as is found in Note 1 on pp. 213–215, is essential.)

The heart pumps blood by a recurring cycle ("heartbeat") made up of two steps: (i) a contraction of the cardiac muscle beginning at the top of the left auricle[3] and passing down to the ventricle (*sistole*), followed by (ii) a short period of rest (*distole*). But the cardiac muscle also exhibits other less regular vibrations, among which the most important are so-called flutter and fibrillation, and it was with these that Rosenblueth and Wiener were concerned. Both types afflict patients with lesions in the left auricle.

As defined by Rosenblueth and Wiener, *flutter* is a continuous periodic wave motion along a closed path, whereas *fibrillation* is an intrinsically irregular vibratory movement comprising several waves following random paths. From the literature they cite, it appears that the last word on these movements had come in 1925, when Dr. T. Lewis claimed that they were intrinsically the same, fibrillation being just a very "fast flutter". Rosenblueth and Wiener held the opposite view, as is evident from the definitions they adopted. They granted that the dispute could not be settled by experiment, since it is impossible empirically to tell a fibrillation from "a fine grained flutter over a multiplicity of complicated fixed paths" [46b, p. 258]. They nevertheless attempted to justify their new view by developing a mathematical theory of flutter and a mathematical statistical theory of fibrillation.

Both theories are abstract and kinematical rather than dynamical. The bioelectrical aspect of the subject, for instance, is ignored. Their theory on flutter is postulational, with theorems appearing, such as "a single stimulus applied to a single point or region can never result in flutter" [46b, p. 231]. That on fibrillation describes a random network of fibers, and establishes the equations of conduction over it. It leans heavily on the ideas introduced in the Wiener-Wintner paper, "The discrete chaos" [43a]. A good portion of the paper is devoted to the propagation of flutter around one or more anatomical or functional obstacles.

Wiener and Rosenblueth felt that their theories would shed light on other "continuous self-sustained transmission of impulses in physiologically connected, non-spontaneously-active networks", such as nerve nets. They pointed to the similarities between the heartbeat and the nerve impulse in their all-or-none and refractory behaviors (see Note 1, p. 214). The flutter corresponds to reverberating nerve impulses along closed chains of neurons, and fibrillation to so-called tonic-clonic corti-

---

3   The *auricle* is the small upper chamber of the heart, and *ventricle* is the larger lower chamber. An interior wall divides each into "left" and "right".

cal responses, i.e. the slow spreading of the local effects, elicited by stimuli, along seemingly random paths without any noticeable periodicity, cf. {R2, pp. 40–42}. It was their hope that some of their theorems could be subjected to experimental tests, and the range of theoretical conjecture thereby reduced.

Although well aware that this 60-page work lacked adequate computational analysis, the paper was presented at the mathematical meeting in Gadalajara, and in the following spring (1946) at the first of a series of semi-annual meetings organized by Dr. Warren McCulloch with the support of the Josiah Macy Foundation. These meetings, limted to a group of about 20 scientists, were for the presentation of fresh research, and informal discussion. The paper which embodied this work [46b], was to serve as a catalyst in the work of Drs. W.J. Osher and T. Cairns in the mid-1960s, and its influence continues even to this day. For details see the comments of Garcia Ramos in *Coll. Works, IV*.

D Muscle clonus

In the summer of 1946 Wiener again went to Mexico, this time with support from the Rockefeller Foundation. He teamed up with Rosenblueth and his colleague, Dr. Garcia Ramos, to analyze experiments on the extensor thigh muscles of a cat in order to study *clonus:* "the familiar spasmodic vibration which many people experience when they sit crosslegged with one knee under the other" [56g, p. 277]. This phenomenon has to do with feedback. Rosenblueth felt that their results were substantial enough to be presented to colleagues, and they were read at the Macy meeting in New York in the fall of 1946 but not published. Their experimental confirmation "did not come up to Arturo's rigid requirements" as Wiener has explained, cf. [56g, p. 278]. Recently, Dr. Garcia Ramos, the only surviving author, has found the supporting evidence, and his updated version [85c] appears in the *Coll. Works, IV*, along with his commentary.

The least technical account of their experiments, which were on the exterior thigh muscles of decerebrated cats under anesthesia, is in Wiener's *Cybernetics:*

> We cut the attachment of the muscle, fixed it to a lever under known tension, and recorded its contractions isometrically or isotonically.[4] We also used an oscillograph to record the simultaneous electrical changes in the muscle itself . . . *The muscle was loaded to the point where a tap would set it into a periodic pattern of contraction,* which is called *clonus* in the language of the

---

4   A muscle is said to contract *isometrically,* when its length is allowed to vary only minimally. It is said to contract *isotonically,* when the tension on the muscle remains constant.

physiologist. We observed this pattern of contraction, paying attention to the physiological condition of the cat, the load on the muscle, the frequency of oscillation, the base level of the oscillation, and its amplitude. [61c, p. 19] (emphasis added)

Wiener's words, "We cut . . ." must be taken with a grain of salt. As Dr. Garcia Ramos explained to the writer, Wiener would come to the laboratory and watch them do their experiments, throwing in an idea or two. Then, following further talk on the results, Wiener would retire to work out a mathematical explanation.

The experiments showed that the greater the mass of the moving system, the greater the tension required to initiate clonus. On the other hand, the frequency of clonic oscillations is nearly independent of the mass. Continued observations, however, revealed slow changes in the basal tension and amplitude. The many other experimental results are too technical to describe.

Theoretically, Wiener and his colleagues approached the neurophysiological circuit involved in clonus from the standpoint of communications engineering. Almost half of their 50-page manuscript is devoted to the application of the mathematical theory of communication to the problem. A preliminary question was the determination of the physiological concepts that should correspond to the different ones such as current or voltage in engineering. And since the circuit in question is highly non-linear, a major concern was the determination of a suitable linearization.

The fact that the rhythmic activity maintains itself, after the stimuli that induce it have ceased, shows that a closed feedback circuit is involved in clonus. The authors assume that it is made up of one afferent nerve and one efferent nerve, besides the muscle and kinesthetic receptors. If the number of impulses transmitted per second by the efferent nerve is taken for the "current", then the circuit is highly non-linear. But it becomes nearly linear if the logarithm of this number is taken for the "current". To quote from their manuscript:

In the clonus studied, the envelope of the motor impulses recorded electrically in the cat's quadriceps, and inferentially in the efferent nerve, has the form:

Thus the system cannot even approximately be linear in the amplitude of such an oscillation. On the other hand, such a curve may very well serve to represent a function of the form $ae^{b \sin ct}$ . . . If then we take as our dependent variable not $x = ae^{b \sin ct}$, but $y = \log(x/a) = b \sin ct$, we shall obtain quantities in which the oscillations could possibly be linear. This is equivalent to considering the nervous processes to be not additive but multiplicative. [85c, p. 501]

With the logarithmic basis, the oscillation frequencies deduced from the data with the aid of the principles of servo-mechanisms turned out to be good approximations to the observed clonic frequencies.

Encouraged by this experimental confirmation, the authors advanced a physiological explanation of clonus in their paper. We may sum it up by using Wiener's words:

> There is a strong suggestion that though the timing of the main arc in clonus proves it to be a two-neuron arc, the amplification of impulses in this arc is variable in one and perhaps in more points, and that some part of this amplification may be affected by slow multineuron processes which run much higher in the central nervous system than the spinal chain primarily responsible for the timing of clonus. This variable amplification may be affected by the general level of central activity, by the use of strychnine or of anesthetics, by decerebration, and by many other causes. [61c, p. 21]

Wiener also felt that the observed slow change in basal tension and amplitude could be studied by the method of secular perturbations in celestial mechanics due to H. Poincare and G. W. Hill. The final part of the chapter on feedback in Wiener's *Cybernetics* [61c] is devoted to this question.

E The spike potential of axons

To regularize their collaboration, Wiener and Rosenblueth induced MIT and the Mexican Cardiological Institute to approach the Rockefeller Foundation for support of an arrangement whereby Wiener would spend one semester every other year in Mexico City, and Rosenblueth a part of the intervening year at MIT. This was readily obtained, thanks largely to Dr. Warren Weaver, and Wiener spent the fall of 1947, 1949 and 1951 in Mexico, and Rosenblueth made several visits to MIT.

A project that Wiener and Rosenblueth undertook in Mexico in 1947 was to determine the shape of the fluctuation, $y = f(t)$, of the *action* (or "*spike*") *potential* at a fixed point x on a stimulated axon, to find out if the shape depended qualitatively on the axon, and how it is influenced by experimental factors such as temperature. This investigation was done in collaboration with Dr. J. Garcia Ramos and Walter Pitts [48c]. It involved considerable experimentation, followed by mathematical description based on the empirical material as well as on theoretical premises.

Wiener and his colleagues removed intradural portions of the dorsal roots, 2 to 4 cms. long, from the lumbo-sacral segments of the spinal cord of cats,[5] and extracted fine strands containing several nerve fibers. They applied rectangular electric pulses to these strands, and obtained electrograms of the response, picked up by electrodes that fed via an amplifier to a cathode ray oscilloscope. Appealing to the all-or-none law (Note 1, p. 214), they concluded that whenever the latency of response varied but the shape of the curve f remained unchanged, only a single fiber was active.

After rectification of various errors, they were led to the hypothesis that the stimulus-response relationship is "piece-wise linear", i.e. that the time-domain can be broken into three consecutive intervals in which the stimulus obeys *different* linear differential equations. This suggested that different factors were at work during the three intervals. The team then tried to account for the observed results by introducing *dynamical partial differential equations* connecting the current along the axon with the potentials inside and outside each cross-section. (For more details, see Note 2 on p. 215.)

This work was superseded four years later by the more revealing (non-linear) partial differential equation for the ionic current due to A.L. Hodgkin and A.F. Huxley, but its pioneering aspect is worth stressing. It was the first to bring in field conceptions, continuous time and PDE's, in axon theory—an area where the overall, all-or-none behavior, is clearly discrete. Even today the subject lacks powerful general principles governing the physiochemical mechanism, and the modus operandi, consisting of educated guesswork based on careful experimentation, is not very different from that followed by Wiener and his co-workers in 1948. Some of their conclusions received experimental confirmation in the work of Dr. I. Tasaki in 1956.

F The statistics of synaptic excitation

A fourth project that Rosenblueth and Wiener undertook in Mexico was on the statistical analysis of the input-output relation in spinal reflexes in dogs and cats [49b]. This too was done in collaboration with Dr. Garcia Ramos and Walter Pitts, and it involved a good deal of experimentation as well as statistical hypothesizing and testing.

In a *reflex action* the nerve impulse engendered by the stimulation of a receptor organ travels along a spinal nerve to the spinal cord,

---

5   The *lumbar-sacral* segment of the spinal cord is the segment that extends from the middle to the lower back, and is thus farthest from the brain. The *dorsal root* of a spinal nerve (which in quadrupeds is located on the upper side of the cord) is composed of incoming sensory neurons. The *intradural* portion is the part inside the Dura Mater, i.e. the outermost of the three sheaths (the meninges) that surround the cord and encase the cerebrospinal fluid.

and then, via one or more association neurons, to a motor neuron, thereby activating a muscle or a gland. (The brain plays at best a secondary role in the action. The activation of the salivary glands by the sight or smell of food in an example.)[6] In the experiments of Wiener and his colleagues, segments of the spinal cord were exposed for study. Stimulating electrodes were placed on the lumbar dorsal root of a spinal nerve, as close to the center of the cord as possible. Recording electrodes were placed on the surface of the nerve at both its dorsal and ventral roots.[7] The recording electrodes were connected in parallel to the input of an amplifier, the output of which was led to a cathode-ray oscilloscope. The stimuli were rectangular pulses, applied at a slow rate with gradually increasing intensity. Since the distances between the point of stimulation and the points of recording on the dorsal and ventral roots were different (as the later root is in another section of the cord), the dorsal-root spike registered on the scope before the appearance of the reflex spike in the ventral root. Thus single pictures with the dorsal spike to the left of the reflex spike were obtained. By plotting the former spikes on the x-axis and the latter ones on the y-axis on a sheet of paper, they got a single input-output curve. This turned out to be an increasing concave curve with the suggestion of a horizontal asymptote as x-increases.

Suppose that in the nervous net just described A afferent (i.e. incoming) fibers branch into N synaptic knobs, which impinge randomly over E efferent (i.e. outcoming) neurons. The authors started with three simple theoretical assumptions:

(a)   Each afferent neuron has about the same number of synaptic knobs;

(b)   Impinging on each efferent neuron is the same number n of independent afferent knobs;

(c)   An efferent neuron fires, if and only if at least m of these n afferents are simultaneously active.

Letting p_k be the probability that at least k afferents are firing, they easily deduced that $E \cdot p_m$ is the average number of efferents which fire in response to an afferent volley comprising m or more firing afferent fibers. But this result did not fit their experimental data, and hence at least one of the assumptions (a), (b), (c) had to be wrong. The authors found cogent reasons for retaining (a) and (b), but rejecting (c) and

6   Such activation by events unconnected with, but associated with the presence of food (e.g. ringing of a bell) is an example of *conditioned* reflex action.

7   The *ventral root* of a spinal nerve is composed of outgoing motor neurons. It is located in the middle (thoracic lumbar) section of the cord. For the *dorsal root* and *lumbar section,* see footnote 5 on p. 208.

therefore treating m as a random variable. They then worked out statistical methods to test the different hypothetical distributions that suggest themselves for the random variable m. They also treated the more complex case of multi-synaptic reflexes.

G Voluntary, postural and homeostatic feedback. The moth-cum-bedbug

Perhaps the most important benefit that Wiener derived from his sustained contacts with Rosenblueth was his understanding of the complexity of feedback in the animal. From these contacts Wiener learned of the classification of such feedback into (1) voluntary, (2) postural and (3) homeostatic. *Voluntary feedback* is the primary one employed in fulfilling a task, and the one by which we gauge by means of sense organs the extent of the task that has not been accomplished. *Postural feedback* is an auxiliary feedback for the maintenance of internal tone (tonus) and involves kinesthetic organs. More fully,

> ... In order to accomplish a purpose successfully, the various joints which are not directly associated with purposive movement must be kept in such a condition of mild *tonus* or tension, that the final purposive contraction of the muscles is properly backed up. In order to do this, a secondary feedback mechanism is required, whose locus in the brain does not seem to be the cerebellum, which is the central control station of the mechanism which breaks down in intention tremor. This second sort of feedback is known as *postural feedback*. [50j, p. 164]

Whereas purpose tremor (which appears only in activity) stems from excessive voluntary feedback located in the cerebral cortex, the complementary disease, Parkinsonianism (in which tremor exists only during inactivity), stems from ineffective postural feedback located in the brain stem. Wiener gave a mathematical explanation of why the stability of a servomechanism composed of many parts requires more than one feedback [61c, pp. 106–110]. He found that "... The sum of different operators each of which may be compensated as well as we wish by a single feedback cannot itself be so compensated" [61c, p. 106]. An important example of an inanimate system controlled by double feedback is the automatic steering of a ship by means of the gyrocompass. The motion of a hand or a finger "involves a system with a large number of joints"; its stability requires both voluntary and postural feedback, the latter for the maintenance of tone in the muscular system.

Finally, there is *homeostatic feedback,* designed to maintain steady metabolism, respiration and the like, and to keep crucial parameters, such as body temperature, within intervals of safety. These feedbacks occur by slow transmission along the (nonmyelinated) fibers of the automonic nervous system. Some of the messages go via non-nervous channels, e.g. hormone transmission by blood circulation. Associated with such feedback are the *homeostatic diseases,* like leukemia, where often

> ... what is a fault is not so much an absence of all internal control over the process of corpuscle formation and corpuscle destruction but a control working at a false level. [56g, p. 292]

The homeostatic feedback is discussed in *Cybernetics* [61c, pp. 114–115], and the mathematical theory of feedback systems, including the role of anticipatory and lagging feedback, is discussed further in the same chapter of the book.

These theories on feedback and its excesses had their inception in the question that Wiener and Bigelow had asked Rosenblueth in the course of their war work in 1942, (cf. § 15B). Although mathematically feasible, their implementation in natural mechanisms was largely conjectural. Wiener wanted the theories put to a practical test. In the late 1940s he teamed up with Dr. J. Wiesner, then in the Research Laboratory of Electronics, and later to become the president of MIT, to build a demonstration-machine with two feedbacks that would play a purposive and postural role, respectively, and exhibit the two types of tremors known to physiologists.

The machine they built, with help from H. Singleton, was a small tricycle cart, with two photocells facing front, one on the right side and one on the left. The output from the cells, after amplification, reaches the tiller controlling the front steering wheel. Depending on the direction of the output voltage, the cart is steered either towards or away from the quadrant with more intense lighting, thus behaving either like a "moth" or like a "bedbug". In either case it acts like a purposive mechanism, either pro- or anti-phototropic. The feedback involved, from light source to cell to tiller, and back, is *voluntary,* for "voluntary action is essentially a choice among tropisms" [56g, p. 166]. When the amplification is increased, however, the feedback becomes excessive, and forces the cart to overdo its movements, thereby getting it into oscillation, as in *purpose* or *intention tremor.*

Also built into the cart was a secondary feedback attached to the tiller, and so arranged that it could be overloaded even with little or no output from the photo cells. In the absence of light this became excessive and the tiller started oscillating. This secondary feedback played a *postural* role, and the second tremor exemplified *Parkinsonianism.*

This tiny "moth-bedbug" somehow drew the attention of the United States Medical Corp. They photographed its tremors in order to compare their profiles with those of human tremors, and so enhance the knowledge of army neurologists.

Norbert Wiener with Drs. J. Weisner and Y.K. Lee

In the early 1950s a number of such mock-zoological creatures, exemplifying various facets of life, were built. Two came from the leaders of the cybernetical movement in England. Dr. W. Ross Ashby built the *homeostat* {A4, A6}. Devoid of any programming, this machine, when disturbed from equilibrium, automatically ("voluntarily") selects from 360,000 possible connections (or paths), randomly laid out in its internal mechanism, one of the few that restores it to some position of internal stability or equilibrium. This behavior, vaguely akin to that of the lazy domestic cat, who when disturbed from its sleep, stirs a little only to find another cozy position to doze off, earned it the nickname *machina sopora*. Dr. W. Grey Walter of the Burden Neurological Institute in Bristol built a machine resembling the bug, but more versatile, called the *tortoise* or *machina speculatrix*. It has a touch receptor in addition to light receptors which can also scan. It is attracted to light of moderate intensity, repelled by strong light, and automatically avoids obstacles in its path. Apart from the phototropism, it continually explores the floor on which it is placed, automatically avoiding obstacles in its way. Interesting social phenomena appear when a few such tortoises are set going.

Note 1: **The nervous system**

The nervous system serves to transmit information collected by both external and internal sensors to other parts of the body, and also to integrate the information from different sensors and, when appropriate, to initiate muscular movement or glandular secretion. By the end of the 1930s a clear overall picture of the histology and physiology of the nervous system was at hand.

In this picture the fundamental element in the nervous system is a specialized cell called the *neuron.* This, like all cells, has a body comprising a nucleus and cytoplasm enclosed in a membrane, but the cytoplasm extends into input-fibers called *dendrites* and into output-fibers called *axons.* The axons are covered by a membrane made up of protein and lipid (*white matter*), while the rest of the neuron is not (*gray matter*). A neuron's dendrites are either in contact with a sense organ or kinesthetic organ,[8] or they are near the terminus of one or more axons of other neurons, the intervening gap being called the *synapse.* The axons of a neuron can also terminate on the sheath of a muscle. The system of neurons inside the brain and the spinal cord is called the *central nervous system* (CNS).

Even when an axon is in its resting state, there is an electrochemical potential difference (PD) of about 70 microvolts between the center and the periphery of any cross-section. This is due to a preponderance of positive ions inside the axon, stemming from the difference in the permeability of its membrane to the outward flow of potassium (K^+) ions inside, and the inward flow of sodium (Na^+) ions outside.

The neuron is highly *irritable:* any stimulus (i.e. energy input) exceeding a very small threshold, applied to its dendrites changes the permeability of the axon membrane, and with it the PD between center and periphery, at the initial end of the axon. This fluctuation in the PD (so-called *action potential*) propagates down the length of the axon until it reaches its far terminus. The neuron is said to *fire* when its stimulation initiates such a propagation, which in turn is called a *nerve impulse.* Upon reaching the terminus of the axon, the nerve impulse produces one or other chemical, which diffuses across the synapse to the dendrite of an adjacent neuron or to a muscular sheath, thereby either encouraging or inhibiting the firing of the second neuron or of the muscle fiber, as the case may be.

---

8   *Kinesthetic organs* (also called *proprioceptors*) are receptors within the body which gather information about the internal environment, much as the sense organs gather information about the external environment.

The nerve impulse has several characteristic features:

(1)   A neuron will fire when and only when the magnitude of its stimulation exceeds the threshold. When it fires, the resulting nerve impulse is independent of the excess. Thus a neuron has only two states: *firing* or *quiet*. (This is referred to as the "all-or-none" law.)

(2)   The speed of a nerve impulse for a given axon is fixed, being dependent on its cross section.

(3)   A nerve impulse along an axon is either *excitatory* or *inhibitory*, depending on whether it tends to encourage or discourage the firing of the neuron or muscle fiber on which its terminus impinges.

(4)   The chemicals produced at the terminus of an axon by the propagation of a nerve impulse take a certain time (the *synaptic delay*) to diffuse across the synapse and reach an adjacent neuron.

(5)   The trans-synaptic excitation of a neuron by such diffusion requires an accumulation over a time period (*temporal summation*), or the simultaneous arrival of the material diffused from more than one axon (*spatial summation*).

(6)   Time is required for an excited neuron to return to normal, i.e. to its resting PD between the center and periphery of cross-sections of its axons. During this *refractory period,* the neuron will not fire, even if subjected to above-threshold stimulation.

There are long axons connecting remote regions of the nervous system. But near the hub—the cerebral cortex of the brain—there are short connections, apparently random in their arrangement. Neurons forming a closed cycle have been demonstrated anatomically. Circulating nerve impulses along such cycles are presumed to serve as short-term memories.

The system of neurons outside the CNS is called the *peripheral nervous system.* The neurons within this system are distinguishable by the roles they perform. *Sensory neurons* transmit information from the sense receptors and proprioceptors to the CNS and the brain. They have long dendrites reaching to the receptors or proprioceptors, and short axons, which connect, via other neurons, to the spinal cord. *Motor neurons* transmit information from the brain to different parts of the body, and bring about muscular movement and adjustment, and glandular secretion. Their cell body and (short) dendrites are in the brain or spinal cord, and their (long) axons lead to glands or to the sheaths of muscles. *Association neurons* are ones connecting sensory to appropriate motor neurons. They belong to the CNS.

Nerves are cords (or bundles) made up of many nerve fibers, i.e. neurons with their dendrites and axons, often as many as 100. Their length can vary from a fraction of an inch to several feet, and their thickness from that of a thread to that of a pencil. The nerves have a distinguished end, called the *center. Afferent nerves* are ones in which the nerve impulse proceeds towards the center. *Efferent nerves* are those in which nerve impulse propagates away from the center. The nerve impulse is carried in each of the separate fibers in the nerve. *Mixed nerves* are those having fibers made up of sensory neurons, as well as fibers made up of motor neurons. Such nerves can transmit impulses in both directions without interference, much as traffic on a two-way street.

The portion of the CNS necessary for carrying out the involutary processes vital to survival (e.g. the beating of the heart) is called the *autonomic nervous system.* It comprises two complete and separate systems of nerves, called the *sympathetic and parasympathetic nervous systems.* A nerve from each system runs to each body organ, and their impulses have a balancing effect on the organ. The sympathetic nerve speeds up the organ in its activity, the parasympathetic nerve slows it down. The advantages are those of a double feedback.

Note 2: Spike potential of axons

Let $y = f(t)$ be the shape of the fluctuation. Then, after correcting for various artifacts, Wiener and his co-workers found that the curve f "rises, inflects, reaches a maximum, and then declines with only one more inflection" [48c, p. 279] (see Fig.). They also found that the second differences that approximate the second derivative f'' were "quite irregular" [48c, p. 283]. It was reasonable to assume that the transformation leading from the stimulus to the response f was only *piecewise linear*, i.e. the domain of f was made up of intervals over which f obeyed *different* linear differential equations, the points t of division being those at which f and f' are continuous but f'' is not.

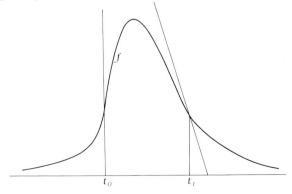

Empirical analysis of the electrograms strongly suggested that f'' "is indeed discontinuous at least at the inflectional point on the ascending branch" [48c, p. 286]. Wiener and his colleagues therefore decided to assume that the curve f for a single fiber is made up of three arcs separated by the ordinates at the points of inflection t_0 and t_1. They then tried to fit single exponential curves to the extreme arcs of f and a modified exponential curve to the middle arc [48c, p. 287]:

$$
\begin{aligned}
y &= Ae^{\alpha t} & t &< t_0 \\
y &= M - Nt + Be^{-\beta t} & t_0 &\le t \le t_1, \\
y &= Ce^{-\gamma t} & t_1 &< t.
\end{aligned}
$$

The eight coefficients A, B, C, M, N, α, β, γ are not independent since f and f' are continuous at t_0 and t_1. The number of independent coefficients is thus at most four. But since the shape of f is independent of the time of experimentation, we can without loss of generality take $t_0 = 0$, and replace t_2 by the time lapse $\tau = t_1 - t_0$ between the inflectional points. Moreover, since the actual values of y reflect extraneous experimental factors such as the mode of electrical contact and degree of amplification, we can take $M = 1$. We then get the equations

$$
\left.
\begin{aligned}
y &= Ae^{-\alpha t}, & t &< 0 \\
y &= 1 - Nt + Be^{-\beta t} & 0 &\le t \le \tau \\
y &= Ce^{-\gamma t} & \tau &> t,
\end{aligned}
\right\} \tag{1}
$$

with the side conditions [48c, p. 288]

$$
A = 1 - B, \quad A\alpha = B\beta - N, \quad C\gamma e^{\gamma \tau} = N - Be^{-\beta \tau}. \tag{2}
$$

These equations are qualitatively consistent with those of the so-called *Hermann model*, an electric network which was proposed to model the action potential. Can the hypothetical curve given by (1), with coefficients subject to (2), be made to fit the recorded electrograms f by assignment of numerical values to the coefficients? The answer turned out to be affirmative for single myelinated fibers of mammals as well as single unmyelinated fibers of certain invertebrates (squids and crabs) [48c, p. 285]. But it was negative for the composite f's recorded from multifibered trunks in which several axons are stimulated. Such an f is *not* the sum of the f's for the individual fibers since the coefficients α, β, γ, τ vary from fiber to fiber in the trunk [48c, p. 290].

The discontinuity of f'' at the inflectional points t_0 and t_1 led Wiener and his colleagues to the view that the fiber is firing only during the time-interval $[t_0, t_1]$, the activity prior to t_0 being attributable to electrotonic spread, i.e. an irregular electrical discharge, and that subsequent to t_1 being due to both electrotonus and recovery. This suggested that the firing threshold is attained at the inflectional point t_0, and the refractory period commences at t_1, (cf. Note 1).

To account for these results, Wiener and his co-workers introduced a dynamics, involving the potentials $v_i(x,t)$ and $v_0(x,t)$ inside and outside the x-cross-section of the axon at instant t, as well as the longitudinal current j (in the direction of increasing x) governed by the differential equations [48c, p. 293]

$$\frac{\partial j}{\partial t} = -A(v_o - v_i), \qquad \frac{\partial v_o}{\partial t} = -Rj, \qquad \frac{\partial v_i}{\partial t} = Zj,$$

where A is the admittance per unit length of the membrane, R is the resistance per unit length of the medium just outside the membrane, and Z is the impedance of the fiber per unit length. The piecewise linearity of the transformation entails that the parameters A, R, Z are not constant, but have three different values A_k, R_k, Z_k, $k = 1,2,3$, over the time-intervals $t < t_0$, $t_0 \leq t \leq t_1$ and $t > t_1$. The dynamics is, however, incomplete since it hypothesizes nothing on the firing conditions.

16 McCulloch, Pitts and the Evolution of Wiener's Neurophysiological Ideas

A McCulloch, the poetic and mathematico-logical neurologist

Wiener first met Dr. Warren McCulloch at the neurophysiological meeting in New York in 1942 at which Dr. Rosenblueth presented their joint work with Bigelow on teleology [43b]. McCulloch was then Professor of Psychiatry in the Medical School of the University of Illinois. In 1917 he had entered Haverford College to honor in mathematics, but in the spring went on active duty with the Naval Reserve for a year, teaching celestial navigation to cadets and learning about submarines. He studied philosophy at Yale and psychology (experimental aesthetics) at Columbia before he entered the Columbia Medical School. He became a serious student of mathematical logic, and investigated the mathematico-logical aspects of schizophrenia and psychopathia while serving at the Rockland Hospital for the Insane. His life's mission is disclosed by his amusing exchange with the Quaker philosopher Rufus Jones at Haverford College in 1917:

> "Warren", he said, "what is thee going to be?" And I said, "I don't know."
> "And what is thee going to do?" And again I said "I have no idea; but there is one question I would like to answer: What is a number, that a man may know it, and a man, that he may know a number?" He smiled and said, "Friend, thee will be busy as long as thee lives." I have been. . . {M13, p. 2}.

It was McCulloch's dream to make an experimental science out of epistemology. By dint of their philosophical learning, predilection for mathematical logic, keen interest in the empirical realm—be it engineering or psychopathy—both McCulloch and Wiener were birds of similar feather; but McCulloch also wrote poetry.

In 1942 McCulloch was interested in the organization of the cortex of the brain. Working with him was Walter Pitts, a 20-year-old student of the logician R. Carnap and of the biophysicist N. Rashevsky, both of Chicago. At the inducement of Dr. Jerome Lettvin, then in the Boston City Hospital, and in order to strengthen his knowledge of mathematics, communication theory and computer theory, Pitts came to MIT in 1943 to study with Wiener. He had the marks of a genius and soon became a very close associate of Wiener during the years

1943–1948. Both McCulloch and Pitts played an absolutely positive role in the evolution of Wiener's ideas in neurophysiology, especially on the problems of logical manipulation, Gestalt or pattern-recognition, gating, brain rhythms and sensory prosthesis. The omission of both names from Wiener's autobiography [56g] is a glaring example of its inconsistency. It arose from exacerbation of some silly dispute.

The influence of McCulloch and Pitts fell on a mind with a unique understanding of the principles of computation, and of their logical foundation in the work of A. M. Turing. To see the effect we must first turn to this understanding.

B The Turing machine

> The *calculus ratiocinator* of Leibniz merely needs to have an engine put into it to become a *machina ratiocinatrix*. The first step in this direction is to proceed from the calculus to a system of *ideal* reasoning machines, and this was taken several years ago by Turing. [53h, p. 193]

To mechanize the operation of addition digitally as on an abacus or desk calculator, we need an *algorithm,* i.e. a routine, or sequence of steps, and an energetic machine or "engine" to execute these steps, all of which e.g. moving beads, pressing keys, cranking levers, pushing buttons, involve mechanical work. The same applies to the digital accomplishment of other mathematical problems considered in §§ 13, A, D, e.g. integration. Such digital devices, which cannot do tasks other than those for which they are designed, are called *special purpose digital computers.*

What Leibniz had in mind for his *calculus ratiocinator,* on the other hand, was a general method in which *all truths of reason* would be reduced to a kind of calculation[1]. Were such an algorithm available, then only an engine designed to execute its steps would be required to arrive at Leibniz's *machina ratiocinatrix,* which would be a universal, *all-purpose* computer. The step to the machine, hard as it is, is nowhere nearly as overwhelming as the problem of finding a universal "all-purpose" algorithm.

What such an algorithm might involve became clear when Charles Babbage adapted in his analytic engine of the 1840s the revolutionary idea of the French engineer V. M. Jacquard to put a controlling unit in his textile loom in 1805. This unit controlled the combination of warp threads that had to be lifted in order to weave a pre-assigned pattern into the fabric. Babbage wanted the controlling unit in his engine to execute the sequence of operations needed to compute a pre-assigned "algebraic pattern", such as an expression $ax^2 + by + c$. As Lady Lovelace (the remarkable Ada Byron, daughter of the poet) put it:

1 In the light of Gödel's 1931 metatheorem (p. 53), for the Leibniz goal to be tenable the word "all" has to give way to "all decidable".

The analytic engine weaves algebraic patterns just as the Jacquard loom weaves flowers and leaves. {G10, p. 22}

Now many different fabrics can be had with the same pattern by altering, for instance, the colors of the warp threads. Likewise, different numbers can be computed from the given algebraic expression by assigning different values to the constants $a, b,$ etc. and the variables $x, y,$ etc. Thus to accomplish a specific algebraic task, the control unit has to prescribe a sequence comprising both *operations,* i.e. steps of the algorithm, as well as the *assignment of values* for the constants and variables.

The Babbage engine, were it constructable, would indeed be a *machina ratiocinatrix* as conceived by Leibniz. This is because any mathematical or logical problem, capable of mechanical solution, can be resolved into a repetition of fundamental steps, and the exact routine to be followed laid out in the control unit as a sequence of instructions. With remarkable insight Leibniz and Babbage visualized this possibility, even though they had no way to render scientific the intuitive idea of *mechanical solvability.* This became possible only in the 1930s after the rise of mathematical logic, and the explication, by K. Gödel, A. Church and S.C. Kleene, of the idea of *recursion* latent in the *Principia Mathematica* of Whitehead and Russell (cf. § 5B). But this idea had to be rendered in a form serviceable to the design of computers. This step, of going "from the calculus to a system of ideal reasoning machines" (Wiener's words) was taken concurrently by E.L. Post and A.M. Turing in 1936.

Computability, Turing's explication of the pre-scientific notion of mechanical solvability is a bit too complex to explain. We shall instead describe a hybrid Turing machine which should give the reader an inkling of Turing's idea. The machine comprises a moveable tape divided into cells, in any of which a single symbol from a fixed finite alphabet can be inserted. Its engine is capable of carrying out just a few simple *operations* $0_1, 0_2, 0_3, \ldots$, such as reading the symbol in the fixed *front location,* or replacing it by another symbol, or moving the tape one step to the right or to the left. The *algorithm* to be used to solve a given mathematical or logical problem is prescribed by a *program,* i.e. a sequence of instructions $I_1, I_2, I_3, \ldots I_n$ to be carried out in some order on the machine at the times $t = 1, 2, 3, \ldots$ Each I_i is of the type:

(1) Perform operation 0_k, and then proceed to instruction I_j, (the k and j depending on i),

(2) perform operation 0_k and depending on whether the symbol in the front cell is a or b or ..., turn to the instruction I_{j_1} or I_{j_2} or ...,

(3) stop.

For instance I_7 might be: "Move the tape a step to the left and then turn to instruction I_4". (For details, see Note, p. 238.)

Success with computers has engendered the faith that given any solvable mathematical or logical problem, a program can be written whereby it can be solved on the Turing machine, i.e. that this machine accomplishes the tenable part of the *calculus ratiocinator* of Leibniz.

The next problem, difficult in its own right, is to engineer the operations involved in the Turing algorithm effectively, and get the *machina ratiocinatrix*. The key step in this, as far as electronic machines are concerned, was C. Shannon's discovery in 1938 that the operations of the Boolean algebra of propositions such as alternation, conjunction and denial (in symbols $\vee$, $\wedge$, $\sim$) are realizable by electrical switching networks {S6}.[2] Let "true" symbolized by '1' be represented by an electric pulse, and "false" symbolized by '0' be represented by its absence. Imagine that a pulse, applied to one of the switches S_1, S_2, ... in a circuit, closes it. Then the truth table of $\vee$, is realizable by a simple network, with switches "in parallel" as the following diagram shows:

$\vee$	0	1
0	0	1
1	1	1

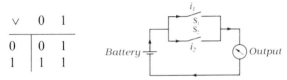

The output indicated by the needle will be 1 only when a closing pulse i_1, i_2 is applied to one or both of the switches S_1, S_2, exactly as in the truth table for $\vee$. Likewise, the truth table for $\wedge$ is realizable by a simple network of switches "in series" as the next diagram shows:

$\wedge$	0	1
0	0	0
1	0	0

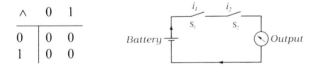

Notice that in the binary scale, the multiplication and addition tables are

$\times$	0	1
0	0	0
1	0	1

and

$+$	0	1
0	0	1
1	1	10

The first matches the table for $\wedge$, and the second differs from the table for $\vee$ only in having 10 in place of 1 in the last place. But the symbol

2 Shannon's basic idea had been outlined by C.S. Peirce in an 1886 letter to A. Marquand, see pp. 197–198 in K.L. Ketner's "The early history of computer design", Princeton University Library Chronicle, XLV, No. 3, Spring, 1984, 187–211.

'10' can easily be had from the symbol '1' by the Turing operations of moving the tape one unit to the left, and then imprinting '0' in the front cell. In the electronic computer more advanced techniques involving semi-conductors are used, but as the reader may guess the logic in the procedure is that of switching. Thus the Turing machine, so equipped, is capable of mechanically performing the fundamental operations of both logic and arithmetic, and is the all-purpose *machina ratiocinatrix* of Leibniz.

With this understanding of the nature of computers, Wiener was able to augment his early ideas on the nature of logic. Important logical concepts such as *proof, contradiction,* and *undecidability* are expressible in machine language. More and more, he came to regard logic as the idealized theory of logic machines. Like our minds, these machines suffer from all the "non-removable limitations and imperfections" affecting all natural mechanisms and organisms [61c, p. 125]. "*All logic is*

Warren S. McCulloch
1897–1969

limited by the limitations of the human mind when it is engaged in that activity known as logical thinking" [61c, p. 125].

This view did not imply an abandonment of the logical empiricist thesis on the analyticity or devoidness of factual content of all logical and mathematical statements, cf. § 5C above, but rather its supplementation by the understanding that our logic reflects the organization of the human mind. The statement '1 + 3 = 4' is devoid of factual content, but our ability to prove the result 1 + 3 = 4, is dependent on the structure of our thinking aparatus. Just as a metallic mechanism will respond only to certain inputs, and black-out when subject to inputs for which it is not designed, so too our thinking mechanism will respond only to certain types of inputs. For both types of mechanisms, the class of possible outputs is limited by the constitution of the mechanism, and the inputs it can deal with. Thus, the logic that the human mind arrives at is limited by the limitations of the neural apparatus. It may also be noted that an assertion such as "our ability to prove '1 + 3 = 4' is dependent on the presence of a bank of switches in the neural mechanism" is not analytic but factual, and therefore subject to the uncertainties inherent in the empirical domain.

C The computing and pattern-recognizing abilities of the brain

The first work of McCulloch and Pitts that affected Wiener was their important 1943 paper on "A logical calculus of ideas immanent in nervous activity" {M14}. The McCulloch-Pitts calculus was a formalization[3] of the intrinsic structure of the neural process, leaving aside its biochemistry. Its atomic sentences are of the form '$N_k(t)$', meaning that *the kth neurons fires at instant t,* and its axioms reflect the all-or-none aspect of the process, the presence of both excitatory and inhibitory neurons, and the occurrence of synaptic delay (cf. Ch. 15, Note 1).

The calculus was very revealing. It suggested that the specific biochemistry of the nervous system is irrelevant to the understanding of its integrative behavior in perception. The integrative behavior of the system is expressible in the notation of the propositional calculus, i.e. in the Boolean notation. When equipped with "a tape, scanners connected to afferents, and suitable efferents to perform necessary motor-operations", the nervous net can compute all and only those numbers which a Turing machine can, cf. {M13, p. 35}. Thus *the concept of Turing's machine is significant not only from the purely analytical standpoint of mathematical logic, but even from the standpoint of the neurophysiological understanding of the human mind.*

3   In the formal language II of Carnap's treatise {C3}, augmented by notation drawn from the *Principia Mathematica.*

Walter Pitts
1923–1969

These discoveries, and their striking and unforeseen parallels to the previous ones of Shannon on electrical networks {S6}, (cf. § 16B), confirmed Wiener's hunch on *the proximity of the brain and the electronic computer*. To use his own words, it became clear that

> . . . the ultra-rapid computing machine, depending as it does on consecutive switching devices, must represent almost an ideal model of the problems arising in the nervous system. The all-or-none character of the discharge of the neurons is precisely analogous to the single choice made in determining a digit on the binary scale. . . The synapse is nothing but a mechanism for determining whether a certain combination of outputs from other selected elements will or will not act as an adequate stimulus for the discharge of the next element, and must have its precise analogue in the computing machine. The problem of interpreting the nature and varieties of memory in the animal has its parallel in the problem of constructing artificial memories for the machine. [61c, p. 14]

The organizational and functional similarities of the brain and the computer came to be increasingly recognized, and in the winter of 1943–1944 a joint meeting of neurophysiologists and computer theorists was held in Princeton to study the implications of this disclosure. Attending were Wiener and J. von Neumann, who was then designing a computer at the Institute for Advanced Study.

The question now arose as to whether nervous faculties, other than reasoning, could also be simulated in the machine. One such faculty, associated with the areas of the cerebral cortex which are connected to the receptor organs, is that of *pattern recognition:* "What is the mechanism by which we recognize a square as a square, irrespective of its position, its size, and its orientation?" [61c, p. 18]. This question was raised in discussions in the spring of 1946, cf. [61c, pp. 17, 18], but it was the work of McCulloch and Pitts, published in their 1947 paper, "How we know universals" {P6}, that gave Wiener some important clues.

This paper offered a theoretical description of the neurophysiological mechanisms whereby the mind recognizes structure. In full accord with Rosenblueth's ideas (§ 15A), *Gestalt* was defined specifically, à la Felix Klein and Hermann Weyl ({W5, pp. 25, 72–74}, {W6, p. 131}), as *an invariant of a group G of transformations acting on a space $\mathscr{X}$ of apparitions.* Both $\mathscr{X}$ and G vary with the sensory modality. But for all modalities, an *apparition* is explicated as the *state of excitation* of an appropriate cluster or manifold $\mathscr{M}$ of neurons inside the cerebral cortex. Since at any moment each neuron in $\mathscr{M}$ is either dormant or firing, the momentary state of excitation can be prescribed by giving the subset A of firing neurons. Thus the *space of apparitions,* or *state space* $\mathscr{X}$ is $2^{\mathscr{M}}$, the set of all subsets of $\mathscr{M}$.

The group G which comprises certain one-one transformations on $\mathscr{X}$ onto $\mathscr{X}$ thus has a finite number N of elements, and $N \leq (2^m)!$, m being the number of neurons in $\mathscr{M}$. Two apparitions $A, B \subseteq \mathscr{M}$ are deemed to be *equivalent* or have the same *Gestalt,* if

$$T(A) = B, \quad \text{for some } T \text{ in } G.$$

For the visual modality, $\mathscr{M}$ is a layer of the visual cortex, and the group G is homomorphic to the group of Euclidean motions in physical space.

McCulloch and Pitts suggested that the nervous system is equipped with neural nets to affect the formation of the transforms $T(A)$ of an apparition A. Let

$$G = \{T_0, T_1, \ldots, T_{N-1}\}, \quad \text{where } T_0 = 1. \tag{1}$$

Then $\mathscr{M}$ is duplicated on $N-1$ cortical sheets $\mathscr{M}_1, \ldots, \mathscr{M}_{N-1}$. The neuron x in $\mathscr{M}$ has N-1 axons leading to the neurons $T_1(x), \ldots, T_{N-1}(x)$ on the sheets $\mathscr{M}_1, \ldots, \mathscr{M}_{N-1}$, respectively. The apparition A, i.e. the state of excitation of the cortical manifold $\mathscr{M}$, will then yield in succession the transformed apparitions $T_1(A), \ldots, T_{N-1}(A)$ having the same Gestalt as

A. The mind recognizes A as a square, for instance, if some $T_k(A)$ overwhelmingly overlaps a standard memorized "square" apparition $\overline{S}$ stored in the brain's memory, i.e. if the symmetric difference

$$\{T_k(A) \setminus \overline{S}\} \cup \{\overline{S} \setminus T_k(A)\}$$

falls below a certain tolerance. How apparitions are stored in the memory is not yet understood, nor in fact is the mechanism of long-time memory.

Another type of net for pattern recognition is a *teleological mechanism* operating on the principle of negative feedback. For any apparition A, let $\overline{A}$ among its replicas $T(A)$ be deemed *standard*. Let $f = (f_1, \ldots, f_r)$ be a multi-parameter which determines the state, i.e. $f_1, \ldots,$ f_r be real-valued functions on $\mathscr{X}$ such that if for each k, $f_k(A) = f_k(B)$, then $A = B$. Let $|\cdot|$ be a norm on $\mathbb{R}^n$ (cf. § 7C). The mechanism computes the deviation

$$\delta_0 = |f(A) - f(\overline{A})|,$$

and unless zero, this deviation induces a movement in the mechanism leading to a new replica $T(A)$ with reduced deviation

$$\delta_1 = |f\{T(A)\} - f(\overline{A})| < \delta_0.$$

This deviation in turn induces a further movement, and this goes on until the deviation falls below a certain tolerance. For instance, if a square appears anywhere in the visual field, such a reflex mechanism activates the muscles which turn the eyeball and bring the square in the center of the field, and so differs from $\overline{A}$ imperceptibly.

The McCulloch-Pitts theory led Wiener to outline a theory of mechanical pattern-recognition by scanning. Let G be the projective group acting on physical space. A finite subset $\{T_1, \ldots, T_n\}$ of G is picked so that any T in G is within a prescribed threshold of some T_k. A figure A in the space is deemed to have the same shape as a standard $\overline{A}$, if and only if for some k, $\overline{A}$ differs from $T_k(A)$ by less than a preassigned tolerance ε. The scanning involved in this method was to be of the same sort that Wiener had recommended in 1940 for the electronic computer (§ 13D). That is, if G is an n-parameter group, then the appropriate region of the parameter space $\mathbb{R}^n$ has to be scanned by the line-by-line procedure indicated for $n = 2$ in § 13D.

These very tentative theories, based on imaginative guesswork, were all that the theoretical neurophysiologist could offer in the 1940s in this most complex area. They re-inforced Wiener's belief that scanning is important to the functioning of the brain, as it is to the electronic computer for solving partial differential equations. The so-called principle of *exchangeability of time and space,* viz. that spatial dimensionality is reducible by augmentation of scanning time[4], cf. {M13, p. 49}, which he had clearly foreseen in 1940, cf. § 13D, now became an important article in his faith. It led him to the view, shared by McCulloch and others at that time, that the period of the scanning cycle might bear a relation to the known (so-called alpha) rhythm in the brain:

> We may suspect that this alpha rhythm is associated with form perception, and that it partakes of the nature of a sweep rhythm, like the rhythm shown in the scanning process of a television apparatus. [61c, p. 141]

This suspicion was to affect much of Wiener's work on electroencephalography, as we shall see (§ 16F below).

D Sensory and muscular-skeletal prosthesis

The ideas of McCulloch, Pitts and Wiener on pattern-recognition found a fruitful and humane application in the problem of sensory and muscular-skeletal prosthesis. Sensory prosthesis involves a problem of pattern-recognition. The blind man, for instance, has lost his visual ability to recognize the gestalts *A, B, . . .* of the alphabet, and he will be able to read only if these universals can be presented to him in a form that he can recognize. To some extent this is true even with non-sensory prosthesis. For instance, for the hand amputee, the gestalt of *roundness* (as in a coin) will have to be presented to him in a non-tactile modality.

Effective prosthesis, as Wiener saw it, thus demands surrogates for certain pattern-recognizing functions of the nervous system. A blind man has lost more than his eyes; he has lost the use of a part of the visual cortex. For the blind, an instrument which can convert light signals diffracted by the letters *A, B, C,* etc. into musical chords of varying durations, "takes over quite explicitly some of the functions not only of the eye but of the visual cortex" [61c, p. 25]. (Wiener was speaking of an instrument that had been designed by McCulloch and Pitts.) It was Wiener's belief that this principle applies to all types of prostheses. The amputee, for instance, has been deprived of both his sensory and motor abilities. He has lost more than the purely passive support of the stump and the contractile power of its muscles. He has also lost all "cutaneous and kinesthetic sensations originating in it".

> The present artificial limb removes some of the paralysis caused by the amputation but leaves the ataxis. With the use of proper receptors, much of this ataxis should disappear as well, and the patient should be able to learn reflexes, such as those we all use in driving a car, which should enable him to step out with a much surer gait. [61c, p. 26]

---

4 To scan an n-dimensional cube of edge-length ℓ is no different from scanning a single edge of length ℓ^n.

Thus the application of Wiener's ideas on pattern-recognition gave a new dimension to the technology of prosthesis. The handicapped, who have lost the use of their limbs or their muscles, have to be aided not only with artificial organs or limbs but also with apparatus to transmit impulses from these to the undamaged layers of the cortex, and thereby enable them to perform the impaired activity with many of the same psychic sensations as a normal individual. On the other hand, those who have lost the use of sense organs have to be equipped with apparatus that act as surrogates for both the impaired organs and the unusable parts of the cortex. This surrogate role involves the use of a second (unimpaired) sensory modality. Incoming signals from the outer world are converted by the apparatus into the second modality and then fed to the subject's sense organs, and he is taught how to translate the messages so relayed to the brain back into the first modality. While the validity of this approach to prosthesis is no longer in doubt, its large-scale technological implementation still remains for the future.

It should be noted that the substitution of one modality for another generally involves a considerable, and naturally imposed, loss of information. Thus hearing is a poor surrogate for vision, since the information received by the eye is from 50 to 100 times that received by the ear. Even if the entire auditory system of a blind person were to be engaged in the transmission of audio-translations of visual signals, only a small fraction of the visual information would get through. But for limited visual tasks, such as reading a page, such substitution is feasible, for a good deal of information supplied to the eye is irrelevant to such a task, and its omission does not impair the activity. This is the basis of the McCulloch-Pitts photocell apparatus to enable the blind to read by converting the shapes of letters into recognizable audio signals. Wiener took a good deal of interest in this apparatus, but his own active involvement was in a project to design an apparatus that would convert auditory signals to tactile signals, and so enable the deaf to "listen" [49d]. This project was undertaken by the Research Laboratory in Electronics at MIT, under the leadership of Dr. J. Wiesner, Wiener's co-worker in designing the "moth" (§ 15G).

This apparatus used the audio-signals of speech to activate vibrators attached to the skin of the palm or the fingers, exploiting the fact revealed by the Vocoder[5] research at the Bell Telephone Laborato-

---

5   The *Vocoder* (from voice-coder) is a system of voice transmission in which not the speech itself but its electrical replication, made by an analyzer, is transmitted to a synthesizer at the receiving end, which then replicates the original speech.

ries that in human speech there are only about five important frequency bands. To use the words of the authors:

> In this method a sound is carried from a microphone to a bank of filters. The output of each of these filters, except possibly the low frequency stage, is rectified and used to modulate a low frequency carrier and the output of these carriers as well as of the possibly by-passed low frequency stage, is carried to a number of stimulators, which may be properly designed electrodes or may be electromagnetic vibrators resting on suitable points of the skin, such as the five fingers. [49e, p. 512]

It was thought that the information-loss in this method of transmission would not be enough to mar the recognition of speech. Thus it would be possible for the subject to learn, by practice, how to reconstruct speech from skin stimulation.

Wiener was especially interested in this project because he realized that the often harsh and grotesque speech of the deaf results from their inability to hear themselves. He was confident that his apparatus would enable the deaf to hear themselves "via their fingers" so to speak, and this would be a psychological asset to them.

Wiener also took an interest in a project to design a better iron lung for paralytics. His idea was to draw electric signals, so-called electromyograms (EMGs), from the undamaged breathing muscles that the paralytic retained, and to use the amplifications of these signals to operate the iron lung. In the models then existing, the patient had to engage in a rigid, unpleasant and unnatural breathing routine in order to use the lung.

Of all the prosthesis projects in which Wiener took an active interest, the one that drew the most attention was an aid for the hand or elbow amputee, which came to be known as the "Boston Arm". Interestingly, this stemmed from a bad accident that Wiener had in the fall of 1962:

> I fell down a staircase and broke my hip and my fore-arm, and naturally I was at the hospital under treatment, . . . in contact with orthopedic surgeons. Also naturally, I could not refrain from talking with them and presenting my hobbies, and I mentioned this idea about artificial limbs. . .
>
> We started in my sick-room discussing plans for the future, and we built up a team of orthopedic surgeons, of neurologists and of engineers to carry this problem further, and we have had the good luck to secure the service and the financial backing of the Liberty-Mutual Insurance Company. This organization is very much interested in rehabilitation problems. [63b, p. 265]

Work in this direction, aspects of which were duplicated in England and the Soviet Union, only reached fruition in 1968, i.e. after Wiener had died. *The New York Times* of September 13, 1968, reported:

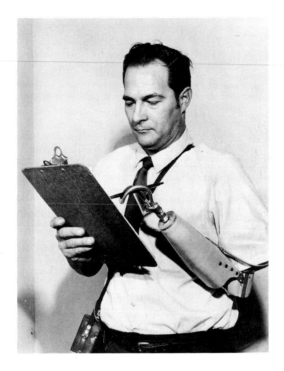

The Boston Arm
(first unveiled in 1968)

It responds to the
signals generated by
the biceps and triceps
muscles in the upper
arm.

Both photographs
courtesy of Mr.
T. Walley Williams,
Liberty Mutual Insur-
ance Company,
Boston, MA

Doctors and engineers unveiled today what they called the "Boston arm", an
electronically operated artificial limb that an amputee flexes simply by willing
it to flex, as is done with a natural arm.
    In part, the Boston arm project grew out of talks in 1960 between
Dr. Glimcher and the late Dr. Norbert Wiener of M.I.T.

A good deal of the work on the arm was actually done under the
direction of Professor Robert W. Mann and his group at MIT. The
"fingers" of this arm are activated by electric stimuli (EMGs) received
from the elbow muscles (biceps and triceps), and as the newspaper
pointed out, the arm was "volitional": its use did not demand any special
initiative by the user beyond the normal appetency.
    Wiener was neither a professional physiologist nor a surgeon nor
an engineer, and his specific recommendations on prosthesis sometimes
proved wrong. Professor Mann has remarked (*Coll. Works*, IV) how
Wiener underestimated the potentialities of the tactile modality for
prosthesis. The best modern aids for the blind rely on this modality
rather than on hearing. Also Wiener's idea to communicate with the
deaf by skin stimulation channeled from a few audio-frequency bands
was oversimplified and did not work out. Some of this specific suggestions on limbs for the amputee were also off-par. His belief that implant-

The Boston Elbow (1987)
The patient can control both the Elbow and the Hand myoelectrically with one
pair of muscles.

ed electrodes are necessary for full utilization of impaired muscles and
that surface electrodes will not do has been disproved by events. Of
course, Wiener could not anticipate the wonders of some of the latest
techniques using microcomputers, lasers and EMG processing.

Wiener's great contribution was rather to the methodology of
prosthesis. He saw clearly, and better than most of his contemporaries,
its fundamental elements: completion of the feedback loop, bilateral
communication between man and prosthesis machine, and learning or
rehabilitation. He clearly understood the commonality of these elements
in all types of prosthesis. Underlying his work was strong faith in the
synergistic symbiosis of man and machine: a nonlinear coupling between
them that can accomplish more than either component can individually.
As Professor Mann remarks:

> . . . I want to emphasize the universal character of Norbert Wiener's broad-
> ranging inquiries. Nowhere in his life's work is this more evident than this
> field of cybernetic prostheses, which after all was a somewhat tangential
> aspect of his primary life's work in mathematics. The comprehensiveness of
> his approach can be illustrated by noting the breadth represented by these six

short papers—the philosophy of science, science, mathematics, brain research, and engineering, all converging on applied prostheses design! {*Coll.
Works*, IV, p. 439}

E The computer as prosthesis for the brain

If machines of this sort can be devised, they will be of particular use in many
domains in which the present theory is computationally so complex as to be
nearly useless. . . . There are many cases where our computational control is
so incomplete that *we have no way of telling whether our theory agrees with
our practice*. Wiener's 1940 memorandum. [85a, p. 134] (emphasis added)

To Wiener, the term prosthesis had a very wide connotation. The human
species suffers from severe neurophysiological and other limitations. A
device, such as the telescope, that enables humans to overcome a limitation, is prosthetic in Wiener's wide sense. As the quotation above
indicates, Wiener visualized a prosthetic role for the electronic computer
even in 1940, i.e. before it came into existence. This vague thought
received a much sharper enunciation in 1946 when John von Neumann
became profoundly interested in the computer from the broadest standpoint. Thus J. von Neumann and H.H. Goldstine wrote:

... really efficient high-speed computing devices may, in the field of non-
linear partial differential equations as well as in many other fields which are
now difficult or entirely denied of access, provide us with those heuristic hints
which are needed in all parts of mathematics for genuine progress. In the
specific case of fluid dynamics these hints have not been forthcoming for the
last two generations from the pure intuition of mathematicians, although a
great deal of first-class mathematical effort has been expended in attempts to
break the deadlock in that field. To the extent to which such hints arose at
all (and that was much less than one might desire), they originated in a type
of physical experimentation which is really computing. We can now make
computing so much more efficient, fast and flexible that it should be possible
to use the new computers to supply the needed heuristic hints. This should
ultimately lead to important analytical advances. {G11, pp. 4–5}

In 1953, Wiener enlarged on this very theme, being guided by the
work of McCulloch and Pitts on form-perception. He suggested that our
inability to visualize geometrical forms (Gestalt) in more than three
dimensions may be mitigated by intelligent use of the computer.
Specifically, he advocated its use in the study of several complex variables, a field in which, in the words of Professor S. Bergman, a prime
difficulty is that "we have no intuitive visualization of the space in which
we are working". {B7, pp. 166–167}

Wiener's proposal (a little too complex for further elaboration)
is now beginning to bear fruit in engineering, and interestingly it is in the
design of robots. The *robot arm* is an assembly of several, relatively

movable, rigid parts connected at "joints". Were the arm to execute certain critical movements, it would break from the stresses created. Thus there are certain "danger zones" which engineers must understand in order to prevent such movements. Now a rigid body has 6 degrees of freedom (§ 12C). Clearly then the robot arm has a much larger number, say n, of degrees of freedom. As we saw in § 12C, its movement can be studied in its *phase space* of $2n$ dimensions. To the "danger zones" correspond certain critical submanifolds in this space, which are beyond our visualization. At the University of Iowa, Professors E. Haug and F. Potra are now using the computer to approximate these submanifolds, and trying high resolution computer graphics to yield insight into their geometry.

F Brain waves and self-organization

> The study of brain waves resembles the detecive story in being an attempt to obtain an organic body of knowledge, leading in a definite and positive direction, from a rather haphazard mass of clues. . . . As in the detective story, the clues seem to the amateur observer misleading and inconclusive; and it is only when they are processed with a very large degree of skill, that the common direction in which they are leading begins to stand out with the evidence of an undeniable fact. The thrill of seeing this undeniable fact gradually emerge from a mass of seemingly uninterpretable data is in no way different in quality from the thrill of reading a first-rate mystery story and of finding oneself in step with the author or ahead of him in arriving at the conclusion. [55e, pp. 5–6]

Wiener had been interested in brain rhythms ever since he had learned of them in the 1930s from Dr. Arturo Rosenblueth. They had been discovered by the British physician R. Caton in 1875, and recorded by the German physiologist Hans Berger in 1928. When a pair of electrodes are placed at two points on the scalp, a faint and fluctuating potential difference (PD) can be detected. This detection is impossible with ordinary instruments, but with the combined use of a powerful amplifier and a cathode ray oscilloscope, a record of this (amplified) PD as a function of time, a so-called *electroencephalogram* (EEG), can be obtained.

When the advent of amplification brought these EEGs into vogue, there was hope that they would tell a useful story about the brain. But the curves turned out to be extremely irregular ("a drunkard's idea of a roller coaster" [57e, p. 115]), and even the most skillful and experienced readers were able to decipher only a few crude rhythms. But Wiener, like some others, was certain that the EEGs were indeed "writings of the brain", worthy of serious study.

Wiener tried to convey the importance of the EEGs to an audience of medical men in New York by asking them what would happen if two electrodes were placed on the wall of the lecture hall:

I would find small differences of potential . . . The actual size of the potential wouldn't mean much to me. I don't know what the walls are made of, I don't know what electric wires are behind them . . . , but I can tell you something that I do pick up . . . Since this is New York City which has sixty cycle alternating current, I should pick up sixty cycle hum . . . This frequency of sixty times a second would be very reliable, as reliable as the regulation of the current of the voltage in the city power stations, and this is quite reliable. . . . Similarly, when we are examining potentials in the brain . . . we cannot expect that the actual size of the observed potentials is very significant. We can expect, however, to get very accurate results from the timing of them. [57e, p. 112]

Such considerations led Wiener to the idea of subjecting the EEGs f to a generalized harmonic analysis (§ 9D). This idea proved to be exceedingly fruitful, for when the autocovariance functions φ of f were plotted (cf. § 9D(12)), remarkable regularities came to light.

The instrument used for the production of autocovariance functions of the EEGs, known as the *MIT autocorrelator*, had a mildly animated pre-history. Unknown to Wiener, Dr. W. Grey Walter had built in England a machine for the spectral analysis of brain rhythms during the war, and its replica was being installed in the Massachusetts General Hospital in Boston in 1946, when Grey Walter came to MIT. The latter on his part had no idea that Wiener himself was much interested in brain rhythms. Consequently during a lecture at MIT, Walter was taken aback when Wiener announced how a spectral analyzer ought to be built, and Wiener was surprised when Walter described the machine at the hospital. Later on they became friends, and as Walter recalls, they "spent many happy and exciting days in one another's homes and at international meetings" {W1, p. 94}.

Dr. Grey Walter has left us with a diagnosis of a well-known Wiener idiosyncrasy: his falling asleep during other people's lectures, and on waking up, asking the lecturer a question or two, often pertinent. Walter first witnessed this during his own lecture at MIT! Speaking of a later day, he writes:

I took several records of his own brain rhythms which proved, amongst other things, that his defensive naps were real deep sleep. He could drop off in a few seconds, but would awaken instantly if one spoke his name or mentioned any topic in which he was really interested. {W1, p. 94}

Actually the MIT autocorrelator was based on better mathematical principles than Grey Walter's machine. It embodied Wiener's ideas from 1930: it was in principle a Michelson interferometer (§ 9F). The EEG f was put on magnetic tape, and played back on an apparatus having two playback heads, thereby creating the output $f(t) + f(t + h)$, the delay

W. Grey Walter
1910–1977

h depending now on the distance between the playback heads. Square law rectifiers, adders and a resistor-capacitor integraph, for evaluating

$$\frac{1}{T}\int_0^T f(t + h)f(t)dt,$$

completed the apparatus. The technique of recording EEGs on tape and the design of the apparatus were developed by Wiener's colleagues at MIT, Drs. W. Rosenblith, Mary A.B. Brazier, J. Barlow, Y.W. Lee and others.

The regularity of the autocovariance curves of the EEGs dispelled the idea that the electrical activity of the brain is chaotic and devoid of significance. Their spectral distribution functions (§ 9D(13)) showed strong oscillations at certain critical frequencies, the most conspicuous oscillation, the so-called *alpha rhythm*, having a frequency somewhere in between $10 \pm (1/50)$ cycles per second, or so Wiener thought. As he stated to his medical audience:

> ... there is a clock mechanism here that can be definitely picked up from the brain; it beats ten times per second and it takes 50 seconds to lose or gain a beat. [57e, p. 116]

This "clock" is affected by the individual's constitution and physiological condition, for the alpha rhythm

> disappears in deep sleep, and seems to be obscured and overlaid with other rhythms, precisely as we might expect, when we are actually looking at something and the sweep rhythm is acting as something like a carrier for other rhythms and activities [61c, p. 142].

Wiener invested a good deal of effort in trying to demarcate the purpose of this physiological clock, and to understand what sustains its steadfast ticking.

As to purpose, Wiener held that the physiological clock serves as a *gating mechanism* to regulate the timing of impulse transmission in the nervous system. Such a gating mechanism is required in an automatic apparatus (such as a milling machine) to control the timing of its different movements in order that it may do its job. The McCulloch-Pitts theory of the nervous system stipulates a short time-interval before the firing of a neuron, during which incoming nerve-impulses can combine to reach the necessary thresholds. For the synapse to work efficiently, there has to be some gating mechanism to ensure that the incoming impulses arrive more or less simultaneously. Wiener hypothesized that the physiological clock regulated this gating mechanism. He found confirmation of this hypothesis in some discoveries of Dr. D. B. Lindsley on the time-intervals between the arrival of visual sensory neurons in the cortical area of the brain and the firing of motor neurons, cf. [61c, p. 198].

Wiener went on to advance an explanation of what keeps the physiological clock ticking steadily. The nervous system is equipped with an assembly of oscillators with frequencies close to 10 cycles per second. These frequencies are perturbed by noise, but the oscillators are non-linearly coupled by negative feedback in a way that tends to pull the frequencies to a common value, and so the nervous system as a whole acts like a virtual oscillator with a slightly varying frequency in the neighborhood of 10 cycles per second. Wiener had in mind the sort of non-linear coupling that exists between the alternators of a large electric power generating network or grid, as in the eastern United States. The grid supplies an alternating current with a steady frequency of 60 cycles per second, despite the fluctuations of the load on individual alternators brought about by the haphazard use and disuse of power by the population. A steady frequency is maintained, because the feedback in the grid speeds up slow alternators and slows down fast ones, and brings into play extra alternators when the load is excessive. Wiener worked out a mathematical theory to account for this sort of phenomenon in [58i] which is too difficult for this book, and is described in {M9, pp. 120–121}.

Wiener attached much importance to this idea of *non-linear entrainment of frequency*. On it was based his understanding of *self-organizing systems*, especially those where a rhythm is organized, as is the case with nearly all living systems. The cells of such a system are both senders and receivers of messages. The cells are non-linearly coupled, so that the frequencies of time-variation of important parameters governing the cells can pull one another together (as in the brain) or push one another apart, thereby creating in either case an organization.

> Such a system as it gathers greater and greater synchronism will emit an impulse which has a greater and greater tendency to synchronize oscillators which have not already been pulled into place, until by a mass action they constitute a definite pulsating organ. [58f, p. 13]

This conclusion clearly suggests that it is impossible to localize the organization in one place. For instance it is meaningless to ask where in the brain the physiological clock is located.

Wiener's conclusions on electroencephalography have not been borne out by subsequent research. In the first place, the alpha rhythm has turned out to be more enigmatic than he realized, and its frequencies range from 8 to 13 cycles per second. Its genesis is still a mystery. Dr. Grey Walter in a paper on this work of Wiener's has pointed out that the gating in the nervous system is considerably more elaborate than Wiener had stipulated. It is more akin to the gating of traffic-flow in a city in which the stop-go lights are activated to a large extent by the traffic itself rather than by a fixed rhythm, cf. {W1, pp. 95–99}. Walter was also critical of Wiener's comparison of the nervous mechanism to that of an electric grid, and he denied the relevance of the underlying mathematical theory. Such a theory demands a spectral density with a sharp peak with a dip on each side. But this sort of profile is not always exhibited by the empirical spectra of actual EEGs.

It should be noted that these critical remarks are confined to the encephalographic applications of Wiener's mathematical theory of self-organization by non-linear entrainment of frequencies. There is evidence that other rhythmic organizations, such as the synchronous flashing of fireflies, seem to function according to his theory.

"The most complicated object under the sun" was von Neumann's description of the human nervous system (cf. letter, p. 243), and of all the areas of inquiry into which Wiener's mind penetrated, this was the toughest. McCulloch admitted that his own theories of the neuro-physiological mechanism were intended largely to suggest fruitful experimentation which hopefully would reduce the need for wide-ranging theoretical conjecture in this difficult subject. Wiener tended to view his theories on electroencephalography in less conjectural terms. If, however, we gauge his contribution in more modest terms, there is little doubt as to its penetrating and wholesome quality.

Note: Structure of the Turing machine

The Turing machine consists of a moveable tape, infinitely long in both directions, divided into equal squares, called *cells* and a scanning and printing *head*. The head can scan only the cell in front of it, i.e. read what symbol it has. For simplicity we assume that at time $t = 0$, each cell has exactly one symbol from an alphabet with two symbols '0' and '1', and that only a finite number of cells have '1's, the others '0'. The machine is equipped with an engine that can execute the following operations:

0_1: Move the tape one cell to the right.
0_2: Move the tape one cell to the left.
0_3: In the observed cell replace 1 by 0, or 0 by 1.
0_4: Read the symbol in the observed cell.

A *program* consists of a finite sequence of instructions I_1, $I_2, \ldots I_n$, to be carried out in some order at times $t = 1, 2, \ldots,$.[6] Each I_i is of one of the following three types:

(1) Perform the operation $0_{k(i)}$; if $k(i) = 1$ or 2 or 3, then go to the instruction $I_{j(i)}$.
(2) Perform the operation $0_{k(i)}$; if $k(i) = 4$, then depending on whether the reading is 1 or 0, go to the instruction $I_{j_1(i)}$ or $I_{j_0(i)}$.
(3) Stop.

Here it is understood that the functions $j(\cdot), j_0(\cdot), j_1(\cdot)$ are specified, the domain of $j(\cdot)$ being the set of i for which $k(i) = 1$ or 2 or 3, and that of $j_0(\cdot), j_1(\cdot)$ being the set of i for which $k(i) = 4$.

---

6   For expository reasons, we are (inaccurately) treating the program as being outside the machine.

17   The Cybernetical Movement and von Neumann's Letter, 1946

A   The Teleological Society

Wiener's intellectual collaboration with Rosenblueth, Bigelow, McCulloch, Pitts and earlier with Lee, Bush and Russell, and his work on electronic computing, predicting and filtering embraced several fields of science. But through this work ran a unifying thread centered on the concepts of message, noise, feedback, communication, teleology and control.

The concept of message was not restricted to messages of human origin—those involved in the flashing of fireflies and those flowing through electrical machinery were also included. The question of *intelligence*, which led back to *logic*, and to the possibility, discerned by Leibniz and Pascal, of *logical automata* and *machine intelligence* were thus included.

The same sort of scientific experience was shared by a group of scientists, many of whom had been in touch with Wiener. Thanks to the generous support of the Josiah Macy Foundation, this group was able to get together from time to time in different places. In 1943 it decided to name itself the "Teleological Society". Led by J. von Neumann and Wiener, the society included eminent engineers, computer theorists and physiologists, among them Rosenblueth, McCulloch, Pitts and Bigelow of whom we have already spoken, and Dr. H.H. Aiken of Harvard, Dr. H.H. Goldstine of the University of Pennsylvania, builders of the Harvard Mark I and ENIAC, and Dr. Lorente de No of the Rockefeller Institute, among others. It gradually dawned on the members of the group that the problems which concerned them, in which message and the other above-mentioned concepts are cardinal, constituted a new interdisciplinary field of scientific activity.

The "cybernetical circle", as the group came later to be known, had made substantial contributions to science during the 1940s. They had analyzed the notion of noise, demarcated the concept of information and demonstrated its negentropic thermodynamic significance, designed filters that could intelligently predict the future values of stationary time-series, discovered the servo-mechanical aspects of voluntary purposeful activity, developed a calculus for the neuronal process and a theory of Gestalt or pattern-recognition, and recognized the common mode of operation of the brain and the computer.

But by the late 1940s there was a growing feeling among many that the movement was losing steam and had not addressed itself to a truly major problem then on the frontiers of science. As Dr. W. Grey Walter was to say later in 1969:

> So often has a cybernetical analysis merely confirmed or described a familiar phenomenon in biology or engineering, so rarely has a cybernetical theorem predicted a novel effect or explained a mysterious one. {W1, p. 94}

In his preface to the second edition of *Cybernetics* [61c], Wiener affirms that "if a scientific subject has real vitality, the center of interest in it must and should shift in the course of years", and again that "it behooves the cyberneticist to move on to new fields and to transfer a large part of his attention to ideas which have arisen in the developments of the last decade". By way of such transfer, he cites his 1958 book *Non-linear Problems in Random Theory* [58i], and the inclusion of fresh chapters on "Learning machines" and "Brain waves and self-organizing systems" in the second edition of *Cybernetics*. This supplementary material embodied, however, ideas that had been brewing in Wiener's mind since the 1940s, if not the 1930s, in his collaboration with Lee and Rosenblueth. Moreover, some of this material is not sufficiently critical as we saw in § 16F, and some on coding and decoding in [58i] is incorrect.

Now a leading member of the teleological circle ever since its informal inception in 1943 was Professor John von Neumann of the Institute for Advanced Study in Princeton. There are many references to him in Wiener's writings, but nowhere does Wiener refer to a six-page letter he received from von Neumann in 1946, in which von Neumann expressed strong misgivings on the almost exclusive preoccupation of cyberneticists with neurophysiology, and went on to advocate a nut-and-bolt study of the virus from a cybernetical perspective . This letter is a remarkable document which must be understood. First, however, we must indicate the cohesive forces that brought on such a close interaction between what were perhaps the two most versatile mathematical minds in America at that time.

B The parallel evolutions of von Neumann and Wiener

There are some remarkable parallels between the intellectual evolutions of John von Neumann and Norbert Wiener. Both began research in the foundations of mathematics — Wiener by way of the *Principia Mathematica*, von Neumann by way of Hilbert's formal set theory — and both made significant fundamental contributions within their respective frameworks. Both felt a compelling attraction to the frontiers of physics, and both created new mathematics to cope with the realities being uncovered: Wiener for the Einstein-Smoluchowski theory of the Brownian motion, von Neumann for the quantum mechanics of the atom {V2}, then fresh from the pen of de Broglie, Heisenberg, Born and Schrödinger. The new mathematics was deep: measures over $C[0, 1]$ and stochastic integration, and the spectral resolution of operators on Hilbert space. Both men pushed the new mathematics to its depths: Wiener's work leading to generalized harmonic analysis and Tauberian theory (foreshadowing Banach algebras), von Neumann's work to rings of operators, appropriately termed "von Neumann algebras" today. Both did cognate work of high magnitude — Wiener on the Hopf-Wiener equation and causality, von Neumann on ergodic theory. During World War II both men were drawn into military projects: Wiener in antiaircraft ballistics, von Neumann in atomic ordnance with the Manhattan Project. Both were intensely interested in electronic computers — Wiener since the days of his association with Vannevar Bush's differential analyzer, von Neumann from the start of the war. And both contributed in their own ways: Wiener, haphazardly, with deep insights, and von Neumann, systematically, with the actual logical designing. Both articulated in mathematical terms the methodological and philosophical insights born of the fire of conflict and war: Wiener by his theory of forecasting and filtering leading to cybernetics, von Neumann by his automata theory and his theory of strategy ("game theory"), {V7} and in this both put to substantial use their early work of the 1920s and 1930s. Both made penetrating assaults on non-linear territory, von Neumann in the more causal, Wiener in the more random.

There is an even more profound interrelation between their work in the different phases. Functional integration, ramifying from Wiener measure, plays a significant role in the quantum field theory ensuing from the von Neumann quantum mechanics, as Professors I. E. Segal, E. Nelson and others have shown. The von Neumann-Stone theory of spectral integral representations of unitary groups provides the best setting for the Wiener-Kolmogorov prediction theory, as Kolmogorov understood. Almost all trajectories of processes governed by the measure-preserving flows of Birkhoff and von Neumann fall in the class S of Wiener's generalized harmonic analysis, as we saw in § 12A, and parts of this Wienerian analysis simplify appreciably when ergodic considerations are brought to bear. Last but not least, if we keep in mind the Heraclitan wisdom that "all things come to pass through the compulsion of strife", then the von Neumann theory of strategy has a place within the field of cybernetics which is concerned with communication and control, and with the breakdown of communication and control.

These two great minds thus had much in common, despite very obvious differences in their thought processes and presentations: von Neumann's being thoroughgoing and immaculate to the last detail, while Wiener's were diffuse and sometimes almost incomprehensible, and despite their widely differing temperaments and administrative propensities. The impression that an intrinsic barrier separated their mathematical attitudes, which a reader of Professor Heims's recent account {H6} may perhaps get, would be quite mistaken. True, von Neumann's great mind had an axiomatic bent, whereas Wiener's relished a "more tighter attitude" (cf. § 7C). True that one could cope with several different problems during a stretch of time and that the other could not. Also true that the words "crazy", "cranky", which might prop up half-affectionately in talking about Wiener, were never applied to von Neumann. But these differences have small significance. Their important ideas overlapped, even to the extent of sensing the prosthetic use of computers. Thus it was natural, if not inevitable, that sooner or later there would be a healthy interaction between the two.

Wiener's active interest in von Neumann's work began with the ergodic theorems in 1931, cf. §§ 9F, 12B, and in the early 1940s von Neumann was struck by Wiener's recognition of the proximity of the brain and the computer, and his teleological viewpoint.

C Von Neumann's letter on the direction of cybernetical research; molecular biology

The correspondence between von Neumann and Wiener on cybernetical issues began around 1943 and came to a close in the 1950s with von Neumann's increasing official responsibilities and deteriorating health. On November 26, 1946, just prior to a projected conversation concerning the direction that cybernetical research should take, von Neumann wrote to Wiener, [MC, 22]. His letter is a sequel to shorter ones in which von Neumann expressed great interest in Wiener's ideas on teleology, and arranged for Wiener to lecture on this at the Princeton University bicentennial celebrations.

THE INSTITUTE FOR ADVANCED STUDY
School Of Mathematics
Princeton, New Jersey

November 29, 1946

Professor Norbert Wiener,
Department of Mathematics
Massachusetts Institute of Technology,
Cambridge, Massachusetts

Dear Norbert:

   This letter represents an effort to do better than I estimated in my letter of November 25, in which I proposed that we might get together for the afternoon or evening of December 4 in Cambridge, and indicated only somewhat vaguely what the subject was that I would like to discuss with you. I am now trying to give you a more detailed advance notice, hoping that this will make our discussion on December 4 more specific.

   Our thoughts—I mean yours and Pitts' and mine—were so mainly focused on the subject of neurology, and more specifically on the human nervous system and there primarily on the central nervous system. Thus, in trying to understand the function of automata and the general principles governing them, we selected for prompt action the most complicated object under the sun—literally. In spite of its formidable complexity this subject has yielded very interesting information under the pressure of efforts of Pitts and McCulloch, Pitts, Wiener and Rosenblueth. Our thinking—or at any rate mine—on the entire subject of automata would be much more muddled than it is, if these extremely bold efforts—with which I would like to put on one par the very un-neurological thesis of A. Turing—had not been made. Yet, I think that these successes should not blind us to the difficulties of the subject, difficulties, which, I think, stand out now just as—if not more—forbiddingly as ever.

   The difficulties are almost too obvious to mention: They reside in the exceptional complexity of the human nervous system, and indeed of any nervous system. What seems worth emphasizing to me is, however, that after the great positive contribution of Turing – cum – Pitts – and – Mc-Culloch is assimilated, the situation is rather worse than better than before. Indeed, these authors have demonstrated in absolute and hopeless generality, that anything and everything Brouwerian can be done by an appropriate mechanism and specifically by a neural mechanism—and that even one, definite mechanism can be "universal". Inverting the argument: Nothing that we may know or learn about the functioning of the organism can give, without "microscopic", cytological work any clues regarding the further details of the neural mechanism. I know that this was well known to Pitts, that the "nothing" is not wholly fair, and that it should be taken with an appropriate dose of salt, but I think that you will feel with me the type of frustration that I am trying to express. (H. N. Russel used to say, or to quote, that if the astrophysicist found a general theory uniformly corroborated, his exclamation should be "Foiled again," since no experimenta crucis would emerge.) After these devastatingly general and positive results one is therefore thrown back on microwork and cytology—where one might have remained in the first place. (This "remaining there" is of course, highly figurative in my case, who have never been there.) Yet, when we are in that field, the complexity of the subject is overawing. To understand the brain with neurological methods seems to me about as hopeful as to want to understand the ENIAC with no instrument at one's disposal that is smaller than about 2 feet across its critical organs, with no methods of intervention more delicate than playing with a fire hose (although one might fill it with kerosene or nitroglycerine instead of water) or dropping cobblestones into the circuit. Besides the system is not even purely digital (i.e. neural): It is intimately connected to a very complex analogy (i.e. humoral or hormonal) system, and almost every feedback loop goes through both sectors, if not through the "outside" world (i.e. the world outside the epidermis or within the digestive system) as well. And it contains, even in its digital part, a million times more units than the ENIAC. And our intellectual possibilities relatively to it are about as good as somebodies vis-a-vis the ENIAC, if he has never heard of any part of arithmetic. It is true that we know a little about the syndromes of a few selected breakdowns—but that is not much.

My description is intentionally exaggerated and belittling, but don't you think that there is an element of truth in it?

Next: If we go to lower organisms from man with 10^{10} neurons to ants with 10^6 neurons—we lose nearly as much as we gain. As the digital (neural) part simplifies, the analogy (humoral) part gets less accessible, the typical malfunctions less known, the subject less articulate, and our possibilities of communicating with it poorer and poorer in content.

Further: I doubt that the "Gestalt" theory, or anybodies verbal theory will help any. The central nervous system is complicated, and therefore its attributes and characteristics have every right to be complicated. Let not our facile familiarity with it, through the medium of the subjective consciousness, fool us into any illusions in this respect.

What are we then to do? I would not have indulged in such a negative tirade if I did not believe that I see an alternative. In fact, I have felt all these doubts for the better part of a year now, and I did not talk about them, because I had no ideas as to what one might say in a positive direction.

I think now that there is something positive to be said, and I would like to indicate in which direction I see it.

I feel that we have to turn to simpler systems. It is a fallacy, if one argues, that because the neuron is a cell (indeed part of its individual insulating wrapping is multicellular), we must consider multicellular organisms only. The cell is clearly an excellent "standard component", highly flexible and suited to differentiation in form and in function, and the higher organisms use it freely. But its self-reproductivity indicates that it has in itself some of of the decisive attributes of the integrated organisms—and some cells (e.g. the leukocytes) are self-contained, complete beings. This in itself should make one suspicious in selecting the cells as the basic "undefined" concepts of an axiomatism. To be more par terre: Consider, in any field of technology, the state of affairs which is characterized by the development of highly complex "standard components", which are at the same time individualized, well suited to mass production, and (in spite of their "standard" character) well suited to purposive differentiation. This is clearly a late, highly developed style, and not the ideal one for a first approach of an outsider to the subject, for an effort towards understanding. For the purpose of understanding the subject, it is much better to study an earlier phase of its evolution, preceding the development of this high standardization-with-differentiation. I.e. to study a phase in which these "elegant" components do not yet appear. This is especially true, if there is reason to suspect already in the archaic stage mechanisms (or organisms) which exhibit the most specific traits of the simplest representatives of the above mentioned "late" stage.

Now the less-than-cellular organisms of the virus or bacteriophage type do possess the decisive traits of any living organism: They are self reproductive and they are able to orient themselves in an unorganized milieu, to move towards food, to appropriate it and to use it. Consequently, a "true" understanding of these organisms may be the first relevant step forward and possibly the greatest step that may at all be required.

I would, however, put on "true" understanding the most stringent interpretation possible: That is, understanding the organism in the exacting sense in which one may want to understand a detailed drawing of a machine—i.e. finding out where every individual nut and bolt is located, etc.

It seems to me that this is not at all hopeless. A typical bacteriophage, which can be multiplied at will (and hence "counted" — by the colonies it forms on a suitable substrate—as reliably as elementary particles can be "counted" by a Geiger-counter) is a phage that is parasitic, I think, on the Bacillus Coli. It has been extensively worked with, e.g. by Delbrueck at Vanderbilt. It is definitely an animal, with something like a head and a tail. Its dimensions are, I think, 60m $\times$ 25 m $\times$ 25 m, i.e. its volume is $60 \times 25 \times 25 \times 10^{-21}$ cm^3 = 3.7×10^{-17} cm^3. The density may be taken to be 1, hence its mass is about 3.7×10^{-17} gr, ie. the same as about 2.5×10^7 H atoms. Since the average chemical composition of these things is usually about C or N or O per one or two H, the average atomic weight of its constituents is about 6. Hence the number of atoms in it is about 4×10^6. Furthermore, it is known from the behavior of physiological membranes, that they are monomolecular—or oligomolecular—Langmuir—layers, which exercise their function in a highly mechanical way. E.g. the so called "active permeability": The peculiar ability to "permit" ions to pass through the membrane against an electrical field—an activity which clearly must, and demonstrably does, require an energy supply from the metabolism—and which is therefore better described as "pushing the ions across": than as "permitting them to pass". I understand that here an ion simply gets seized by the opposite—polarity end of one of the rare (charged) radicals in the membrane.[7] Very similar things can be said about the functioning of the "phosphate bond", which seems to be the main physiological device for localized, short time energy storage—i.e. the equivalent of a spring. Thus one can really talk of "mechanical elements", each of which may comprise 10 atoms or more. Thus the organism in question consists of six million atoms, but probably only of a few hundred thousand "mechanical elements". I suppose (without having done it) that if one counted rigorously the number of "elements" in a locomotive, one might also wind up in the high ten thousands. Consequently this is a degree of complexity which is not necessarily beyond human endurance.

The question remains: Even if the complexity of the organisms of molecular weight 10^7–10^8 is not too much for us, do we not possess such means now, can we at least conceive them, and could they be acquired by developments of which we can already foresee the character, the caliber, and the duration. And are the latter two not excessive and impractical?

I feel that the answer to these questions can already be given, and that it need not be unfavorable. Specifically:

I am talking of molecular weights 10^7–10^8. The major proteins have molecular weights 10^4–10^5 (The major proteins have molecular weights 10^4–10^5—the lowest one known actually appears to be only about 7,000) and the determination of authorities like Langmuir and Dorothy Wrinch consider it promising. Langmuir asserts, that a 2–4 year efforts with strong financial backing should break the back of the problem. His idea of an attack is: Very high precision x-ray analysis, Fourier transformation with very massive fast computing, in combination with various chemi-

---

7 which then turns around and deposits the ion on the other side of the membrane.

cal substitution techniques to vary the x-ray pattern. I realize that this is in itself a big order, and that it is still by a factor 10^3 off our goal—but it would probably be more than half the difficulty.

In addition there is no telling what really advanced electron-microscopic techniques will do. In fact, I suspect that the main possibilities may well lie in that direction. The best (magnetic) electron-microscope resolutions at present are a little better than 10 Å = 1 mμ. With 4×10^6 atoms in a volume of 3.7×10^{-17} cm^3 = 3.7 $\times 10^4$ mμ^3, the average atomic volume is 10^{-2} mμ^3, and hence the average atomic distance about 1/5 mμ. Hence the 1 mμ resolution is inadequate—but not very far from what might be adequate. A resolution that is improved by a factor of 10–20 might do. It is dubious whether electron lenses can be improved to this extent. On the other hand, the proton microscope need not be more than 2–4 years in the future, and it would certainly overcome these difficulties.

Besides all these developments might be pushed and accelerated. Of course, everybody knows what a 1–1/2 Å resolution would mean: One could "look" at an H atom, and with a little more, say 1/5 Å, one could "see" the Schroedinger-charge-cloud-of the orbital electrons. But the physiological implications are even more extraordinary, and they should receive a great deal of emphasis in the immediate future.

At any rate, I think that we could do these things:
Study the main types of evidence: Physiology of viruses and bacteriophages, and all that is known about the gene-enzyme relationship. (Genes are probably much like viruses and phages, except that all the evidence concerning them is indirect, and that we can neither isolate them nor multiply them at will.)

Try to learn a reasonable amount about the present state of knowledge and opinions concerning protein structure.

Study the methods of organic-chemical structure determination by x-ray analysis and Fourier-analysis with their necessary complement of manipulations.

Study the principles and methods of electron-microscopy, both in the direction of electron optics and in the direction of object-manipulation. Try to get oriented to the possibilities of proton-microscopy.

Finally: Compile for our common use two lists: (1) Relevant publications, with the main emphasis for the immediate future, considering our lack of education, on books and survey articles. (2) Persons from whom we might learn most about the state of affairs and the outlook in these fields.—

I did think a good deal about self-reproductive mechanisms. I can formulate the problem rigorously, in which Turing did it for his mechanisms. I can show that they exist in this system of concepts. I think that I understand some of the main principles that are involved. I want to fill in the detals and to write up these considerations in the course of the next two months. I hope to learn various things in the course of this literary exercise, in particular the number of components required for self-reproduction. My (rather uninformed) guess is in the high ten thousands or in the hundred thousands, but this is most unsafe. Besides, I am thinking in terms of components based on several rather arbitrary choices. At any rate, it will be necessary to produce a complete write-up before much discussing is possible.

Certain traits of the gene-enzyme relationship, of the behavior of some mutants, as well as some other phenomena, seem to emphasize some variants of self-reproductivity, which one would be led to investigate on purely combinatorial grounds as well. E.g.: Self-reproductivity may be symbolized by the schema A — A. What about schemata like A → B → C →A, or A → B → C → C, or A → B → C → D → E → C, etc.?

John von Neumann
1903–1957

   This is as far as my ideas go at this moment. I hope you will not misinterpret the anti-neurological tirade at the beginning of this letter. Of course I am greatly interested in that approach and I have the greatest respect for the important results that have been obtained in that field, and in our border area with it. I certainly hope these efforts will continue. I wanted to point out, however, that I felt that the decisive "break" was more likely to come in another theater. I was trying to formulate and to systematize my motives for believing this, and the simplest literary mode to do this is the controversial one. I hope therefore, that I have not given you a false impression of the spirit in which I am starting a "controversy".

   I am most anxious to have your reaction to these suggestions. I feel an intense need that we discuss the subject extensively with each other.

   Hoping that this letter has not been unbearable just by its sheer length, and hoping to hear from you and to see you again soon, I am, with the best regards,

                    Yours, as ever,

                    John von Neumann

The dialectic in von Neumann's letter is both penetrating and compelling. The very success of the coarse-grain axiomatization of nervous nets precludes the *experimentum crucis* needed to go deeper, and necessitates study at the finer microscopic cytological level. But even the ant with its 10^6 neurons is too complex, let alone the human being with 10^{10} neurons. Indeed, even the simplest cell is over-complex by virtue of the structure acquired during millions of years of evolution. Something more primitive and formative was needed: a subcellular organism like the virus—in the twilight zone between the living and the non-living, as Wiener used to say.

Prophetic was von Neumann's suggestion to investigate the "physiology of viruses and bacteriophages and all that is known about the gene-enzyme relationship". Such study, with revolutionary ramifications on biology, was indeed undertaken a few years after the letter, though not unfortunately by the cybernetical circle. In 1952 the experiments on bacteriophage by Hershey and Chase showed that the hereditary code is carried in the nucleic acid in the phage and not in its protein shell; and in 1955 Frankel-Conrat showed that it was an enzyme on the phage shell which cracks the bacterial wall. Later the nucleic acid in the phage was identified as DNA.

Wiener himself was interested in microorganisms, especially those in the "twilight zone", such as the virus, cf. [50j, 31–32]. In J.B.S. Haldane he had a friend who was a great geneticist with a keen perception of the quantum mechanical bearing on life, and Wiener's interests included questions such as the organization of protein chains of amino acids in the chromosome, cf. [61c, pp. 93–94].

Unfortunately, this writer has no record of what transpired at the putative meeting on December 4, 1946, between Wiener and von Neumann. He learned of the letter and of the meeting at MIT only after the deaths of Professors McCulloch, Pitts, Haldane and Rosenblueth. Professors J. Garcia Ramos and H. von Foerster, whom he did consult, were unaware of both letter and meeting. One ostensible result was an invitation to Professor Delbrueck to participate in the Macy Foundation meetings. He attended one meeting, but as Professor Heims reports, he found the discussion "vacuous in the extreme and positively inane" {H6, pp. 205, 475}. There is no record we know of that von Neumann's letter was ever brought to the attention of the cybernetical group. One can only surmise that this did not happen, and that in all probability Wiener's ineptitude was the cause. Like the Wiener memorandum on computers, the letter did not get the attention it merited.

An unfortunate ramification of the neglect of von Neumann's letter was the absence of a nexus between the cybernetical group and the researchers in the important field of molecular biology, which was on the verge of emergence. This area, on the borderlines of physics, chemistry and biology, is ideal cybernetical territory: central to it are the ideas of purpose, functionality, self-organization, corruption and reproduction, cf. {M16}. As Rosenblueth has pointed out, the very conception of such a subject came from the cybernetical ideas nascent in Schrödinger's incisive book *What is Life*? {S4}, cf. {R2, p. 22}. Second, cybernetics was left out of the applications of molecular genetics, i.e. out of genetic engineering. This new technology marks the beginning of a "third" Industrial Revolution, to continue Wiener's interesting classification, § 19A. We know from history that shortsightedness and rampant greed can run such ventures afoul of human welfare. It was Wiener's hope that cybernetical ideas could serve as an enlightening force during these industrial transitions.

D Wiener's thoughts on molecular biology

Only toward the end of his life did Wiener take cognizance of molecular biology, which by then had come into its own. His motive, however, was to see what light it shed on certain neural activities and certain problems of reproduction.

In a short paper [65b] Wiener suggested that the template principle of gene and cell reproduction be studied further by the spectral analysis of the far infra-red molecular radiation, and the role of radiation in the formation of new molecules be investigated. In a lecture prepared for the Central Institute for Brain Research in Amsterdam, posthumously published, but not delivered because of his sudden death in March 1964, he compared the state of neurophysiology to that of genetics at the time of Darwin and Mendel. Like genetics, neurophysiology had to move in a molecular direction. In that lecture Wiener made a gallant attempt to apply the new microbiology to the unresolved question of the mechanics of long-term memory. Wiener conjectured that these might be "nucleic acid memories":

> There may be nucleic acid complexes other than genes or viruses which have a familiar mechanism of multiplication. Such substances are beautifully adapted to carry memories, not only in the race but in the individual. It is an observation which has been made within the last few years, that nucleic acid complexes are to be found in the nervous system more readily than in other tissues. Not only are they to be found in the nervous system but they seem to be largely located in the neighborhood of synapses, in other words, in the places where the nerve fibers communicate with one another and where we should naturally suppose that some of the most critical nervous activity takes place.
>
> Thus it is quite reasonable to suppose that the stable memory which is not so instantaneously available, is carried not by reverberating circuits in the nervous system but by nucleic acid complexes. Our pattern is not a

switching pattern but a pattern of some molecular activity, let us say of molecular vibrations. This is the sort of dynamics which we study in solid state physics. And if we wish to tie up nervous activity with statistical mechanics this may well be the phenomenon which we should investigate. [65b, p. 403, 404]

In the same undelivered lecture, Wiener advocated Ashby's view that "the basis of nervous organization is statistical" [65b, p. 407]. His idea was to supplement the all-or-none aspect of neurophysiology, stemming from the nerve impulse along axons, by a statistical theory of micro-radiation stemming from the core and the dendrites of the neuron. These micro-radiations are pulled into phase at certain frequencies by the non-linear coupling of nucleic acid complexes. Wiener wanted to use the theory of entrainment of frequencies by non-linear coupling, which he had worked out in [64a], his very last mathematical paper.[1] He had in mind a kind of random process "closely related to that by which the molecular vibrations of the particles of the air combine to produce organized sound waves". He felt that the study of such sound "with long range forces such as occur in plasmas", would provide the statistical mechanics required for the study of nucleic acid complexes, cf. [65b, pp. 405–407]. An evaluation of this work by Dr. J.S. Barlow is given in *Coll. Works* IV.

1   There he attempted to show that a Hamiltonian or even low dissipative dynamical system, excited by random turbulence, would under certain circumstances generate non-linear oscillations confined to narrow frequency bands.

18   Cybernetics

A Cybernetics and its historical origins

In the summer of 1947, Wiener was on one of the more interesting of his many travels. On the way to Nancy, France, to attend a conference on harmonic analysis organized by his collaborator S. Mandelbrojt, he stopped off in England to visit many friends. He saw Haldane in London, A. M. Turing at the National Physical Laboratories in Teddington and Grey Walter in Bristol. He also looked up friends in Cambridge and Manchester. The conference itself was interesting. Apart from it leading to a new paper with Mandelbrojt [47a], he got to know first-rate mathematicians of a younger generation such as L. Schwartz, and to reunite with old friends such as Harald Bohr.

But the most eventful aspect of the trip occurred in Nancy, when a representative, M. Freyman, of the publishing house Hermann et Cie called on him with a request to write a short book on the fundamental aspects of his work in communication and control. He turned out to be, in Wiener's words, "one of the most interesting men I have ever met". He was the founder of the group of French mathematicians who publish under the pseudonym of "Bourbaki", and to give them a residence he had just "created" the "University of Nancago" (from "Nancy" and "Chicago") that very year. Wiener felt that "it would be fun to get in with so interesting a group", and readily signed a contract for the book "over a cup of cocoa in a neighboring patisserie" [56g, pp. 316–317]. In Mexico, late that same summer, Wiener, the traveler, wrote the book, dedicating it to Arturo Rosenblueth [48f]. So began a journey of great scientific, philosophical and social interest that sometimes landed Wiener in strange and unforeseen circumstances: at one point the Soviet press was to call him a "cigar-smoking slave of the industrialists". [XII, p. 16]

The subject matter of Wiener's impending book was to be the *theory of messages*, and he felt that a Greek word signifying "messenger" would make a good title. But the only such word he knew was *angelos*, and as this had the connotation in English of *angel* or *messenger from God*, he abandoned the idea [56g, p. 322]. For the title he then picked the word *Cybernetics*, which he coined from the Greek "kubernetes" for steersman, and began using the term to refer loosely to the theory of communication and control in the animal and the machine. Wiener was surprised to learn later that over a 100 years earlier the French physicist Andre-Marie Ampère, an early leader in electromagnetism (after whom the unit of electric current the "amp" is named) had coined the term "cybernetique" to refer to the "art of government" {A3}, (1838, 1843), and that in the same year, 1843, the word "cypernetyki" appeared in a Polish work on management by S. Trentowski in Poznań.

Ampère's cybernetique and Wiener's cybernetics are far apart, however, as Ampère's Chart on the classification of the sciences and the place allotted to cybernetics testifies.

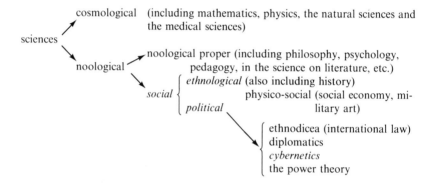

Ampère wrote, cf. {D7, p. 47}:

> . . . I call it cybernetics from the word κυβερνήσις which from its first narrow acceptation of the art of managing a vessel, eventually developed the far broader meaning—already with the Greeks—of *the art of governing in general.*

Although James Watt had patented the centrifugal governor in 1784, Ampère showed no awareness of the existence of regulatory processes outside politics, and of the feedback of information as a key element in regulation. In his time, even the concept of energy had not taken hold, let alone information. While military art and diplomatics are listed in his chart, engineering is conspicuous by its absence.

What Wiener never found out, however, was that the term κυβερνητική, the art of steersmanship, occurs in "Gorgias", a dialogue of Plato, in a passage on the arts of navigation and rhetoric. In both activities the purpose is to control, and in both the feedback of information is central: the impact of sea waves on the ship's hull, applause and jeers from the audience. Was then Plato a precursor of cybernetics? The answer is unclear, for we know of no place where he affirms that the back-and-forth flow of messages is what unites navigation and rhetoric. We may tend to read too much in his use of the term, forgetting that it had a long career:

> The word "κυβερνάω" and its derivatives are fairly frequent in Homeric poems as well as in Ancient literary and philosophic Greek writings. The primary meaning of this family of words denotes the steering of a vessel, or guiding the vessel in a specified direction (*Odyssey*, 3, 283) and also driving chariots (Plato, *Theages*, 123 c). Several writers use this verb with a metaphorical tint to denote "guiding" or "governing". {G4, pp. 144–145}

In this century, a clear awareness of the centrality of "reversibility", i.e. what we all call feedback, is shown in the book *Psychologie Consonantiste* (1938, 1939) written by Stefan Odobleja, a military physician in Romania, {O1}. He attempted to build the science of psychology around the concept of *consonance*, i.e. a continuing confluence of factors maintained by feedback. This psychological starting point, and Odobleja's incognizance of engineering feedback, resulted in his blurring of the distinction between positive and negative feedback, and in his interpretation of feedback coupling (or reversibility) as being energetic rather than informational. Notwithstanding these limitations borne of time and place, Odobleja asserted the important role of consonance and reversibility in phenomena outside psychology, for instance in biological processes, bringing a fresh light to bear on the work of French physiologist Claude Bernard (1879). Where the informational aspect in a feedback is negligible, his descriptions are weighty. For instance, his theory of *attention*, i.e. of bringing and holding an object in mental focus, is rather like Wiener's treatment of visual feedback (cf. § 16C).

Odobleja's role in the history of cybernetics is well described by the title of a recent collection of essays: *Odobleja Between Ampère and Wiener* {D7}. Ampère emphasized regulation or control, Odobleja the feedback loop, and Wiener and Shannon the message, as well as the connection between all three elements.

The cybernetical attitude carries with it much more than an appreciation of control by informational feedback, however. Among its other central ideas are:

(1)   a universality that assigns only a secondary status to the autonomy of the different sciences;

(2)   a recognition that intelligence extends into the inanimate world;

(3)   an indeterministic causality in which the universe is a cosmos, but of an incomplete order that permits teleology and freedom;

(4)   an ontology-indifferent functional or "black-box" approach to inquiry.

The scientists who pioneered these attitudes must be recognized as progenitors of cybernetics.

With regard to (1), Leibniz must be accorded a first place by virtue of the very wide spectrum of phenomena embraced in his monadic doctrine. A place must also be found for those who stressed this unifying perspective in a modern context, such as J.B.S. Haladane (1934) and G.D. Birkhoff (1938, 1941), cf. § 10D. As for (2), we must again include Leibniz for his remarkable foresight that cerebration is an enlarged calculation, and that the latter can be put in-the-metal. Mention must also definitely be made of the builders of computers, small or large, analogue or digital: R. Llull (c. 1275), W. Schickard (1592), B. Pascal (c. 1642), G. Leibniz (c. 1673), C. Babbage (c. 1840), J.C. Maxwell (1855), Lord Kelvin (1876), and later workers. Nor must we leave out Frederick Taylor, and Frank and Lillian Gilbreth who, with camera and stopwatch, revealed the labor process as a patterned sequence of basic steps (1910s), thus foreshadowing the concept of an *industrial program* and the coming of the automatic factory, [50j, p. 189].

With regard to (3), we cannot leave out the founders of the concept of entropy, S. Carnot and W. Thompson (Lord Kelvin), and the fathers of statistical mechanics, J.C. Maxwell, L. Boltzmann and J.W. Gibbs. A special place must be assigned to the philosopher H. Bergson for his discerning, if hazy, reflections on time, purpose and creativity. Especially significant is the role of scientists who brought indeterminism and its Bergsonian aspects to bear on biological evolution, such as Sir Ronald Fisher, the statistician. In his 1934 paper "Indeterminism and natural selection" {F2}, he has left us with an especially clear and succinct description of the attitude (3):

> The historical origin and the experimental basis of the concept of physical determinism indicate that this basis was removed with the acceptance of the kinetic theory of matter, while its difficulties are increased by the admission that human nature, in its entirety, is a product of natural causation. An indeterministic view of causation has the advantage (a) of unifying the concept of natural law in different spheres of human experience and (b) of a greater generality, which precludes the acceptance of the special case of completely deterministic causation, so long as this is an unproved assumption. It is not inconsistent with the orderliness of the world, or with the fruitful pursuit of natural knowledge. *It enriches rather than weakens the concept of causation.* It possesses definite advantages with respect to the one-sidedness of human memory, and to the phenomena of aiming and striving observable in man and other animals. Among biological theories it appears to be most completely in harmony with the theory of natural selection, which in its statistical nature resembles the second law of thermodynamics.

Henri Bergson
1859–1941

In an indeterministic world natural causation has a creative element,
and science is interested in locating the original causes of effects of special
interest, and not merely in pushing a chain of causation backwards ad infinitum.
{F2, p. 99} (emphasis added.)

The importance of the attitude (4) in cybernetics has been stressed by Dr. W. Ross Ashby, (cf. §§ 18B, C.). Nature presents us with a lot of organisms, the internal structure of which is inaccessible to us. The cybernetical attitude is to examine them behaviorally, i.e. to study their response to different stimuli, and to gain insight by fabricating a model organism exhibiting the same behavior. The great importance of such actual synthesis was emphasized by Giambattista Vico (c. 1700), and it was reiterated by C.S. Peirce. The latter's theory of relations contributed immensely to such study of functionality.

The theory of the subject that Wiener christened thus has a long history extending all the way back to Plato. When we turn from this theory to cybernetical craftsmanship, we have to go to the very dawn of human history. The trap, one of the earliest of human contraptions is a fine example of a teleological control mechanism. (With a bait it resembles in a way a guided missile; it attracts the quarry instead of being attracted to it.) The mechanisms described by Heron of Alexandria (c. 100 A.D.?) are feedback systems in the strict sense, and many of them were used as irrigation or other devices in the river-valley civilizations of antiquity, cf. A. Giuculescu {G4}.

B The subject-matter of cybernetics

A pithy characterization of cybernetics as a scientific discipline was given in 1956 by Dr. Ross Ashby in his book, *An Introduction To Cybernetics* {A6, pp. 1–7}. Cybernetics is the *general study of mechanism from the standpoint of functionality and behavior rather than internal structure and material*. What the word "mechanism" is to mean is stated by Wiener:

> For us, a machine is a device for converting incoming messages into outgoing messages. A *message*, from this point of view, is a sequence of quantities that represent signals in the message. Such quantities may be electrical currents or potentials, but are not confined to these, and may indeed be of a very different nature. Moreover, the component signals may be distributed continuously or discretely in time. A machine transforms a number of such input messages into a number of output messages, each output message at any moment depending on the input messages up to this moment. As the engineer would say in his jargon, *a machine is a multiple-input, multiple-output transducer*. [64e, p. 32] (emphasis added).

Wiener's concept of transducer subsumes Ashby's definition of a *machine*, as a function on $S \times E$ to S, where S is the set of *internal states* and E the set of *external conditions* {A5, p. 242}. For, such a machine is merely a transducer for which "incoming messages" and "outgoing messages" are members of $S \times E$ and S, respectively. Ashby's conception is very fruitful. As he explains:

> A machine is that which behaves in a machine-like way, namely, that its internal state, and the state of its surrounding, defines uniquely the next state it will go to {A5, p. 251}.

Irrelevant is the stuff from which it is made. An angelic assembly need only behave in a "machine-like" way to become a machine.

Conceived exclusively in these broad terms, cybernetics would include almost any subject, and could hardly have an agenda of its own. However, cybernetics investigates the machine from the standpoint of Bergsonian time, cf. §§ 12F and 18A. Of special interest are the issues centering on information, noise, entropy, feedback, causality and purpose. *Signal detection* (more accurately, message detection), i.e. the recovery of the incoming message by the filtration of noise from the incoming signal, is a fundamental problem. Especially important is the demarcation of the category of *purposive mechanisms* and the subcategory of *teleological mechanisms*. [43b, 50d], cf. § 15B.

Tagged on with these qualifications, the term *cybernetics* ceases to refer to a particular empirical subject-matter such as geology, and begins to refer rather to a *method for viewing and solving problems* irrespective of subject-matter, i.e. to a methodology. The concepts most important to it are *signal*, *message* and *noise*, related by the rough formula

$$\text{signal} = \text{message} + \text{noise},$$

as well as *feedback*. The laws governing the signal are more complicated and harder to handle than those governing the message. The efficacious way to solve a practical problem is to extract the message, i.e. to filter out the noise, to deal exclusively with the uncontaminated message (even though it occurs nowhere in nature in isolation), and finally correct the error due to idealization by taking into account the presence of noise and of transients. There are, of course, many classical problems in which the noise is so small that it can be regarded as absent. For purposive systems the analysis of the effect of the signal on the receptors of the system and the internal feedback of the outputs within it are important. Another task is the *coding* and *decoding* of messages so as to minimize the influences of noise and maximize the flow of information.

Cybernetics thus brings to the forefront concepts and ideas relevant to Bergsonian time such as noise, entropy, feedback and purpose, message and control, one or more of which have appeared in specific fields such as thermodynamics or servomechanism engineering, but which have been absent from the vocabulary in which the scientific method is ordinarily presented. To this extent, *cybernetics is the extension of the scientific methodology necessitated by the existence of processes for which time is Bergsonian.*

C Is cybernetics a science?

In the preface to the 1961 edition to his book *Cybernetics* [48f], Wiener speaks of cybernetics as no longer a program but an "existing science". There is some uneasiness among scientists as to the nature and domain of this science. (See, for instance, J.R. Pierce {P5, pp. 208–209}. Its status and domain are discernible, however, from the discussion in the previous section. Cybernetics is a branch of mathematics, with its own specific primitive terms, *machine*, *signal*, *noise*, *information*, etc., much like hydromechanics with its own specific terms, mass, pressure, vorticity, etc. Although the axiomatization of the subject may be premature today, it is possible, and the axioms can be changed from time to time to accommodate the growth of knowledge of mechanism and communication from new observation and experiment (much as the postulates of fluid flow can be replaced by new ones). In regard to applications, cybernetics is like the theory of differential equations in that these belong to many different areas. Such an application will be fruitful, only if it incorporates specific knowledge about the subject in addition to the cybernetical principles. At an advanced stage the cybernetical postulate-system may permit the prediction of results unknown (as did the Maxwell and Einstein theories), thereby obviating costly experimentation, as well as yield theorems relating to the efficient design of machinery and the efficient use of hardware. Cybernetics will then be able to contribute to the economic life of man more thoroughly than at present.

That cybernetics should be conceived in these mathematical terms was eloquently argued by the eminent medical cybernetician, W. Ross Ashby:

> Cybernetics stands to the real machine—electronic, mechanical, neural, or economic—much as geometry stands to a real object in our terrestrial space. {A6, p. 2}

Just as the objects of geometry are idealizations of real objects, and just as a strange geometry with no immediate application to real objects may merit serious consideration and prove later to be of great practical value, so too with the study of mechanisms. This thought is again masterfully conveyed by Dr. Ross Ashby:

> Cybernetics, then, is indifferent to the criticism that some of the machines it considers are not represented among the machines found among us. In this it follows the path already followed with obvious success by mathematical physics. This science has long given prominence to the *study of systems that are well known to be non-existent*—springs without mass, particles that have mass but no volume, gases that behave perfectly, and so on. To say that these entities do not exist is true; but their non-existence does not mean that mathematical physics is mere fantasy; nor does it make the physicist throw away his treatise on the Theory of the Massless Spring, for *this theory is invaluable to him in his practical work*. The fact is that the massless spring, though it has no physical representation, has certain properties that make it of *the highest importance* to him if he is to understand a system even as simple as a watch. The biologist knows and uses the same principle when he gives to Amphioxus, or to some extinct form, a detailed study quite out of proportion to its present-day ecological or economic importance. {A6, pp. 2, 3} (emphasis added)

This trenchantly expressed view—that the key to practical efficacy lies in the study of the non-existent — is Pythagorean-Platonism

William Ross Ashby
1903–1972
Courtesy of
Professor N.N. Rao
Electrical & Compu-
ter Engineering De-
partment
University of Illinois

at its militant best, i.e. it retains Plato's emphasis on abstraction but not his disparagement of experimentation, precisely as Roger Bacon advocated (c. 1260). To obtain a closer fit, perhaps the word "non-existent" should be replaced by "transcendent", in the sense of being a replica of what exists, but a replica that transcends natural limitations and is, therefore, ideal. In writing the formula, signal = message + noise, and in singling out the message as the significant term despite its non-existence in nature, the cybernetician is following the Pythagorean-Platonic method. From this methodological standpoint, the origins of cybernetics indeed go back to Plato. So do the origins of most sensible moral and aesthetic activity. The music-lover, listening to a short-wave broadcast of Mozart on a rainy day, seeks the (non-existent) pure Mozart in the signal, and ignores the accompanying krrr . . . sound. The good battle-commander concentrates on the battle-plan, not on the carnage around him, cf. A.N. Whitehead: "Mathematics and the Good" {W10, pp. 666–681}.

We are referring here to the *epistemological* aspect of Platonism, not to its ontological aspect. The latter is of little interest to the cybernetician for whom the behavior of the machine, not its materiality, is central. It is the knowledge of behavior that helps "in *changing* the world", and not just "in *interpreting* it in various ways" (Karl Marx, Thesis 11 on Feuerbach, {M6, p. 84}. The prediction of the existence of the positron by Dirac four years before its discovery in the laboratory was by adherence to the time-tested Pythagorean-Platonic method. Likewise for the meson and the neutrino. These examples of the use of Plato's methodology to advance the atomistic standpoint, at variance with his ontology of the world as a plenum, shows the total irrelevance of the Platonic ontology to Platonic epistemology. This, Marx, and more so Engels and Lenin, failed to realize (next section).

D Soviet views on cybernetics

As conceived by Wiener and Ashby (§ 18C), the science of cybernetics has several branches, among which are *data-processing* and *automata theory*. In the Soviet Union the tendency still current, is to narrow the scope of cybernetics by redefining it as data-processing and automata theory. For instance, Academician V.M. Glushkov writes in his 1969 article entitled "Contemporary Cybernetics":

> It is usual nowadays to define cybernetics as the science of the general laws of data transformations in complex control systems and systems of information processing.
>   When defining the subject of cybernetics it is important to avoid two extremes. These are, first, including in cybernetics everything which concerns control; and secondly attempting to reduce cybernetics to a comparative study of the relation between control systems in engineering and those in living beings. {G7, pp. 47, 48}.

Replace "complex control system" and "data" by Wiener's "multiple-input, multiple-output transducer" and "message" respectively, and we are, in essence, back to the Wiener-Ashby definition. Furthermore, Soviet cyberneticians have been able to incorporate a good deal of Wienerian thought in their own framework. It is rather the overtones of the word "data", the near-exclusive concern in Soviet cybernetical literature with digital data and algorithmic considerations, and the omission of analogue devices that go against the attitude of universality (cf. § 18A (1)) and are confining.

For Soviet cyberneticians the narrower interpretation of cybernetics offered certain advandages. Because of the officially imposed ban on mathematical logic in the Soviet Union until the late 1950s, they had a good deal of catching up to do in the subject, and so an extra emphasis on automata theory was advantageous. Indeed, Glushkov's book {G6} is the only one known to the writer that both bears the title *Introduction to Cybernetics*, and presents the second order predicate calculus together with a proof of Gödel's incompleteness theorem. The book is also conspicuous by its limitation "to digital methods of representation of information and to digital information processors" {G6, p. v.}, so much so that Wiener's name appears nowhere in it. Secondly, Kolmogorov and other Soviet mathematicians brought in profound new insights on the nature of "algorithm", and it was natural for Soviet cyberneticians to put these to use. Thirdly, the Russian definition, by giving the subject a more sharply defined boundary and a much smaller philosophical range than the Wiener-Ashby definition, protected it from ontology-infested officialdom, suspicious of revolutionary methodology. For in the 1950's there was considerable official hostility to the cybernetical movement, witness the following excerpt from the official Soviet *Short Philosophical Dictionary* (*Kratkii filosofskii slovar*):

> Cybernetics: A reactionary pseudo-science arising in the U.S.A. after the Second World War and receiving wide dissemination in other capitalistic countries.
>
> Cybernetics clearly reflects one of the basic features of the bourgeois worldview—its inhumanity, striving to transform workers into an extension of the machine, into a tool of production, and an instrument of war. At the same time, for cybernetics an imperialistic utopia is characteristic—replacing living, thinking man, fighting for his interests, by a machine, both in industry and in war. The instigators of a new world war use cybernetics in their dirty, practical affairs. {K11, pp. 236–237}.

Less ontology-bonded Marxists did not fall into this trap. Wiener informs us that in 1947 Professors J.D. Bernal, J.B.S. Haldane and H. Levy, all distinguished British Marxists, "certainly regarded it (cybernetics) as one of the most urgent problems on the agenda of science and scientific philosophy" [61c, p. 23].

In the late 1950s the official Soviet attitude towards cybernetics sobered considerably. Wiener was invited to the Soviet Union in 1960, treated with respect, and was requested to address the philosophical section of the Soviet Academy of Sciences, and expound his views in the periodical *Voprosy Filosofii*, [61b]. Cybernetics came to be actively pursued in the Soviet Union, but with the narrower conception we mentioned. The narrowness runs the danger of fostering over-specialization, reducing communication between cyberneticians and workers in cognate fields (excluded from cybernetics by definition), retarding the application of cybernetics to industry, and inhibiting the synthesis of a cosmical or even global cybernetical perspective, commensurate with the needs of contemporary civilization.

E Ontogenetic learning

Cybernetics, in the full sense of Wiener and Ashby, has wide scope, for definable within its framework are conceptions of great consequence such as purpose, learning and self-organizing. Of these the concept of purpose, broached in § 15B, is fundamental.

Wiener realized that once the concept of a purposive mechanism is defined (cf. § 15B), that of a *learning mechanism* can be explicated. Every machine has an input-output transformation T defined by setting $T(f)$ equal to the output when the input is f. In a purposive mechanism there is, moreover, an ideal transformation T_o, which is cherished but unrealized, i.e. $T(f)$ only approximates $T_o(f)$ to some degree. A purposive mechanism is said to *learn*, if it has the ability not only to perform the transformation T, but also to change T from time to time so as to bring it closer and closer to the ideal transformation T_o, i.e. to reduce the difference $|T(f) - T_o(f)|$. Wiener visualized such a learning mechanism as a system comprising a *performing filter A* coupled by feedback to a *non-linear filter B* [61c, p. 173; 64e, pp. 14, 20–21]. *A* carries out the routine of transforming input f into output $T(f)$. *B* keeps a record of past inputs, outputs, and errors of performance of *A*, and has devices for the re-evaluation of the parameters governing this performance. Periodically, the system takes "time-out" to make this re-evaluation. The results are automatically fed back to *A*, and the routine is resumed with improved efficiency. For instance, the component *B* of a learning anti-aircraft battery would record the long-time trajectories of incoming airplanes, and compute therefrom estimates of the covariances of the hypothetical underlying time series, cf. § 14C. For this, *B* would have to include non-linear devices such as square law rectifiers. When the improved estimates of the covariances are fed into component *A* of the battery, the new $T(f)$'s produced by *A* will be closer to the ideal predictor $T_o(f)$ than before, i.e. the performance will improve.

Wiener believed that although our understanding of learning machines is in its infancy, chess playing automata embodying the principles just outlined would be realizable in the future [50j, p. 176–177]. These automata, of "higher type" (cf. p. 52), would be superior to the existing machines based on Dr. Shannon's designs {S8}.

Wiener compared this ability of a mechanism to learn, i.e. to improve its performance by appropriately modifying its response transformation T in the light of environmental realities, to *ontogenetic learning* among biological organisms. He believed that in the higher animals the parameters which determine T are synaptic thresholds (which control the firing of neurons) and that the "re-evaluation" or "learning" consists in changing these thresholds for the better. He also felt that in effecting such a change, the neurophysiological system in the animal acts more like an analog computer than a digital one.[1] The thresholds are governed by continuous quantities such as temperature, and the changes are affected through non-nervous channels such as hormonal flow through the blood stream. But Wiener admitted that these views were largely conjectural.

Wiener's theory of the mechanical simulation of ontogenetic learning also included situations in which the ideal transformation T_0 is not held fixed. He envisioned *learning mechanism capable of modifying the ideal T_o itself*, whenever the sequence of A's errors of performance (as evaluated by B) do not diminish with a sufficient rapidity. Such a judgmental or opportunity-seeking mechanism has to possess extra non-linear components $C, D, \ldots$ to carry out the requisite new evaluations. It will be higher in (Russellian) type in relation to the mechanisms for which T_0 is fixed, cf. § 5B. This point of view is an extension of Wiener's very early idea of a "highest good" that evolves as new vistas open up, which he expounded in his Bowdoin Prize essay [14C], cf. § 5D above. The use of Russellian types has the same cogency here as it does in military theory and in the theory of game-playing automata, which we shall examine later in § 20C.

F Phylogenetic learning and non-linear networks

In later years Wiener thought about phylogenetic or racial learning in biological evolution, and about the possibility of its mechanical simulation.

Imagine a species $\mathbb{S}$ of organism $S_1, S_2, \ldots$, which are both adaptive and asexually reproductive with a Lamarckian genetics, in which improved thresholds acquired in learning are inherited. The performance of the *assembly* itself will improve with the passage of time, for each individual organism will be more learned, i.e. better designed, and will pass on its improved design to its offspring. We may say that the species $\mathbb{S}$ itself "learns". Thus we have an asexual Lamarckian analog of the phylogenetic learning of a biological species.

The interest of such asexual Lamarckian phylogenetic learning for Wiener lay in the possibility of its simulation by non-linear electrical networks. This, as Wiener saw, has far reaching consequences. It becomes possible to so design a teleological machine that it can both learn, i.e. "adapt" itself to its environment, and reproduce, i.e. create an "offspring" in its own image. Given sufficient access to external energy, the assembly $\mathbb{S}$ of such teological reproducing, learning machines would

---

1   This again points to the shortsightedness of the Soviet policy of excluding analogue machines from the domain of cybernetics.

then exhibit a behavior akin to phylogenetic learning in biological species. Such phylogenetic learning would be of different types depending on the goals of the teleological units. Were the assembly $\mathbb{S}$ made up of machines eager to learn chess, the learning would occur in the field of chess. There are several other such alternatives.

The possibility of such simulation rests on Wiener's theory of non-linear filters, a work of genius involving some of the deepest mathematics ever done. It originated from fairly obvious attempts to extend the linear theory, but then Wiener's revolutionary idea to use random probes as inputs, shifted the theory into much deeper waters. This deeper theory is touched on in Note 2 on p. 270 at the end of this chapter. Let us see how it arose, and how it allows the simulation of Lamarckian reproduction.

> Electrical engineering ... having dealt with substantially linear networks throughout the greater part of its history is now rapidly introducing into these methods elements, the non-linearity of which is their salient feature, and is baffled by the mathematics thus presented and requiring solution. {B23, p. 448}

So wrote Vannevar Bush in 1931. Thus Wiener was probably aware in the 1930s of the limitations of the linear network from the standpoint of general circuit theory, and of the need for remedying this situation. But the first evidence of any serious thinking on his part on this matter appears only in his 1942 report, *Response of a non-linear device to noise.*[2]

In this report Wiener started with the plausible supposition that the network's non-linear input-output transfer operator T is of the form

$$T(f) \;=\; W_1*f \;+\; W_2*(f \times f) \;+\; W_3*(f \times f \times f) \;+\; \ldots \qquad (1)$$

in which the first term

$$(W_1*f)(t) \;=\; \int_0^\infty W_1(t - \tau)f(\tau)d\tau$$

gives the "linear part" of the network, the second term

$$\{W_2*(f \times f)\}(t) \;=\; \int_0^\infty \int_0^\infty W_2(t - \tau_1, t - \tau_2)f(\tau_1)f(\tau_2)d\tau_1 d\tau_2$$

gives its "quadratic part" and so on. Thus, the single convolution weighting W of the linear network, cf. § 9B (11), gives way to a sequence of weightings W_1, W_2, ... of higher and higher orders. Such expansions had already been studied by Volterra in the 1900s. The problem now was to synthesize the higher order convolutions W_2*, W_3*, ... by extending the type of circuitry used in the Lee-Wiener network for synthesizing the simple convolution $W*$.

---

2 MIT Radiation Laboratory Report, V-186, April 6, 1942.

From the very outset, however, Wiener was dissatisfied with this scheme. By a remarkable intuition, he replaced the f by a Brownian movement input, and took the convolutions of W_1, W_2, etc. with the chaos determined by the Brownian movement, § 12E(6). Thus he replaced the last two integrals by the *chaotic integrals*

$$\int_0^\infty W_1(t - \tau)dx(\tau, \alpha), \qquad \alpha \in [0, 1],$$

$$\int_0^\infty \int_0^\infty W_2(t - \tau_1, t - \tau_2)dx(\tau_1, \alpha)dx(\tau_2, \alpha),$$

which he had first introduced in the early 1930s, $x(\cdot, \cdot)$ being the Brownian motion, cf. § 7D(3). The response of the network to such a Brownian input signal $x(\cdot, \alpha)$ is now itself the signal $g(\cdot, \alpha)$ of stochastic process. This type of study of the non-linear network is made in Wiener's 1942 report, referred to above. Wiener wanted to take the *shot effect* as the random input since it offers the best physical realization to the ideal Brownian motion, cf. § 7D.

What prompted Wiener to use shot noise as a probe for the filter? It was his belief that to "know" a black box it suffices to observe its response to the most chaotic signal conceivable, to wit, the Brownian motion. Wiener seems to have had this insight at least by 1930, for in his memoir [30a] we read:

> Imagine a resonator—say a sea-shell—struck by a purely chaotic sequence of acoustical impulses. It will yield a response which still has a statistical element in it, but in which the selective properties of the resonator will have accentuated certain frequencies at the expense of others. [30a, p. 215]

Thus the response to "purely chaotic" impulses will "reveal" the resonator or filter, i.e. it willl give us enough clues about its character, perhaps enough to construct its duplicate. Indeed, this is the case with time-invariant linear filters, for one very important result in [30a] asserts that if

$$g(t, \alpha) = \int_{-\infty}^\infty W(t - \tau)dx(\tau, \alpha), \qquad \alpha \in [0, 1],$$

then for almost all α, the response signal $g(\cdot, \alpha)$ has the spectral density $\sqrt{2\pi}|\hat{W}(\cdot)|^2$. Thus we can determine $|\hat{W}(\cdot)|^2$ by bombarding the filter with shot noise, and passing its response through a spectrophotometer. For a causal filter (§ 9B) we can then retrieve $\hat{W}$ from $|\hat{W}(\cdot)|$,[3] and thence W.

For linear filters, however, it is not necessary to use such random probes, for the character or sinusoidal inputs are just as revealing, as we

---

3   This requires a factorization of $|W(\cdot)|$ into optimal holomorphic factors. For details, see {M9, § 20}, especially (20. 4). Wiener became especially conscious of the importance of such factorization when he started working on the prediction of airplane trajectories during World War II, and obtained Hopf-Wiener integral equations, cf. § 14C.

emphasized in § 9A. But for non-linear filters we are left with the Brownian motion as the most effective probe. This thought, which dictated Wiener's choice of a probe, is stated lucidly in Wiener's post-humously published book:

> The output of a transducer excited by a given input message is a message that depends at the same time on the input message and on the transducer itself. Under the most usual circumstances, a transducer is a mode of transforming messages, and our attention is drawn to the output message as a transformation of the input message. However, there are circumstances, and these chiefly arise when the input message carries a *minimum of information*, when we may conceive *the information of the output message as arising chiefly from the transducer itself*. No input message may be conceived as containing less information than the random flow of electrons constituting the shot effect. Thus the output of a transducer stimulated by a random shot effect may be conceived as a message embodying the action of the transducer.
>
> As a matter of fact, it embodies the action of the transducer for any possible input message. This is owing to the fact that over a finite time, there is a finite (though small) possibility that the shot effect will simulate any possible message within any given finite degree of accuracy . . . That is, if we know how a transducer will respond to a shot-effect, input, we know ipso facto how it will respond to any input. [64e, p. 34] (emphasis added)

Thus Wiener felt that random probes have a revelatory role in electrical engineering that is similar to that of probes such as ink-blot tests, random interrogation, etc., in human psychiatry.

This radical viewpoint suggested to Wiener the abandonment of the Volterra scheme (1), in favor of a much deeper one, latent in his own work on the homogeneous chaos in the late 1930s. However, some systematic work by R.H. Cameron and W.T. Martin {C1} and by S. Kakutani {K3} had to occur before Wiener could spot the latency, and take off. The basic ideas are those of harmonic analysis, but the term "harmonic" has now a much wider connotation than that in § 9A.

Wiener considered a *black box* or *filter* B with an input terminal and an output terminal, the internal structure of which is unknown. We assume, however, that if at instant t it receives the shot noise (or Brownian motion) signal $x(t, \alpha)$ at its input end, then its response $g(t, \alpha)$ at the output end can be read. (Here $0 \leq \alpha \leq 1$, cf. § 7D.) In short, we know only the transfer operator T of the filter B. Wiener showed that if the filter B is time-invariant and causal (in the sense of §§ 9A, B) and stable (in the reasonable sense explained in Note 2 on p. 270), but otherwise quite non-linear, then T is expressible in one and only one way as a series:

$$T = \Sigma a_p W_p, \tag{2}$$

where $p = (p_o, p_1, \ldots, p_m)$, m is a positive integer and $p_o, p_1, \ldots, p_m$ are any non-negative integers, the a_p are real numbers dependent on T, and *the W_p are standard white operators*, i.e. they are transfer operators of a one-input, one-output filter, the circuitry of which is fully known and standardized. The summation is over all the p. The equality in (2) is to be taken in the sense of root-mean-square error. Roughly speaking, not for all inputs f, but only for all *typical* ones, will $T(f)$ be equal to $\Sigma a_p W_p(f)$.

Wiener showed that for any such black T, the coefficients a_p can be determined. Once the a_p's are found, we can get the circuit for $a_p W_p$ by merely juxtaposing multipliers (amplifying/attenuating devices) to the circuit for W_p. We then build the huge circuit A by hooking up all the $a_p W_p$ circuits, according to the recipe $\Sigma a_p W_p$. This A *will be a white offspring of the black box B*. That is, A will have the same transfer operator as B, and thus be identical to its parent in overt behavior, but it will have a fully blueprinted circuitry that may be quite different from the internal structure of B, concerning which, we know nothing.

All this must be taken with many grains of salt, for obviously not all the infinitely many $a_p W_p$ can be hooked together. Not only must the series in (2) be conveniently truncated, but the formidable formulae for the a_p impose more approximations. Thus different offspring of the same B will in practice show variations much as in a biological population.

This process, leading from B to A, can be represented by a block diagram with a unit, corresponding to the analysis that yields the crucial coefficients a_p, and the other that corresponds to putting the pieces together:

If from the diagram we omit B and A, we are left with the diagram of a *genetic scheme* or *incubator* for transducer reproduction.

To return to phylogenetic learning, we now consider an assembly $\mathbb{S}$ of teleological learning mechanisms $S_1, S_2, \ldots$. Each of these automatons S_i is capable of the ontogenetic learning considered in § 18E. But we now imagine that they are also offspring to a common ancestor, and can themselves reproduce via the agency of the incubator. We then get a growing geneology $\mathbb{S}$ of automatons that itself learns as time flows, i.e. we have mechanically simulated phylogenetic or racial learning.

G The distinction between scientific inquiry, and stratagem or contest

Raffiniert ist der Herrgott, aber boshaft ist er nicht. (A. Einstein)

Let A and B be purposive filters with contravening goals. Were B absent, then A would execute a sequence of moves $a_1, a_2, a_3 \ldots$ towards achieving its objective. But B is present; after A has made the move a_1, B makes a counteracting move b_1. The second move a'_2 that A will now make will depend on b_1 and will differ from a_2. Likewise the third move a'_3 that A will make will depend on the second move b_2 of B, and so on. Thus by virtue of B's presence, A is obliged to adopt a new sequence of moves a'_1, a'_2, a'_3, ...

We may describe such a sequence of moves of a purposive mechanism, which depend on the countermoves of another such mechanism with a contravening goal, as a *fight* or *contest*. A rather common type of contest occurs when A's objective is to *control* the purposive mechanism B itself, and one of B's objectives is not to be so controlled. We then have a fight or contest between two adversary mechanisms.

The process of adopting increasingly efficient policies for winning a contest is called *strategic learning*. As to the concept of *strategy* itself, it has been axiomatized in the important work of J. von Neumann and O. Morgenstern {V7}. Wiener, guided by Russellian typology, classified strategic policy into tactics, stratagem, superstratagem, etc., cf. §§ 20C, D.

A purposive mechanism A makes an *inquiry* of another mechanism B, purposive or not, when its objective is to gain information concerning B. Such an inquiry may just be part of the larger objective of doing research, i.e. of determining the laws governing B.

A most vital aspect of scientific inquiry and research is that, barring a very few insignificant exceptions, it involves no contest with the mechanism being studied. Nature is not an adversary in the relevation of its structure, and there is no room for stratagem in such a study. This is the thrust of Einstein's aphorism "Subtle is the Lord God, but capricious He is not". This sentiment goes back to Heraclitus, 500 B.C.:

> The Lord, whose oracle is at Delphi, neither reveals nor conceals but gives tokens.

Nature neither assists nor restricts investigation. The Pythagorean-Platonic tradition, to which cybernetics belongs, rests on this firm faith.

The few exceptions to this rule are well known. A person may resist giving out data on his health. A secluded tribe may resist an anthropological investigation. Laboratory animals may behave abnormally in captivity, and to this extent foil the scientists' objectives.

The contrast between inquiry and contest is reflected in our use of language. We adopt one mode of discourse to facilitate communication, another to administer contests and conflicts. The types of confusion to which the two forms of discourse are subject are quite different [50j, p. 193]. That affecting the first stems from human error and/or inanimate randomness such as atmospherics or other forms of noise; that in the second often comes from secret coding, jamming and/or bluffing. We may look upon scientific discovery as "decoding the code of Nature", but this would be a mistake without the prior recognition that *the code of Nature is open*, that decoding difficulties stem exclusively from the limitations of the decoding apparatus, i.e. the scientist and his language, and that stratagem is irrelevant, [50j, p. 124].

Nevertheless, the tendency persists to look upon research as a "game" to be played against Nature with the purpose of outsmarting it! From the cybernetical standpoint, these attempts to intellectually foray with Nature are quixotic, and well merit Wiener's description, "the shadow boxing with ghosts" [50j, p. 150].

Note 1. **Comparison with life processes** (cf. § 18F)

Transducer-reproduction bears some interesting similarities with physical, biochemical and biological reproduction. The reproduction scheme or incubator of § 18F cannot be used, because the large modifications required change it beyond recognition. This is inevitable since cellular tissue has a fine structure that is attuned for reproduction to a much higher degree than the metallic pieces of our incubator. Nevertheless, certain similarities subsist.

In gene reproduction, for instance, nucleic acids determine the laying down of a chain of amino acids in the form of a pair of helices. When these helical pairs split, each gathers to itself the molecular residues needed to form its partner, and we get two genes instead of one. Here the substrate, the nutrient medium, which is capable of assuming a large number of forms, is made to assume the particular form of a gene, by the presence of one such gene. Thus, the indeterminate substrate and the single gene play a role that clearly resembles that of the *indeterminate white box A* and the *black box B* of transducer theory, respectively.

When it comes to biological reproduction, the deviations from our reproduction-scheme are even more pronounced. Apart from sexuality, which is absent in transducer-theory, biological reproduction is "Mendelian", i.e. acquired characteristics, such as responses learned ontogenetically, are not genetically transmitted; but they are so transmitted in our "Lamarckian" transducer-reproduction. Furthermore, *social heredity*, i.e. vital responses learned *after birth* by imitation of or training from live parents, and memorized, which is so important to the life of the higher animals, is absent in our transducer-reproduction scheme.

Note 2: **Mathematical representation of non-linear filters** (cf. § 18F)

In the representation § 18F(2) of the black box B, the white operator W_p, corresponding to the index $p = (p_o, p_1, \ldots, p_m)$ is defined by its response to the shot noise signal $x(\,\cdot\,, \alpha)$ at instant t:

$$[W_p\{x(\,\cdot\,,\alpha)\}](t) = \prod_{k=0}^{m} H_{p_k} \{\int_{-\infty}^{t} L_k(t - \tau)dx(\tau,\alpha)\}. \tag{3}$$

Here the L_k are Wiener's favorite Laguerre functions, cf. § 13B(4), and the H_{p_k} are the well-known Hermite polynomials.

The W_p filter is indeed "white", i.e. it is synthesizable. For each k from 0 to m, a cascade of lattice circuits, as in the Lee-Wiener network, will deliver L_k*. To form the polynomials H_{p_k}, all we need are square-law rectifiers (which turn $f(t)$ into $f(t)^2$), scale-amplifiers and summing circuits. To form their product we can again fall back on square law rectifiers and adders, by dint of the equation

$$f(t) \cdot g(t) = \frac{1}{2}[\{f(t) + g(t)\}^2 - f(t)^2 - g(t)^2].$$

Let the black operator T be time-invariant, causal and also *stable* in the sense that for each t,

$$\int_0^1 | T\{x(t, \alpha)\}|^2 d\alpha < \infty. \tag{4}$$

Then the coefficient a_p in its expansion (2) is given, for almost any α in [0, 1], by the time-average:

$$a_p = \lim_{A \to \infty} \frac{1}{A} \int_{-A}^{0} [T\{x(\,\cdot\,, \alpha)\}](t) \cdot [W_p\{x(\,\cdot\,, \alpha)\}](t)dt. \tag{5}$$

This is arrived at from a phase-average expression for a_p, by applying the Birkhoff Ergodic Theorem to the flow of the Brownian motion, § 12B(4).

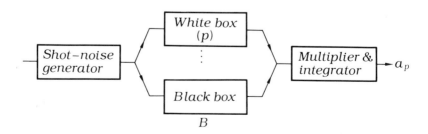

The formula (5) offers a mechanical evaluation of the coefficients a_p. By simultaneously bombarding B and the white box (b) with transfer operator W_p described above, by the same shot noise signal, and passing the outputs through a square law rectifier and integrating device, we get by (5) an approximate value of a_p. (See Fig.) The synthesis A of the black box B is then given by the circuitry discussed in § 18F.

The mathematical theory from which these remarkable expressions are derived is discussed in Wiener's book [58i, lectures 10, 11] and in a setting, employing the concepts of Cameron, Martin and Kakutani in {M9}. See also the Comment on [62b] in *Coll. Works*, IV. The representation obtained in the general theory is often called the *Hermite-Wiener representation*. An equivalent representation, called the *Fock representation*, appears in quantum mechanics. More on the Volterra and Wiener series can be found in Schetzen {S3}

19 The Second Industrial Revolution and its Educational, Economic and Social Challenges

A Automatization and the labor movement

Wiener's thinking on automatization and its socio-economic consequences was far ahead of its time. Guided by his clear perception of the
electronic computer (cf. his 1940 memorandum, § 13D), and his sharp
understanding of feedback, derived from his war work and physiological interests (§ 14E, 15B), he had concluded by about 1942 that a
high-speed computer, properly programmed, could be made to run an
automatic factory. In his words:

> ... *the modern ultra-rapid computing machine was in principle an ideal central
> nervous system to an apparatus for automatic control*; and ... its input and
> output need not be in the form of numbers or diagrams but might very well
> be, respectively, the readings of artificial sense organs, such as photoelectric
> cells or thermometers, and the performance of motors or solenoids. With the
> aid of strain gauges or similar agencies to read the performance of these
> motor organs and to report, to "feed back," to the central control system as
> an artificial kinesthetic sense, we are already in a position to construct
> artificial machines of almost any degree of elaborateness of performance.
> [61c, pp. 26, 27] (emphasis added)

By 1944 Wiener had surmised the social consequences of the "assembly
line without human agents" that the automatic factory would entail.
Lucid descriptions of the engineering of such a factory and the mode of
its operation are given in [50j, Ch. IX]. In that chapter Wiener also
traces the engineering involved to fundamental discoveries in physics,
and compares the modern situation in this regard to what prevailed
during the early phases of the Industrial Revolution 200 years ago.

Initially, such a factory would be able to perform automatically
tasks that involved the repetition of a short sequence of easily prescrib-
able operations requiring only very low level judgments, e. g. the labeling
of cans. Its operation would demand skilled technicians, trouble shooters
and maintenace crews having a higher level of engineering judgment
than possessed by the vast work force. Wiener also foresaw that such
"type 1" automatic factories would be followed by those of "type 2", in
which many of the tasks carried out by the lower echelons of supervisors
of the type 1 factory would be automated, and these type 2 factories

would in turn be followed by those of "type 3", which would displace some of the supervisors of the type 2 factory, and so on.

Wiener aptly characterized this process of successive automatization as a *second industrial revolution*. Just as the first Industrial Revolution involved the increasing use of *power-amplifying* machines (e.g. the pneumatic drill), the second had to involve the increasing use of *intelligence-amplifying* machines, from the pocket calculator up. Obviously the human intelligence needed to run an automatic factory of type n would have to increase with n, i.e. supervisors with more and more advanced engineering knowledge and judgment would be needed.

Wiener saw at once that smooth automatization demanded a steady advancement in educational standards. Serious unemployment and social dislocation would result from a disparity between a high rate of shift to automatic factories, and a low degree of systemic effort to upgrade the level of the work force. In the hope of sparing the workers the suffering and ugliness that accompanied the first industrial revolution, Wiener decided to alert the public to the changes that lay ahead, and to this end he tried to contact the leaders of American labor in the mid-1940s. But he failed to evoke any response from them until 1949 when Walter Reuther, president of the United Automobile Workers, began to show an interest.

The Wiener-Reuther correspondence is punctuated by telegrams from Reuther such as "Much interested. Would like to meet you after negotiations with Ford", or "Regret schedule will not permit meeting you on . . . ". The two finally met on March 14, 1951, and drew up a plan for a Council on Labor and Science made up of scientists from different fields. But then came the Chrysler strike, and the implementation of the plan was delayed.

A more deadly delay in going ahead with Wiener's plan came from the outbreak of the Korean War on June 25, 1950. Reuther's interests now shifted to the geopolitical scene. In a letter to Wiener dated July 18, 1950, he wrote:

> In the sincere belief that our best prospect of avoiding war and winning the peace lies in our pursuit of a plan of action that combines a fully adequate military defense with a positive peace offensive, I have submitted the enclosed proposal to President Truman for his consideration.
>
> I have not suggested an alternative to provisions for fully adequate military defense, but rather a supplementation and strengthening of our military defense by a positive program of social and economic action directed at elimination of poverty and social injustice which are the source of communist power. [MC, 121]

Reuther's proposal is entitled "A total peace offensive", and he requested comments on it from Wiener. Wiener responded in a five-page letter dated July 26, 1950. In this letter Wiener did not really comment on the Reuther proposal but expounded his own geopolitical viewpoint. What effect, if any, this letter had on Reuther is not known. (It is reproduced in full in § 20B, where Wiener's geopolitical thought is discussed.)

The Wiener-Reuther correspondence resumed on February 5, 1952, when Reuther wrote:

> Our Union would be honored to have you speak at the session of the Conference during which we will discuss the subject "Achieving and Maintaining Full Employment in a Free Society".
>
> Leon Keyserling, Chairman of the Council of Economic Advisors will share the platform with you and will speak on likely economic developments in the future. [MC, 146]

But Wiener declined, stating that he had had a hectic time in France and Mexico and was "under doctor's orders to take it easy and rest".

So ended, it would seem, the contacts between Wiener and

Walter Reuther
1907–1970

Courtesy of
Ms. Norma Bogovich,
UAW.

Photo by
Alexander Archer
(UAW Solidarity
Magazine)

Reuther. They had only a marginal impact on the UAW, and nearly none on the labor movement.

Neither Wiener nor Reuther foresaw the moral disability latent in a "trade-union consciousness" that operates without a "revolutionary consciousness", to use Lenin's phraseology. It imposes a cringing short-sightedness on both laborers and their leaders. All aspects of the employer-employee relationship (even production of dubious utility) are held sacrosanct, as long as the worker can enhance his short-term economic gains. Lenin described its effect on the Russian worker around 1900 in the words: "A kopec added to a ruble is nearer and dearer to him than any socialism or any politics" {L7, p. 69}. For the UAW workers of the late 1940s (who some years later were to flood the market, without a qualm, with flashy tail-finned gas guzzlers), one could say in the same vein: "A quarter added to a dollar was nearer and dearer to them than an understanding of the dislocation that might hit them 15 or 20 years later as a result of the ongoing automatization".

Apropos of Wiener's efforts to alert labor on the socio-economic consequences of automatization, a letter he received from a migrant laborer, J. Printise Womack, is of considerable interest. Unlike the situation we just described in which Wiener had to go after labor leaders, here it is a laborer who on his own initiative asked Wiener to intervene. The letter indicates on awareness of the limitations of isolated trade-union consciousness and of the impingement of cybernetics on religion, at a time (1950) when Wiener had hardly begun to think in these terms. Here is the letter [MC, 122]:

15 Aug. 1950

Dr. Wiener,

In the introduction to *Cybernetics* you indicated that you made an effort to reach the labor unions and interest them in the problems which cybernetics poses.

As a farmer child, young man farm worker in this highly industrialized farming area (the San Joaquin Valley), I wish it were possible for you to be heard by the President's Commission on Migratory Labor, which is now holding open hearings in various parts of the country.

It seems rather obvious that the industrial revolution on the farm is making most farm workers and many farms obsolete. And the shadow of cybernetics, etc. hangs over them.

I believe the particular situation of the migrant in our success-ethic society affords us a unique opportunity to get a hold on our selves. As a migrant worker or producer he's on his way out. As a Migrant, as an individual, an indefinable, an aloneness, he is valuable, if we choose to value ourselves as such.

Of course, from migrant to Migrant requires a transcending motion—requires us to go beyond our identifications with religion, nationality, politics, race, etc. to an open attitude toward man.

In any case, in view of cybernetics it would seem futile to try to integrate the displaced farm workers and farmers into our success ethic way of life.

Too the promises of Stalinism to the peasants and workers of the world are apt to back fire on them. Industrialization will spread to the farms; cybernetics, etc. to the factories, etc. Shall we continue to work along "integrating" lines, me-tooing the Communists, or should we pose a vision of man as Migrant, or some such, hold out the hope that individual Migrants of the world shall be relatively free to devote themselves to their true work of finding out (trying to find out) who they are?

The address of the President's Commission on Migratory Labor is: 1400 Pennsylvania Avenue, N.W., Temps V, Washington 25, D.C. The Commission was directed to inquire into social, economic, health and educational conditions, among migratory workers, both alien and domestic, in the U.S. etc. Your work is relevant to these problems. I hope you will be heard.

J. Printise Womack

201 Park Way
Dinuba, Calif.

Wiener responded as follows [MC, 122]:

September 20, 1950

Mr. J. Printise Womack
201 Park Way
Dinuba, California

Dear Mr. Womack:

I am very interested in your letter of the 15th of August. Since my *Cybernetics* has come out, I have actually got in touch with the Labor Unions through Mr. Walter Reuther. I appreciate very much the problem of itinerant farm labor, but I have not sufficient direct experience of them to be in a position to take action by contacting Washington authorities in the matter myself.

I suggest very strongly that you see my new book, and particularly the chapter on the Second industrial revolution[1]. It may give you ideas for going ahead yourself.

Very sincerely yours,

Norbert Wiener

This correspondence suggests that Wiener might have had better luck with a trade union, perhaps The United Mine Workers of America, grounded on more biblical soil than the UAW. Of course, Wiener's interest in the UAW was based on the belief that the automobile industry would very likely be an early victim of haphazard automatization, and recent history has borne him out.

Wiener also tried to get his concerns known to management, and here he was pleasantly surprised to find considerable interest. He was invited to speak at meetings by several groups, among them the Society

---

1 Wiener was referring to Chapter IX of [50j].

for the Advancement of Management, and the American Society of Mechanical Engineers. His audiences, which often included leading executives, were very sympathetic to his views on the necessity of up-grading large sections of the labor force, and on the dangers of "de-valuation" of the human being and "glorification" of the gadget on the sole basis of work-efficiency. They conceded that the tasks ahead would require considerable cooperation of both labor and management. As Wiener told the cadets of the Armed Forces College, some of the executives sounded "like straight William Morris" [XII, p. 11]. Thus Wiener's educative efforts brought about a general awareness among the leaders of management and labor, and the public at large, of the problems facing the impending age of automatization, but with imper-ceptible practical effects.

B The educational challenge

Wiener's way of averting dislocation and agony during the second Industrial Revolution called for labor and management to "take time-out" to sustain an educational effort in which the upgrading of the work force would keep pace with automatization. Initially, this effort could be confined to worker retraining within the factory or union, but soon new specialized schools would be needed to retrain as well as to re-educate. With the automatization of more and more tasks of higher and higher skills, and the growing need for human labor with more and more engineering know-how, changes in the public school curriculum would be necessitated: more and better courses in languages and sciences, elimination of the spurious and the frivolous. Wiener was fully for the American public school system, but he had been an early observer of its poor performance. In 1935 he had written:

> Our secondary schools are characterized by a slipshodness and amateurish-ness of which we can only think with pain. [35d]

Thus, he realized that the labor-management educational effort would have to extend into the domain of public education, and he was aware of the considerable challenge that this posed.

Holding sway in the public schools was a distorted version of the educational outlook of Professor John Dewey. Dewey stood apart from the great line of philosophers, Plato, Aristotle, Roger Bacon, Aquinas, Descartes, Leibniz, Spinoza, Vico, Kant, Mach, Peirce, Bergson, Russell and Whitehead, all of whom contributed to the advancement of the scientific methodology. Like Rousseau, Marx and Engels, and some other social philosophers, he had at best a hazy idea of the scientific methodology, of the propaedeutic role of mathematics, the importance of strange analogies, and of the distinction between inquiry and stratagem, cf. § 18G. His unhistorical attitude debilitated his efficacy as a reformer. Unappreciative of the important role of moral evil in shaping history, he tended to assume uncritically the contemporary presupposition on the achievability of unlimited human progress by education.

Dewey's faith was inspiring. He wrote:

> I see no ground for criticizing those who regard education religiously. There have been many worse objects of faith and hope than the ideal possibilities of the development of human nature, and more harmful rites and cults than those which constitute a school system. Only if all faith that outruns sight is contemptible can education as an object of religious faith be contemned. This particular form of faith testifies to a generous conception of human nature and to a deep belief in the possibilities of human achievement, in spite of all its past failures and errors. {D4, p. 148}

On the importance of holding onto this faith in the course of action, no cybernetician of Wiener's persuasion could disagree. The disagreement

John Dewey
1859–1952

came in giving the devil his due, and on the nature of the actual outcome of action and reaction. "The paradox of homeostasis is that it always breaks down in the end" Wiener wrote, and T. S. Eliot has reminded us that "sin grows with doing good" {E6, p. 44}. Dewey did not realize the havoc that vested interests could cause by claiming his discipleship. The religion he so eloquently described, brought on precisely what he did not wish:

> ... a mass of dogmas called pedagogy and a mass of ritualistic exercises called school administration. {D4, p. 149}

According to Dewey, the schools "must represent present life":

> —we violate the child's nature and render difficult the best ethical results by introducing the child too abruptly to a number of special studies, of reading, writing, geography, etc., out of relation to this social life.
> —the true science of correlation on the school subjects is not science, not literature, nor history, nor geography, but the child's own social activities. {D4, p. 8}

Science, or the study of nature, must not be the center of education, for,

> ... apart from human activity, nature itself is not a unity; nature itself is a number of diverse objects in space and time, ... {D4, p. 8}
> . . . .
> —the study of science is educational in so far as it brings out the materials and processes which make social life what it is. {D4, p. 11}

Dewey laid stress on expressive and constructive activities in the school, such as "cooking, sewing and manual training", cf. {D4, p. 10}.

The outcome of the actual enforcement of these pragmatic ideas in education was harmful, as indeed Dewey himself realized. It led to (1) the loss of the rigorous core of classical training — Latin, Greek and Euclidean demonstration; (2) a happy-go-lucky approach to the English tongue, and the neglect of foreign languages; (3) shabby treatment of mathematics and the sciences; (4) the insertion of a wide range of over-specialized courses, some vocational but many frivolous.

Wiener expressed strong dislike for the system of electives that took the place of genuine education:

> In general our mathematics teachers do not know mathematics, our language teachers do not know languages.
>   There are a large number of rather non-descript school subjects which combine the greatest possible consumption of time with the least possible acquisition of any permanent tangible training. [57g]

Indeed, the product turned out by the public schools often failed to meet minimal Deweyian or cybernetical standards. Those who crossed the hurdle of the three R's, often tended to view science as an unedifying game played against Nature in order to wrench out its secrets for personal gain. Weak in geography and history, largely ignorant of human culture, and peer-pressed, their picture of life often tended towards cynicism. Their entry into the adult world was marked by the senior prom, rather than by the dignity of a traditional initiation or bar mitzvah, and this carried a social cost.

Wiener underscored the observation of the linguist Otto Jespersen that usage, not norms, should guide the development of language [50j, p. 191]. But in an aberrant educational environment, usage might buttress confusion and reduce linguistic ability. This has indeed happened with many high school graduates, who have to stud their sentences with senseless verbiage such as "you know", "OK", "just", "I mean", etc. This disability in trenchant articulation also carries a social cost, for the clear delivery and reception of messages is intrinsic to the life of man:

> For if the trumpet gives an uncertain sound, who shall prepare himself to the battle? So likewise ye, except ye utter by the tongue words easy to be understood, how shall it be known what is spoken? For ye shall speak into the air . . . Therefore if I know not the meaning of the voice, I shall be unto him that speaketh a barbarian, and he that speaketh shall be a barbarian unto me.
> I Corinthians, 14:8–14:11

Some normative control is clearly called for. Science instructors are rather helpless in this regard.[2] Perhaps extensive reading from the King James edition of the Bible in our English courses, starting with the story of Babel, might be the right antidote.

Wiener knew that this stranglehold on education was being maintained by the teachers' colleges with their licensing privileges. They insisted on teacher training that focused on a "neo-pedagogy" which intellectually belittled and neglected the subject to be taught. Wiener's sentiments were unwittingly expressed in the contents, and especially in the title, "The Emperor's New Clothes or Prius Dementat"[3] of an article written by Professor H. J. Fuller {F6} in the early 1950s. Following it came an interesting book, *Quackery in the Public Schools*, by A. Lynd {L10}.

2   In the 1960s, university instructors in mathematics began calling tensors and matrices "gadgets" instead of "abstractions", presumably to create "rapport" with their students. Nowadays, the term "guys" has replaced "gadgets" and "O.K."s abound. All this has injected a bit of malarkey in the mathematics lecture room that was not there in the writer's student days at Bombay and Harvard.

3   From "Quos Deus vult perdere prius dementat" (Whom the Gods seek to destroy, they first make mad). "The Emperor's Clothes" was Wiener's favorite caption for many an inane painting. He was no more taken in by avantgardism than by neo-pedagogy.

For so fettered an educational system to upgrade itself at an adequate speed to cope with growing automatization in the economy and attendant worker-displacement, was a stupendous undertaking. Ordinarily, the teachers' trade unions would have joined this battle. But Wiener was aware of the unwritten alliance between these unions and teachers' colleges to turn out mathematics teachers who "do not know mathematics" and language teachers who "do not know languages", and to hold on to the educational status quo. Abetting the malaise were also the Parent Teacher Associations, as became clear when they greeted the protests of the colorful Admiral Hyman Rickover against education-al gimmickry with the advice to "stay submerged". (The admiral's views were close to Wiener's.) Wiener offered no concrete suggestions as to how this enormous educational challenge could be met. All he could do was to try to create an enlightened public opinion.

But Wiener's warning fell on deaf ears. Since his death the rate of educational decline has soared, and the present situation borders on the chaotic. The havoc is not just from a slipshod curriculum. Television stupor, drug addiction, pornography, and other paraphernalia, let loose in both home and school by unbridled mammonism, take a heavy toll. During the very period in which the crying need has been to upgrade the population at large educationally and culturally, the actual trend has been down. With us are ailments more severe than those of the first Industrial Revolution against which Wiener had warned more than forty years ago.

C Anti-homeostatic aspects of the American economic system

Wiener questioned the veracity of the belief, held by most Americans and attributed to Adam Smith, that

> in a free market the individual selfishness of the bargainers, each seeking to sell as high and buy as low as possible, will result in the end in a stable dynamics of prices, and will redound to the greatest common good. [61c, p. 158]

He pointed out that the market is an n-person game with no unique choice of an optimal line of play for any individual player, and conse-quently a game with an indeterminate outcome. To begin with:

> The individual players are compelled by their own cupidity to form coali-tions; but these coalitions do not generally establish themselves in any single, determinate way, and usually terminate in a welter of betrayal, turncoatism, and deception, . . . [61c, p. 159]

In short, there is no stability or homeostais. This analysis is based on the premise that the players are completely intelligent and totally selfish. In fact, they are often confused and gullible:

> Where the knaves assemble, there will always be fools, and where fools are present in sufficient numbers, they offer a more profitable object of exploitation for the knaves. Instead of looking out for his ultimate interest, after the fashion of von Neumann's gamesters, the fool operates in a manner which, by and large, is as predictable as the struggles of a rat in a maze. [61c, p. 159]

The actual game is thus a highly volatile and unpredictable process without any trend towards stability.

Wiener was not a socialist, however. He realized that human survival rested on technological mutations that revolutionized the mode of production, and that such occur only where entrepreneurs have both the freedom and the resources to strike out on their own. In such ventures into the uncharted seas of invention and technological possibilities, institutionalized bureaucracies are a hinderance. The short-term economic disequilibrium brought on by such entrepreneurship is an inevitable part of the social life of man, and must not be confused with the volatility stemming from the avaricious conflicts or games discussed in the previous paragraph. Indeed, Wiener knew that the momentum that capitalism imparted to entrepreneurship was its prime virtue, one that offset in part the havoc caused by erraticism in its markets. He saw nothing wrong in financing industrial production, geared to short-term human needs, by the promise of a return on investment.

Wiener, however, clearly saw the absurdity of such investment-for-profit for industrial undertakings meant to last several generations, such as the building of dams or cities or universities, and therefore most definitely did not regard capitalism as a panacea to man's economic problems:

> Suppose, now, that a sum of money at the time of Christ had been left at 2 per cent compound interest; for example, the thirty pieces of silver of Judas. By what factor would it have been multiplied up to the present time? We are approaching the year 2,000 and in order to express our result in round numbers, let us suppose that we are at the year 2,000. Then one dollar at the time of Christ would amount, at 2 per cent, to a quantity with over ninety-seven zeros. At any conceivable scale of evaluation, one cent at the time of Christ, put in a bank at 2 per cent compound interest would amount to something like 10^{84} times all the value of the goods in the world at the present time. This is ridiculous, but it still has a meaning. [62c, p. 30]

The meaning is simply that:

> The sums earned by money put out to interest have been wiped out time and time again by wars, famines, plagues, and other catastrophes. These cata-

strophes have been great enough to wipe out every single commercial undertaking of an antiquity of thousands of years, and *if they had not taken place, the rate of interest for a long-time investment could scarcely have been two-tenths of a per cent.* [62c, p. 30] (emphasis added)

It follows that modern capitalism is able to offer attractive returns on private investment in long-time undertakings only by its condescension to bankruptcies during the down-phases of its periodical trade cycles. For the well-off the resulting losses are often on paper, but they are painfully real to poorer people thrown out of work. Thus the system is not socially homeostatic.

Wiener also pointed out a non-homeostatic feature in American marketing. The American market system depends heavily on so-called *hard-sell promotion.* This consists in the ad nauseam repetition of irrelevant, untrue and sex-slanted blurbs. Such repetition can only influence the gullible. The powerful marketing sector of capitalism, the members of which make their fortunes from such hard-selling technique, and this includes actors, sportsmen and a host of less known spouters of banality, thus has a stake in maintaining mass gullibility, and with it a system of public education, with next to no standards, that caters to it. The marketing system thus exerts a downward pull on educational levels, and with it on the national economy, the national culture, and the national defense, while augmenting iniquity in the distribution of wealth (see p. 93 for illustration). Thus, there is a strongly ingrained anti-homeostasis.

A long-term anti-homeostatic trend in the life of capitalism stems from its unenduring linkage to the so-called Protestant[4] work-ethic. Capitalism is indeed a product of the ethic of hard work and frugality. But soon this ethic meets with competition. The marketing section starts inducing the public to consume and consume, and enthusiastically promotes a "buy now, pay later" mentality, that results in a credit-card sponsored, self-centered hedonism, antithetical to both the work ethic and to frugality. This anti-homeostasis, operating through the moral domain, weakens the foundations of capitalism. The true disciples of the hardwork ethic, the free-enterprising Benedictine monks, the Shakers, the Amish, and others managed their economies without promoting hedonism and without marketing acrobatics.

To sum up, capitalism is anti-homeostatic in four respects:

1. The capitalism market is a volatile *n*-person game, the outcome of which is unpredictable;

---

4   Protestantism has no unique relationship to the hardwork ethic, witness Benedictine monks, Buddhist craftsmen and others.

2. High profits on long-time investments are wiped out during periods of bankruptcy;

3. Hard-sell promotion, dependent on public gullibility, prevents the upgrading of general intelligence required to sustain automatization;

4. Promotion of credit-card prodigality destroys the hard-work ethic and frugality, on which the life of capitalism depends.

Indeed, among the economic systems practiced by mankind, capitalism is the most affected by its bondage to avarice. In the words of Lord Keynes:

> . . . Modern capitalism is absolutely irreligious, without internal union, without much public spirit, often though not always, a mere congeries of possessors and pursuers. {K4, p. 131}

There is however, a more serious inhumane aspect of assembly line production, viz. the deprivation caused by the alienation of work from self. K. Marx and scholars on human civilization have dwelt on it. Wiener certainly sensed it, although he never specifically articulated it. Indeed, one of his hopes was that increasing automatization would mitigate this alienating aspect of the machine. Professor Lewis Mumford has described the problem succinctly:

> Wherever tools and muscle power were freely used, at the command of the workers themselves their labors were varied, rhythmic, and often deeply satisfying, in the way that any purposeful ritual is satisfying. Increase of skill brought immediate subjective satisfaction, and this sense of mastery was confirmed by the created product. The main reward of the craftsman's working day was not wages but the work itself, performed in a social setting.
> . . . .
> Machine culture in its original servile form did not share these life-enhancing propensities: it centered, not on the worker and his life, but on the product, the system of production, and the material or pecuniary gains therefrom. Whether kept in operation by the task-master's whip or by the inexorable progression of today's assembly line, the processes derived from the megamachine worked for speed, uniformity, standardization, quantification. . . The compulsions produced by this system were more insidious than outright slavery, but as with slavery, they finally debased the controllers as well as the working force so controlled. {M18, p. 238, 239}

Forebodings on the moral outcome of capitalism had been voiced as early as the 17th century by Christian thinkers, cf. Tawney {T2}. Its alienating and exploitative character and its economic instability were known to Marx. But Marx's materialism narrowed his intellectual vision. He predicted that the system would collapse from the excessive misery it would inflict on the working class. The mills of God grind by more subtle feedbacks. Some of these Wiener was able to discern, more by shrewd observation, clear thinking, and a humane disposition than by any special expertise. He also realized that all economic systems are within the devil's reach, socialism being no exception.

D Wiener's economic ideas

To a scientist, and especially a humane one, who saw through the pitfalls of both capitalism and socialism, it was an intellectual challenge to work out a socially-responsible, free enterprise, devoid of the dislocative, hedonistic and anti-intellectual propensities of capitalism, a system whose strong homeostasis could absorb technological transitions.

Wiener's non-linear theory of self-organizing systems met with difficulties in its applications to brain-wave encephalography, and unfortunately he made no attempt to apply it in the economic realm. To go by his conversations, he had in mind an economic set-up vaguely resembling the power station of an electric grid, in which "the total generating system acts as if it possessed a virtual governor, more accurate than the governors of the individual generators . . . " [61c, p. 201]. That is, the economy should consist of a number of semi-autonomous units (some of which may be work-motivated rather than profit-motivated, or publicly owned) coupled to a central agency which continually corrects, or even intervenes more directly, to reduce the volatility of the market game, and so approximates a homeostasis conducive to steady economic growth.

Wiener did engage himself a little more concretely on economic questions while at the Indian Statistical Institute in Calcutta during 1955–1956. Social and economic planning were among the prime concerns of the Institute. Wiener felt that he could give a rationale to these problems, and exchanged ideas with Professor Jan Tinbergen, the distinguished econometrician from Holland, who was also staying at the Institute. It is best to let Wiener speak:

> In any economic situation there are certain factors beyond our control which are given statistically. These include the weather, the fertility of the crops, and other factors of the sort. In addition there are certain factors which we can control. For example, the amount of seed grain to be planted, the rate of interest on agricultural loans, etc. The problem of planning is to optimize, or in other words minimize, some quantity depending on the controllable and the uncontrollable statistical factors in such a way that this minimization is maintained on the average. Such a problem is of a statistical character, and therefore of an informational character.
>
> From the economic point of view, this is what we may call the statistically stable planning problem. If we have any planning situation which is meant to continue, it must of necessity lead to a statistically stable planning situation. On the other hand, for a planning situation to be statistically stable, it is not necessarily true that we can arrive at it from existing

conditions in a statistically stable manner. In other words, this concept of statistically stable planning is only part of the social planning problem and must be supplemented by other conditions which enable us to make an effective planning of transient states and which arrive at a statistically stable situation. [58f, pp. 11, 12]

At the Institute Wiener expounded these ideas more fully to a small audience by considering only one factor beyond control, and only one under control. He then reduced the problem to one of bivariate prediction, and to the factorization of a 2×2 matrical spectral density. (To some of his listeners, this came as no surprise.)

E The control of conflict in civil society

Human conflicts and contests also of course occur outside the market place, and lead to games of the von Neumann-Morgenstern type. Judicature and diplomacy are attempts to control them according to the rules of justice operative in the society. As Wiener observes:

> ... the problems of law may be considered communicative and cybernetic —that is, they are problems of orderly and repeatable control of certain situations. [50j, p. 110]

He points out that the language of jurisprudence has an ingrained equivocation, to offset which a large body of legal precedents is necessary. Even so, there is scope for chicanery. Wiener cites the treaty negotiations with the American Indians and criminal law litigation as examples. In fact, much litigation is a three-person game, in which the first move of the two litigants is to win over the third party, to wit, judge and jury [50j, pp. 106–111]

Another area rife with conflict, contest and game-playing is the realm of bureaucracy, governmental or otherwise. Here the equivocation is such that the very nature of the contest and the identities and positions of the players are in doubt. Do president A and his vice-presidents B and C operate as a team, or are B and C rivals, both out to get A's job? Should department A trust or distrust committee b of department B in seeking a certain objective? Often, the answers are buried under cloaks of secrecy, and fluctuate erratically with the passage of time.

Not surprisingly, the writings of C. Northcote Parkinson on public administration {P1} express the cybernetical standpoint on the issue of bureaucracy. His concern with communication and control in bureaucratic organization, his demarcation of bureaucratic diseases, and his direct style all help.[5] Wiener's brief articulation on this subject ignores details and points to a more lasting upshot of the bureaucratic process. He invoked the "Circumlocation Office" from the novel *Little Dorrit* by Charles Dickens. (This Office gives the run-around to Daniel Doyce, the innovative and honest craftsman who wants to get a patent for his invention.) The reference to *Little Dorrit* occurs in the course of Wiener's discussion of patent law [50j, pp. 113–115], but it serves to show up one of the more lasting drawbacks of the bureaucratic process, viz. its stifling of initiative. This strong anti-homeostatic factor is ingrained in existing systems of public administration.

The mechanical engineer prescribes definite forms of lubrication for his machinery to compensate for friction, and the communication engineer incorporates elaborate filters in his amplifiers to dampen noise. The public manager, however, leaves equally significant dissipating factors in the economic and political process unattended. Everyone knows that the execution of a major economic or political undertaking can be distorted by bureaucratic dissipation. Concerning such dissipation is a world literature ranging from the Hindu book of animal fables, the *Hitopedesa* (500 B.C.), and the treatise, the *Arthasastra* (300 B.C.), to the present day humorous writings of Parkinson. Yet as of now no social scientist has tried to represent this persistent and significant factor in his equations and flow diagrams. The measurement of bureaucratic dissipation, and the design of filters to eradicate it, is an outstanding problem in modern political economy that continues to get pushed under the carpet.

While the games mentioned in the previous paragraphs are genuine, there exists a tendency to see game-playing where there is none. For instance, in his 1976 book, D. Bell writes:

> Life in pre-industrial societies . . . is primarily a *game against nature.* . . .
> Industrial societies . . . play a *game against fabricated nature.* . . .
> A post-industrial society . . . is a *game between persons.* {B4, pp. 147, 148}
> (author's italics)[6]

Very important events in the life of early man, such as the discoveries of fire, the wheel and writing, were not games at all but results of inquiry

---

5 His occasional humorous use of symbolism, for instance, $I_3 + J_5 \rightarrow I_3 J_5$, for the bureaucratic-chemical reaction in which the elements, incompetence I, and jealousy J, when mixed in the right proportions, yield the compound *injelitance*, $I_3 J_5$, {P1, p. 102}, adds to the cybernetician's delight.
6 Bell's classification, viz. pre-industrial, industrial and post-industrial, is less intrinsic than Wiener's classification: zeroth, first, second, . . . industrial revolutions, based on whether the tools of production exemplify the amplification of force, power, intelligence, . . ., cf. § 19A.

devoid of stratagem, cf. § 18G. The stochastic nature of the cosmos does not turn natural processes into games. The uncertainties of weather, for instance, do not make the struggle of farming a game. Nor is the process of manufacturing a car a game against anything. The only players against whom man can play games are the animals whom he domesticates or kills or pushes around, and other men with whom he has conflicting relations. (Friendly games such as bridge are another matter.) The latter conflicts, social, economic or political, were as much a part of the life of early man as they are of modern life.

F Communal information

The existence of Social Science is based on the ability to treat a social group as an organization and not as an agglomeration. Communication is the center that makes organizations. Communication alone enables a group to think together, and to act together.

What is true for the unity of a group of people, is equally true for the individual integrity of each person. The various elements which make up each personality are in continual communication with each other and affect each other through control mechanisms which themselves have the nature of communication. N. Wiener, cf. {D2, p. 77}

Of all these homeostatic factors in society, the control of the means of communication is the most effective and the most important. [61c, p. 160]

Wiener was speaking of societies, too large for direct contact between individuals, for which the press (books and newspapers), the radio, the television, the theater, the schools and the church are essential for communication. The last is what holds a community together, its *cohesiveness* being dependent on the amount of easily retrievable *communal information*. Information is *communal,* if it so affects the interrelationship between two or more members, that this minority behavioral change is noticed by the rest of the community, and spreads to the majority. See [61c, pp. 157, 158].

Wiener pointed out that *information* is not a commodity, i.e. an entity easily passable from hand to hand without loss of value, and that this value is not an extensive (i.e. finitely additive) quantity such as the volume or the weight [50j, p. 116]. In a closed system information decreases spontaneously (Entropy Law). It is not storable without "overwhelming degeneration of its value" and is useful only when it can freely circulate, and assist the growth of knowledge [50j, p. 120].

It is dangerous, Wiener claimed, to leave the apportioning of the channels of communication to the anti-homeostatic mechanism of the market. For then, the ownership of a channel will fall in the hands of those who operate it for its *secondary* benefits, viz. monetary gain, and not for its *primary* benefit: the furtherance of socially conducive communal information. "The man who pays the piper, calls the tune." The channels will be cluttered by the repetition of half-true, sex-symbolic blurbs; in short, with misinformation. Worse still, as we saw in § 19C, the channel-owners will have a stake in the maintenance of mass gullibility and in the continuation of an indifferent educational set-up that caters to it, cf. [61c, pp. 161, 162]. The result will be not informative communication, but noise and misinformation; not homeostasis, but conflict.

On the other hand, the considerably greater homeostasis of the small communities such as those in New England, where Wiener grew up, and which he came to love, cf. [53h, Ch. VII], stems from the long-time sharing of uncontaminated communal information, and the uniform levels of intelligence, behavior and altruism that this engenders. In such a community:

> The average man is quite reasonably intelligent concerning subjects which come to his direct attention and quite reasonably altruistic in matters of public benefit or private suffering which are brought before his own eyes. In a small country community which has been running long enough to have developed somewhat uniform levels of intelligence and behavior, there is a very respectable standard of care for the unfortunate, of administration of roads and other public facilities, of tolerance for those who have offended once or twice against society . . . in such a community, it does not do for a man to have the habit of overreaching his neighbors. There are ways of making him feel the weight of public opinion. [61c, p. 160]

In these communities the situation is not all favorable for the "merchants of lies" and "exploiters of gullibility", for even were a resident foolish or knavish, the prospects of his acting wisely are considerably enhanced by the forces of cohesiveness and communal wisdom.

G The grammar of the social sciences

> . . . While human and social communication are extremely complicated in comparison to the existing patterns of machine communication, they are subject to the same grammar; and this grammar has received its highest technical development when applied to the simpler content of the machine. N. Wiener, cf. {D2, p. 77}

The complexity in the social fields, to which Wiener refers, comes from the strong coupling therein between the observer and the observed, and from the proximity of the periods of their natural rhythms.

The methods of precise science have worked only in fields characterized by a very *loose coupling of observer and observed,* i.e. in which observation does not significantly disturb the observed object. Our astronomical observations have no influence on astronomical phenomena, and the same applies to a large extent to our observations of terrestrial macroscopic phenomena. On the other hand, our observations of atomic phenomena do disturb them irrevocably. But because the spatio-temporal scale on which we operate is so much larger than that on which the atomic individuals do, our macro-physics deals with the *average effects* caused by an enormous number of atomic events, and so cancels out the random disturbances brought on by observation. Statistical physics takes over.

> In short, we are too small to influence the stars in their courses, and too large to care about anything but the mass effects of molecules, atoms and electrons. [61c, p. 163]

The situation is different in sociology. First, observation often exerts a considerable influence on what is observed. For instance, the observation and publication of the price of wheat in the commodities market in Chicago affects the behavior of traders, which in turn affects the future price of wheat. The anthropological investigation of a secluded primitive community may affect it irrevocably, were its seclusion an essential concomitant to its very structure and survival. Secondly, the period of the natural rhythms of many sociological phenomena are long and their spatial ranges are large. But over these large tracks of space-time, the relevant (sociological) conditions are *not* constant, as they are in meteorology, for instance. Thus a sociological time series, sufficiently long for statistical analysis, is (unlike a meteorological time-series) liable to be non-stationary.

Briefly, in sociology, both the *strong coupling of the observer and the observed* and the *relative spatio-temporal scales* conspire against the use of the methods of deterministic physics based on differential and integral equations, as well as against the methods of statistical physics based on stationary stochastic analysis. The methods of *precise science* will not work because we are in the realm of what Wiener called the *semi-exact sciences*.

To be of use, the mathematical theories have to be intrinsically adapted to the social fields, with new concepts specific to the subject matter. One such is the game theory of von Neumann and Morgenstern {V7}. Another is P. F. Lazarsfeld's probabilistic explication of sociological terms such as "cohesiveness" by means of the covariance matrices and covariance tensors of their so-called indicators. A third is the theory of psychological tests of the random ink-blot variety. Note that it was Wiener, who in 1930 pioneered the use of "coherency matrices" in optics [30a, § 9], and also alluded to the usefulness of "a purely chaotic sequence of acoustical impulses" in black-box analysis, cf. § 18F.

Wiener was convinced that the social sciences were *semi-exact,* and while limited areas within them could be handled by such appropriate mathematics, the main mode of discourse had to be largely narrative as in the study of history. Wiener was especially skeptical of attempts to treat economics as an exact science. Speaking of "elusive quantities" such as supply, demand and unemployment, he wrote:

> Very few econometricians are aware that if they are to imitate the procedure of modern physics and not its mere appearances, a mathematical economics must begin with a critical account of these quantitative notions and the means adopted for collecting and measuring them. [64e, p. 90]

Wiener argued that many an economic quantity, such as "the demand for steel" x, is brought in without being given a clear operational definition. Even granting that it is well-defined, its time series will be far from stationary. Disturbing the stationarity of the time series of x, for instance, are all changes in the technique of steel production or of steel usage, all inventions affecting the supply of competing materials, as well as certain changes in national policy. As for the time series of commodity prices, e.g. wheat, Wiener cited the work of Dr. Benoit Mandelbrot which indicated that

> ... the intimate way in which the commodity market is both theoretically and practically subject to random fluctuations arriving from the very contemplation of its own irregularities is something much wilder and much deeper than has been supposed, and that the usual continuous approximations to the dynamics of the market must be applied with much more caution than has usually been the case, or not at all. [64e, p. 92]

Here Wiener felt that the appropriate economic analysis had to embody the fractal concept of Mandelbrot. Dissipative factors also pose a problem in the social fields as we saw in § 19E.

Such considerations, rooted in the concepts of complementarity, time scale discrepancy and the persistency of dissipation (§ 19E), led Wiener to reject systematic social engineering as a possible solution to our social ills. Such engineering demands more precision and a looser coupling between the social engineer and his object than is possible in the social fields. Recall that Wiener had learned the Complementarity Principle the hard way, while grappling with the difficulties of generalized harmonic analysis and quantum mechanics in the late 1920s, and that his awareness of time scales and chaotic time series, acquired from his studies in the early 1920s, underwent severe testing in his war work (Ch. 14). Wiener's thoughts on sociology thus bore the impress of ideas he had acquired in the course of his deepest and hardest mathematical research.

H Wiener and Marxism

According to Wiener, science is a means by which mankind maintains its rapport with nature.[7] The rapport is dynamic: it demands a continual man-nature interaction in which both the human community and its environment get transformed. The cybernetical epistemology itself stresses this active element: as Ashby put it, cybernetics asks not "what is this thing?" but "what does it do?" {A6, p. 1}. Thus, sensible action geared towards the improvement of human institutions and the human environment, in which scientists participate, is a practical imperative of cybernetics. Wiener would have readily affirmed Karl Marx's thesis: "The philosophers have *interpreted* the world in various ways, the point however is to *change* it" {M6, p. 84}. He would have also echoed Marx when he asserted:

> The question whether objective truth can be attributed to human thinking is not a question of theory but is a practical question. In practice man must prove the truth, i.e. the reality and power, the "this-sidedness" of his thinking. {M6, p. 82}

For Wiener, like most scientists, believed in the unity of theory and practice.

While Wiener was not acquainted with Marx's theories of surplus value and its expropriation, he was aware of the exploitation and alienation of labor. {§§ 19A, C} He knew that technological advances such as the introduction of gas and electric lighting had led to the imposition of night work in stuffy factories, and often worsened the worker's life. To the question asked at the turn of the century by the French Marxist, Karl's son-in-law, Paul Lafargue:

> When Science subdued the forces of nature to the service of man, ought she not to have given leisure to the workers that they might develop themselves physically and intellectually; ought she not to have changed the "vale of tears" into a dwelling place of peace and joy? I ask you, has not Science failed in her mission of emancipation? {L1, pp. 25, 26}

Wiener would have replied affirmatively.

Thus Wiener was in sympathy with the enlightened principles of Marxism. But Wiener, unlike many Marxists, realized that to attain effective contact with reality, it is necessary to have faith in the rational Logos—Spinoza's God, as Einstein would say, and to adhere to the Pythagorean-Platonic tradition of viewing reality as a mixture of message and noise, then dealing with the (non-existent) message, and finally bringing these ideal considerations to bear on reality, cf. § 18C. He could

---

7   Prolegomena to Theology, 1961, unpublished manuscript MC, 877–881. The relevant text is quoted on p. 358.

not forget what he had learned from Perrin that "the logic of the mathematicians has kept them nearer to reality than the practical representations of the physicists".

Furthermore, Wiener could not accept the dialectic materialist position according to which the "paramount question of the whole of philosophy" is: "Which is primary, Spirit or Nature?" (Engels {E7, p. 21}), and that "the answers which philosophers gave to the question split them into two great camps" {E7, p. 21}. To Wiener and Ashby, this question, far from being paramount, was largely irrelevant. The line between spirit and nature is fugitive as the great Marxist, J.B.S. Haldane understood, cf. § 10D. The ontological differentiation, idealism and materialism, is inconsequential as Wiener noted:

> I see no essential difference between a materialism, which includes soul as a complicated type of material particle, and a spiritualism that includes particles as a primitive type of soul. [34c, p. 480].

In short,

dialectic materialism = dialectic spiritualism.

The first requirement for a sound philosophy is that it shake off the shackles of ontology, and seek a Leibnizian universalism.

As to the actual implementation of dialectical materialism in the Soviet Union, Wiener could hardly accept as enlightened a philosophical position that condemned the *Principia Mathematica* till the 1960s, settled for a pseudo-scientific hodgepodge in genetics in the 1950s, and called cybernetics "reactionary" until 1960—a position, moreover, that tended to confuse the wisdom of the myth with superstition, and had difficulty imbibing peaks of human endeavour such as the epics of Homer, the *Divine Comedy* of Dante, the *Paradise Lost* of Milton, Michelangelo's non-secular frescoes and Bach's *St. Matthew Passion*. Wiener was not gullible enough to accept all that Marxism had come to mean, and was skeptical of those who did so uncritically, and this included most communists. Today, of course, many communists acknowledge the distortions their movement underwent under the fire of experience. The solution, however, lies not in a *glasnost* (openness) applied to message and noise alike, but in a *tchestnost* (honesty) that recognizes the efficacious aspects of Pythagorean-Platonism.

Professor Heims's description of Wiener's attitude on the cold war (involving some words of Struik {S10, p. 37}) as being one of "a plague on both your houses", cf. {H6, p. 311} is grossly inadequate. Wiener understood the anti-homeostasis latent in capitalism (§ 19C). But he could hardly hail a political economy that trailed in advanced technology and automatization, suffered food shortages, and had to keep importing food and advanced technology from capitalist markets.

Nor could Wiener swallow the claim that the cold war represented a class struggle between oppressors and oppressed, when what was hampering the struggle of the oppressed was dissension among communists and chronic bureaucratism. Thus he had little to choose between the two "isms". However, Wiener had great respect for the humane aspects of Russian Socialism—such as free housing for the poor—as he had for the freedom and democracy he loved in American towns. He was a humanist. If he ever declared blight on the two houses, it could only have been on the bureaucratic diseases that have corrupted their high ideals.

Wiener had an abiding love for American democracy and was ardently patriotic. He tells us that from the start "he was repelled by the totalitarianism of the communists", but recognized that "the appeal of communism" felt by his younger colleagues "was an appeal to their humanitarian instincts" [56g, pp. 220–221]. With the humane side of Marxism he had no quarrel. His objections to other aspects of Marxism came not from adherence to a fixed liberal democratic position, but from plain clearheadedness and objectivity, and from an honest disbelief in groups that are afraid to trim the beards of their favorite prophets.

20 Wiener on Global Policy and Military Science; von Neumann's Position; Science and Human Welfare

A Wiener's thoughts on geopolitics

The calamity of Hiroshima pulled Wiener and von Neumann in the same general direction: both became public figures who endeavored to influence the course of events. But the paths they followed were guided by differing conceptions of what constituted good governmental policy.

Wiener's strategic conceptions were molded by his understanding of the impending age of automatization and were at variance with those of the Truman and Eisenhower regimes. He felt that reliance on nuclear superiority offered too flimsy a foundation for a sound foreign policy and defense policy. He was critical of the big money spenders in Washington, for he felt that much of the money would go down the drain, and while possibly corrupting its dispensers, would not materially improve the security position of the United States vis-a-vis its adversaries. He decided to devote himself to the dissemination of cybernetical ideas, especially those bearing on public policy, cf. §§ 19A, B.

Wiener's thoughts on the complex issues of national and global policy are not easy to gauge. They are scattered in bits and pieces over several articles, books and letters. This difficulty is augmented by Wiener's occasional lapses into immature articulation.

Wiener's desperate attempts to enlist in World War I, his happy days at the U.S. Army Proving Grounds in Aberdeen, his prompt and affirmative response to Bush's 1940 memorandum on war work, and his involvement with military projects during World Wars I and II, not to mention his very concept of duty (Chs. 6, 14) suggest, if anything, a militaristic stance. Nor did he become a pacifist after Hiroshima, an impression one might get from Professor Ulam's book {U2, p. 137}. He continued to collaborate on military projects for the Office of Naval Research and for the Industrial College of the Armed Forces, witness the Reports ATI 162232, A3-011433 of November 1947, and February 1953. Free from anti-soldier prejudice, his relations with the armed forces were cordial until the end. He readily accepted an invitation to lecture at the Industrial College of the Armed Forces as late as September 1963, and the thank-you letter he received from the Deputy-Commander, Major-General T. R. Stoughton, was profuse:

> Your lecture was exceptionally well received and there have been many favorable comments regarding it. The students were highly impressed with your enthusiasm, the vitality and comprehensiveness of your presentation and your candid responses to their questions.
>
> We are most grateful that you found time in your busy schedule to address the College, and we consider ourselves honored to have had the benefit of your wide experience. [MC, 330]

The wrong impression that Wiener was hostile towards the U.S. military after the war came from the attention attracted by his exaggerated verbal reactions, and by his two immature articulations [47b, 48d], and too little attention to his more mature judgments. The reactions sometimes took the form of his turning down requests for de-classified reports of his defense work, on the presumption that they would be used to make new weapons of destruction. Such was his response, for instance, to requests from Dr. G. E. Forsyth (of December 3, 1946 [MC, 72] and Professor Oskar Morgenstern (of January 3, 1947 [MC, 75]). In his replies Wiener pointed out that the reports could be procured directly from Washington.

The reply to Forsyth had more, however, and its content was reprinted in the *Atlantic Monthly* and the *Bulletin of Atomic Scientists,* cf. [47b]. In it Wiener spoke of "the tragic insolence of the military mind" and wrote:

> I do not expect to publish any future work of mine, which may do damage in the hands of irresponsible militarists. [47b]

These are among the worst of Wiener's articulations. We are left in the dark as to who these "military minds" and "militarists" are. Are they primarily the civilians who decide defense policy at the highest levels, or are they senior officers in uniform who advise on the implementation of policy?

The story of the atomic bomb shows the difficulty. Let us begin at the beginning. What initiated the Manhattan Project was a letter from Einstein and Szilard to President Roosevelt, and the recommendation to drop the atomic bomb without warning came from a committee (appointed in spring 1945 by H. L. Stimson, the U.S. Secretary of War) that was headed by Wiener's distinguished colleague and friend Dr. Vannevar Bush. It was accepted by President Turman, whose military career was confined to World War I, and who can hardly be described as a "military mind". This decision had been opposed by his chief of staff, the Fleet Admiral W. D. Leahy, a "military mind", but supported by General George C. Marshall, another "military mind". Before the first Soviet nuclear explosion in 1949, when the option of preventive nuclear strikes on the Soviet Union was in the air, among the supporters of such strikes were scientists of the caliber of J. von Neumann (cf. Blair {B12}), and Wiener's teacher, Bertrand Russell {R8, vol. III, pp. 7, 8}, a World War I pacifist and, all told, one of the humane philosophers of this century. Perhaps "the tragic insolence of some *civilian* minds", or "the impudence of wealth and arrogance of power", would fit the facts a little better. Wiener's article has several such ambiguities. Some of these positions he reiterated in another short, but noisy article [48d]. (These articles as well as an exposure of their misleading nature appear in the *Coll. Works*, IV.)

In the early 1950s, however, Wiener the philosopher took over. He sensed the futility of the position he had taken:

> I tried to see where my duties led me, and if by any chance I ought to exercise a right of personal secrecy parallel to the right of governmental secrecy assumed in high quarters, suppressing my ideas and the work I had done.
>
> After toying with the notion for some time, I came to the conclusion that this was impossible, for *the ideas which I possessed belonged to the times rather than to myself*. If I had been able to suppress every word of what I had done, they were bound to reappear in the work of other people, very possibly in a form in which the philosophic significance and the social dangers would be stressed less. I could not get off the back of this bronco, so there was nothing for me to do but to ride it. [56g, p. 308] (emphasis added)

Wisely, pursuing his mathematical interests, Wiener went on to publish work in multivariate prediction [50b, 55b, 57d, 58b, 59a, 59c], even though its applicability to seismic and underwater signals gave it a defense potential. Indeed it had this potential, as is clear from the paper "Mathematical development of discrete filters for detection of nuclear explosions" by E.A. Robinson, *J. Geophys. Res.* 68 (1963), 5559–5567.

From this evidence we have to conclude that the letter to Forsyth and its imprinting in [47b] reflected an immature and exaggerated reaction. It must therefore be discounted in determining Wiener's true attitude on military matters.

Wiener's authentic stance towards the military can only be gleaned from his more mature judgments. One such concerns the atomic bombing of Japan. Speaking of the sudden end this brought to the Japanese war "without heavy casualties on our part", he wrote:

> Yet even this gratifying news left me in a state of profound disquiet. I knew very well the tendency (which is not confined to America, though it is extremely strong here) to regard a war in the light of a glorified football game, at which at some period the final score is in, and which we have to count as either a definite victory or a definite defeat. I knew that this attitude of dividing history into separate blocks, each contained within itself, is by no means weakest in the Army and Navy.

> But to me this *episodic view of history* seemed completely superficial. The most important thing about the atomic bomb was, in my opinion, *not the termination of a specific war without undue casualties on our part, but the fact that we were now confronted with a new world and new possibilities* with which we should have to live ever after. [56g, p. 299] (emphasis added)

Here Wiener asserts that (1) the football game concept of war is silly; (2) such a concept is popular in our army and navy; (3) of the two effects of dropping the atomic bomb—reduced American casualties, and the commencement of a future conflict—the latter is just as important.

The assertion (1) is supported by the doctrine in General Karl von Clausewitz's 1832 classic on war {C10}[1], and today it should be obvious to a competent global strategist that modern wars, unlike football games, form one type of weave that intermittently appears within the growing fabric of world history. The assertion (2) was true when Wiener wrote, and still is, it would seem. As for (3), the fact that the United States chose to annihilate a couple of cities to save its troops from combat triggered off a set of military responses from other powers. The costliness of future struggles that might ensue from these responses should have been carefully weighed in making the decision to exercise the atomic option. Indeed, a responsible body of opinion (including Admiral Leahy, Chief of Staff to Presidents Roosevelt and Truman) has argued that a frontal attack on the mainland could have been averted, American casualties significantly reduced, and the civilian population spared, by a sustained naval blockade and by strategic air-strikes solely against industrial targets. Captain Sir Basil Liddell Hart has pointed out that by July 1945, the United States was made aware of Japan's desire to end the war, by the appointment of a new government headed by Prince Konoye {L9, pp. 694–695}. The atomic bomb was dropped on August 6.

In short, if we ignore the tempests of the late 1940s, Wiener's judgments reveal a good grasp of military realities.

Another of Wiener's serious concerns was the destructive effect of war on scientific morale:

> Before the war, and particularly during the depression, positions in science were not easy to get. The requirements for these positions had become exceedingly high. During the war, this situation had changed in two respects. First, there were not enough men to carry out all the scientific projects which the war involved. Secondly, in order to carry out these projects at all, it became necessary to organize the work so as to use those with a minimum amount of training, ability, and devotion.
>
> The result was that young men who should have been thinking of preparing themselves in a long-time way for their careers lived in a light-

1 See Note on pp. 316–317.

hearted way from hand to mouth, confident that the existing boom in scientists had come to stay. Such men were in no state to accept the discipline or hard work, and they evaluated whatever intellectual promise they might have as if it had been already realized in performance. With the older men crying out for assistance and manpower, these boys would shop around for those masters who would demand least and grant them the most in indulgence and flattery.

This was a part of a general breakdown of the decencies in science which continues to the present day. In most previous times, the personnel of science had been seeded by the austerity of the work and the scantiness of the pickings. There is a passage in Tennyson's "Northern Farmer: New Style" which says "Doänt thou marry for munny, but goä wheer munny is!" [56g, p. 271]

Wiener also believed that more harm than good came from insistence on undue secrecy in defense work, since it tended to lower quality by inhibiting the flow of ideas as well as the flow of criticism. His belief was backed by his experiences in his own defense work in World War II, cf. § 14G.

There is nothing in all this that is "anti-militaristic" or to which a good battle commander could object. No commander worth his salt would settle for troops with a "minimum amount of training, ability and devotion", who "shop" for easy and flattering masters, or who would let his channels of communication be clogged by bureaucratic interference.

As for Wiener's non-association with and public criticism of the leadership in Washington, this reflected a disapproval of policy, not of the constitutional form of the American government nor of the Department of Defense per se. One can be pro-defense without being pro-Pentagon, witness Admiral Hyman Rickover. Wiener had an abiding love for American democracy—especially as it affected the small town, cf. Ch. 3, and was ever ready to rally to its defense, as we know from his conduct during the two wars.

Wiener's views on atomic bomb diplomacy and his iconoclastic views on the American economy (§ 19C) were not in tune with those of the Truman and Eisenhower regimes. Finding official channels uncongenial, Wiener decided to devote himself to alerting the public to the socio-economic problems that the impending age of automatization would pose, and this led to his contacts with Walter Reuther, as we saw (§ 19A). With the advent of the Korean War in 1950, however, Reuther's attention shifted to geopolitical issues, and Wiener's followed suit.

B Wiener's letter to Walter Reuther on geopolitics, and von Clausewitz's principles on war

Wiener's geopolitical views appear in paragraphs 2 to 12 of his five-page reply to Walter Reuther's letter of July 18, 1950. In his letter (which is quoted in § 19A) Reuther sought comments from Wiener on a proposal that Reuther had submitted to President Truman, entitled "A total peace offensive". Wiener's reply, to which paragraph numbers have been added for ease of reference, reads as follows [MC, 121]:

Mr. Walter P. Reuther
President UAW-CIO

July 26, 1950

Dear Mr. Reuther:

1. I have received the galley proofs of your article, A TOTAL PEACE OFFENSIVE. I do not pretend to be a good judge of practical policy, and you are far better able to judge than I am, how likely it is that you can muster the force and influence to get your policies carried out, and what the ways and means may be. As to the importance of your objectives, and the necessity of stressing them, I have no doubt whatever.

2. I do not think that the average American has much idea of the difference between the present conflict and all others in which we have been engaged. Let us begin with the purely military aspects. In all previous foreign wars, except those two ridiculously unequal conflicts, the Mexican war and the Spanish war, we have not been the main object of the hostility of our enemies. In the Revolution, even before France and Spain became our allies, they were more important to England as possible enemies than we could ever be. In the War of 1812, we were merely a side issue to Napoleon. We did not enter into the first World War until, to put it mildly, the French and British picadores and bandilleros had sapped the strength of the German bull, and it was not a matter of excessive difficulty to give the home thrust with our estoque. This was almost equally true of the European part of the second World War; while Japan, our other antagonist, was too far removed from us in comparison with its own weapons of attack to threaten our homeland seriously at any time.

3. Only once before have we faced the main strength of a serious enemy. This was in the Civil War, which we only won at the cost of five years of bleeding and bitterness, and which we very nearly did not win at all. Like the present conflict, it was an ideological one: the freedom of the slave against states' rights. We could never have won it if the dogged fervor of the North had not matched the elan of the South; and even then, we should not have won it if the population and the industrial resources of the North had not been immeasurably greater than those of the South.

4. We are now in a conflict with the forces of the Soviet, call it Cold War, Korean War, or what you will, which is unlike any situation in which we have previously found ourselves. We are not facing Russia across the buffer of a nearer ally. More than half the world has been devastated by two wars, and is sick and tired of war. Another war means for them a final exit from the stage of history. Their cities are no more blasted than their will, as ours would be and will be after two succesive generations who have been through five years of Hell. If we are to fight Russia, we may be sure that we shall have to carry nine tenths of the burden of war against Russia and its satellites on our own backs.

5. Yet the good will of the remaining countries of the world is not indifferent to us. If they decide to stand aside, regarding Russia and the United States as two amoral collossi fighting it out in their own back yards, or if they see more hope in Russia than in the United States, let us remember that we are inferior to Russia in population, not their superiors in martial fervor and strategic competence, and well matched by them in the ability, to equip and maintain an army. As to the vaunted

American "knowhow", it is our own special way of bearing our breasts and scaring our enemies, like the French "panache" and the British "doggedness". The quality is there, but it is not an exclusive American possession.

6. As to the atomic bomb, it is in the long run our greatest liability. Asia associates our use of it in Japan with a point of view in which yellow women and children are worth less than white ones, and such soldier terms as "gook" do nothing to dissipate this impression. To use the bomb again will put the seal of permanency on our growing reputation for stupid brutality. It makes no difference that the Russians are delighted for us to have acquired this reputation. This is all the greater reason for not playing into their hands. I say that the use of the atomic bomb in the last war was the work of a fool, and that its repeated use can only be the work of an enemy of the United States.

7. Apart from this, from the purely technical point of view, we have over-reached ourselves in trusting to the bomb as a weapon. As industrial countries go, we are old, and have established an economic system in which a vital part of our factory potential is placed in our great cities: in exactly those places where the first killing effect of an atomic bomb would be supplemented by the inevitable secondary disorganization and panic which it will produce. Our very military authorities, in expressing their confidence in our defenses, explicitly do not undertake to give us the assurance that no Russian atomic bomb will land on its target.

8. Unlike our own, the Russian industrial potential has come late to the scene, and has been developed in an era in which mass bombing, if not atomic bombing, has been continually contemplated. An appreciable part of it has been developed under the potential threat of the atomic bomb. In a country of vast distances, it is unthinkable that this new order has identified the centers of industry with the great cities to the extent to which we have done it. Thus in comparing the Russian atomic strength with ours, it is a vast underestimation of the Russian situation to count bomb against bomb.

9. Even now, a full-scale third world war will be for us a fight against odds. The first conclusion is that if it comes, we should do nothing to hasten it, and that our civilian defenses need as much bolstering as our military defense. Strictly speaking, as long as we have not decentralized our cities, we are not on a war footing, no matter how strong our armed forces are. Nevertheless, this decentralization is so long and expensive a job that it is not practical politics in less than a decade or so. The conclusion is obvious: if we have to fight at the periphery of our strength, let us do so, let us gather strength, and let us abide the issue. However, let us do nothing to hurry on Armageddon, hoping, as is very probably true, that the Russians do not yet care to involve themselves in the protracted and mutually destructive conflict that this would be.

10. The preparation for war and its expenditures mean at least a slowing up of social progress, and perhaps a reversal in its tide. We must not forget that there are elements in this country which regard this slowing up, and this reversal, with sardonic glee. It is the chance of a certain type of business man, and of a certain type of military man, to get rid once for all of the labor unions, of all forms of socialization, and of all restrictions to individual profiteering from below. It is a trend which may easily be turned into fascism. This we do at our own peril, for it is only the personal and moral advantage of the American way of life over what the Soviets can offer us, which makes it worth while for the average man, the body of the country, to undergo the hardships and dangers of years of prospective conflict, rather than to surrender at once.

11. Even though we can look to the democratic countries only for a limited immediate military help, it is far from a matter of indifference to cultivate their good will, as well as the ultimate good will of countries which are at present Soviet satellites,

by offering them substantive financial and organizational help, as well as the even more important intangible help which consists in the support of their developed and inchoate liberal institutions. Certainly there is nothing to be said for the sabotage which we have applied to all countries which like England, have accepted any form of socialism. We must also avoid the backing of discredited and —[2] regimes, such as that of Chiang Kai Shek in China. We must show enough cooperative interest in the problems of economically backward and primitive countries, such as India or China, to make these countries feel that we can give them a more promising and more secure future than can Russia. We must be alert to all plans in Europe to lay aside old hatreds, and to unite.

12. If we do these things, and do not merely attempt to "sell" America to Europe by a propaganda as shallow, as stupid, and as lying as any cigarette advertising campaign, we shall find a strong underground to work for us, not only in the countries at present neutral, but even in some of the Russian satellites. We shall eventually find divisions of soldiers rallying to our aid, and we shall have a good fighting chance to survive in a world fit for us to live in—for this is the maximum of what winning can mean with modern weapons. If however we fail to realize that we can win the world only by accepting as ours the interests of the world, moral as well as material, then we shall perish, as we shall deserve to perish.

13. This is my comment. Use it as you will, provided that if you publish it, you keep it essentially together, and that you consult me.

Sincerely yours,

Norbert Wiener

One of the important principles in General Karl von Clausewitz's classic on warfare {C10}, asserts that the "most decisive act of the statesman or general is to understand the *kind of war in which he is engaging, and not to take it for something else*"[3]. In paragraphs 2, 3, 4 of his letter to Reuther, Wiener (who probably never read Clausewitz) devotes himself to an exposition of "the kind of war" that the United States would be facing in the 1950s. Wiener's letter is Clausewitzian in other respects. Absent in toto is the episodic interpretation of war as a "game", which starts when a gun is fired and ends when a flag marked "victory" or "defeat" is waved. War is seen in both its military and political phases as an ongoing contest. Wiener's perspective is global, and does not (as in bad military planning) leave out or misread salient factors such as the needs, attitudes and aspirations of the populaces and the state of their morale. This Clausewitzian tone of the letter is not surprising, for some of Clausewitz's ideas, such as war being a trinity, one term of which, to wit its chaotic aspect, "makes it a free activity of the soul" come rather close to Wiener's ideas, such as the dependence of freedom on contingency. As the letter shows, Wiener instinctively veered

---

2   Illegible word.
3   Cf. Note on pp. 316–317, and the principle No. 9. To appreciate what follows, the reader should glance through the list of principles.

towards the *real* as opposed to the *absolute* side of war, in Clausewitz's terminology.

Noteworthy are the paragraphs 11 and 12 of the letter. Wiener stresses the importance of defining the terms "enemy" and "ally" intelligently, and of resisting the attempt to "sell" America by silly propaganda. Rather we must take a worldview and cultivate the goodwill of the people of Europe on both sides of the curtain, and win the United States a strong underground within the far side and also perhaps "divisions" of their soldiers. In short, "We can win the world only by accepting as ours the interests of the world".

In gauging this letter we must remember that it was written 37 years ago, drafted within a few days of the receipt of Reuther's proposal, and that Wiener was not privy to classified information on the disposition of Soviet economic and military might. Quite possibly some of his premises were wrong and with them the conclusions he drew. Furthermore, our comment with regard to his work on prosthesis (§ 16D) of profundity in-the-large but not necessarily in-the-small, is applicable to other aspects of his work. The letter does reveal, however, a penetrating appreciation of geopolitical realities, and exemplifies Wiener's intellectual integrity and courage.

C Type-classification in military science

Parts of Wiener's understanding of war, not covered in his letter to Reuther, are broached in his pithy address to the American Association for the Advancement of Science in December 1959 [60d] on the game-playing propensities of learning automata. One such part concerned the attribution of types, à la Russell, to the military categories, tactics, strategy, super-strategy, etc. and another part concerned the location of the von Neumann-Morgenstern theory of games within this hierarchy. Underlying these ideas was Wiener's acute understanding of the mechanism by which game-playing machines improve their performance by taking "time-out", cf. § 18E.

> The game theory of von Neumann and Morgenstern may be suggestive as to the operation of actual game-playing machines, but it does not actually describe them. . . .
> The von Neumann theory of games bears no very close relation to the theory by which game-playing machines operate. The latter corresponds much more closely to the methods of play used by expert but limited human chess players against other chess players. Such players depend on *certain strategic evaluations,* which are in essence not complete. While the von Neumann type of play is valid for games like ticktacktoe, with a complete theory, the very interest of chess and checkers lies in the fact that they do not possess a complete theory. Neither do war, nor business competition, nor

Chess playing automaton, built by Tores y Quevedo in 1890. Dr. Quevedo's son
and Norbert Wiener at the Cybernetical Conference in Paris, 1951.

any of the other forms of competitive activity in which we are really interested. [60d, pp. 1355, 1356] (emphasis added)

Wiener alluded to the battles of Napoleon and Nelson to show
how great commanders exploit their enemies' limitations, not just in
manpower and material but also in "experience and military know-
how". Thus a realistic theory of games (more accurately, of strategic
contests) has to take cognizance of the element of *experience*, i.e. of a
record or sequence of previous performances. Such a contest involves
moves at three levels at least:

(1) Actions according to a chosen local and short-term policy;
(2) Changes in this policy of action on the basis of the records of
     one's own performance, and of the adversary's performances
     against oneself and against his other adversaries;
(3) Changes of the policy of action on the basis of the knowledge of

Chess playing automata, old and new.

Chess playing automaton, the Belle, designed by Dr. Kenneth Thompson, Bell Telephone Laboratories, 1980.

the adversary's theoretical shortcomings: his ignorance or prejudice or misunderstanding of history, and consequent inability to foresee impending developments.

Hence, as Wiener put it:

> . . . in determining policy . . . there are several different levels of consideration which correspond in a certain way to the different logical types of Bertrand Russell. There is the level of tactics, the level of strategy, the level of the general considerations which should have been weighed in determining this strategy, the level in which the length of the relevant past—the past within which these considerations may be valid—is taken into account, and so on. Each new level demands a study of a much larger past than the previous one. [60d, p. 1356]

As in the theory of sets, a mix-up of types will result in contradictions, cf. § 5B.

Thus in all strategic contests, except the simplest like ticktacktoe, we cannot operate entirely by the rules of the von Neumann-Morgenstern theory. We indulge in *strategic evaluations,* arrived at on the basis of our past experiences in such games, and our past experiences of our opponent's weaknesses in playing it: rigidity, ignorance of certain move-configurations, scientific paucity, etc. It is only in playing short-term subgames within the game that we make our moves according to the von Neumann-Morgenstern theory, by looking at the "board" and seeing where we stand.[4] The over-all strategic evaluation is based on a record or *time series* of previous actions and exhibitions of enemy strength and weakness, and the longer the series the better. This, in brief, was Wiener's view.

Influenced by A.L. Samuel's checker-playing machine {S2}, Wiener brought in the Russellian theory of types in his analysis of game-playing automata. A 1960 lecture of his began with the words:

> The problem I wish to discuss concerns the programming of programming and machines that are of higher logical types according to Bertrand Russell's division of word types. [60b]

In that lecture Wiener discussed a chess game between two players, A and B, which is played by correspondence. B, however, lets a computer do the playing and acts merely as a messenger between A and the computer. The question is whether A will be able to tell that he is playing against a machine and not a man. Wiener's answer: if the computer has been programmed only to be the "first order", A will be able to tell; but if the programming is of a "higher order", A will have a hard time deciding whether his opponent is a human being or an automaton. Thus Wiener assigned a very important role to the type structure of game-playing machines. The mind-brain problem was thus addressed in indisputably clear empirical terms.

D The pitfalls in computerized atomic war games. Time scales

These considerations point to the pitfalls in playing atomic war games with computers and the even greater pitfalls inherent in the heavy reliance on computers at the executing levels in waging actual atomic war. A good part of Wiener's Terry Lectures at Yale University in 1962

---

4   For instance, during a battle within a global war, the soldiers and their immediate officers make their move according to the von Neumann-Morgenstern theory. The generals behind the scene do not worry if the "board" keeps getting worse and worse in a certain sector during a certain period, if this is in accord with the overall war plan.

(which appeared posthumously in his book [64e]) as well as his address [60d] to the AAAS are devoted to this subject.

Suppose that we are planning for an atomic war. Such a war has never been waged before; our time series of previous actions and exhibitions of strengths and weaknesses has zero-length, and we have no experience on which to make a strategic evaluation. To wage the war according to old rules would be a flagrant (and possibly fatal) violation of Clausewitz's Principle 9. Hence the only way to make the necessary strategic evaluations is by *artificial experience,* i.e. by learning from trial and error by playing a large number of atomic war games.

It is tempting to play the atomic war games by computer. But Wiener pointed out that computers are *literal minded*:

> If you are playing a game according to certain rules and set the playing-machine to play for victory, you will get victory if you get anything at all, and the machine will not pay the slightest attention to any consideration except victory according to the rules. If you are playing a war game with a certain conventional interpretation of victory, victory will be the goal at any cost, even that of the extermination of your own side, unless *this condition of survival is explicitly contained in the definition of victory according to which you program the machine.* [64e, pp. 59, 60] (emphasis added)

It is extremely hard, however, to program the computer so that "victory" for it means the same as it does for us. From a series of steady "victories" in atomic war games played according to a certain strategy on the computer, we cannot therefore conclude judiciously that the strategy will be successful if adopted in an actual atomic war. In short, *atomic war games begin to lose their educative value when played by computer.*

The pitfalls of relying on the computer in actual atomic warfare are, of course, much greater. The computer operates at incomparably higher speeds than our central nervous system, and the way in which it is programmed hides from us the many intermediate stages and parameter values by which it makes its quick decisions. A corrective action that we may take may be off substantially, because a parameter governing it was slightly off. The familiar accident of a car crashing into a wall, because the brakes were applied a bit too late or the wheel turned through a slightly wrong angle, has ominous analogues in the realm of atomic warfare. This danger is especially great in atomic wars in which the computer is expected to press the atomic button. The biblical tone in Wiener's warning: "Render unto man the things that are man's and unto the computer the things which are the computer's" [64e, p. 73], is not misplaced, when judged from the sound standpoint of the disparate latencies in the human and computer nervous systems.

In Wiener's way of thinking, the dangers to which we have been referring fall under the heading: *failure in communication between con-*

catenated transducers operating at different speeds. "Operating on different time-scales" was Wiener's expression. He believed, as we saw in § 16F, that many an active transducer (e.g. the firefly) has its own internal rhythm, and that this is controlled by its own internal "clock". The rhythm of this clock determines its "time-scale". Wiener was wont to analyze in these terms many a conflict, which most of us would consider in ad hoc ways. It is worth noting that he addressed in this way the very question of science and the promotion of human welfare that engaged Professor Heims {H6}. Wiener's thoughts on this question are articulated in the last section ("time-scales") of his address [60d] to the AAAS. He poses the issue as follows:

> What are the moral problems when man as an individual operates in connection with the controlled process of a much slower time scale, such as a portion of political history or—our main subject of inquiry—the development of science? [60d, p. 1356]

The scientific enterprise is a *process* (i.e. a transducer evolving in time) and the scientist's activity is another such process in concatenation with the first. They operate at very different speeds; fifty years for the latter "are as a day" for the former. The scientist can comprehend only a small sector of the enterprise. Action, such as a project in genetic engineering, that to him may seem conducive to the enterprise may in fact prove harmful to it. There being no central commander of the enterprise to turn to for appraisal of the action or for advice, all the scientist can do is cast "an imaginative forward glance at history". Not abandoning in the least his faith in the superiority of knowledge over ignorance, and in the relentless pursuit of fundamental research, be should, according to Wiener, question

> the naive assumption that the faster we rush ahead to employ the new powers for action which are open to us, the better it will be [60d, p. 1356]

In practical terms this would mean devoting a fraction of his time and energy to the kind of societal thinking that the members of the Federation of Atomic Scientists engaged in.

What would Wiener's suggestion amount to if in the last paragraph we replace "the scientific enterprise" by "Clausewitzian war" in the sense of an *ongoing contest,* and replace "scientist's activity" by "execution of defense"? It would mean a smaller allocation of defense departmental funds to weapons development and production, and a larger allocation for the cybernetical analysis of weaponry and global strategy in broad and far-sighted terms. This substitution is of course misleading in that war is waged against an adversary and involves the use of stratagem, whereas the scientific enterprise does not, cf. § 18G. Nevertheless, in waging war the defense organization is faced with some of the same dilemmas that confront the scientist in his activity. There is no war-god it can turn to for a long-term evaluation of its strategic decisions, and here again "an imaginative forward glance at history" is all that is available. To this extent the substitution is useful. It suggests that at least some of the savings arising from decreased weapons production may in fact prove beneficial, and actually strengthen the defense posture.

E Synopsis of Wiener's military thought

By combining all these different considerations, Wiener was able to round-off his ideas on global and local policy into a whole that was at once scientifically well-founded, patriotic and economical, and as humane as a world prone to war would allow. The main points of his military doctrine are:

1. War is not a "glorified football game" which starts with the firing of a gun and ends with the waving of a flag, but one phase of an ongoing geopolitical process.
2. The wisdom of a military act depends on its immediate effects as well as on its long-range ramifications on future contests.
3. Moves, tactics, strategies, superstrategies ... form a hierarchy of types 0, 1, 2, 3, ... (a tactic is a sequence of moves, a strategy a sequence of tactics, and so on). As in set-theory, there will be trouble if type lines are uncritically transgressed.
4. The von Neumann-Morgenstern game theory applies only in military operations of low type n.
5. Great generals win by going up the type-ladder. Their moves are based not on an evaluation of the board (as in checkers), but on *strategic evaluations* that involve not just the enemy's limitations in manpower and material, but also in his experience, military know-how and political understanding.
6. Each new level demands a study of a much larger past than the previous one.
7. Game theory is useful for low type military operations. At the higher levels, a time-series analysis of the enemy's experience is necessary.
8. For a concatenation of transducers to function smoothly, there must be a free flow and feedback of information between them

and continual self-correction. In particular, for the successful execution of a military operation there must be "continual reconnaissance", even with weapons.

9.   The A-bomb is a bad weapon by virtue of the impossibility of corrective action, apart from self-damage by radioactive fallout.

> One-shot weapons don't combine well with reconnaissance and are exceedingly dangerous for everybody. They are guns which kick almost as much with the butts as they do with the bullet. [XII, p. 6]

10.   It is dangerous to play atomic war games by computer, because the latter are literal minded and it is very hard to program them so that "victory" means what we want it to mean. The lessons learned from such games can be very misleading in forecasting the consequences of a real atomic war.

11.   A very large defense budget will improve the defense posture, only if a large part of it is spent on cybernetical education inside and outside the armed forces.

F Von Neumann's position; science and human welfare

The views on global stratagem of scientists whose conceptions were in tune with the Truman and Eisenhower doctrines were often in stark contrast with those of Wiener. Perhaps the most distinguished among these scientists was Professor John von Neumann of the Institute for Advanced Study in Princeton, New Jersey, of whom we have already spoken at length (§§ 5B, 9F, 10B, 12A, 16E and Ch. 17). Von Neumann continued his active association with the Department of Defense, which had begun even prior to the Manhattan Project, and became a major advisor on weapons development. In 1951, Mr. Gordon Dean, Chairman of the AEC, described von Neumann as "one of the best weapons men in the world", {H6, p. 251}. He eventually left Princeton for Washington, D.C. to become a Commissioner of Atomic Energy.

   Von Neumann's awesome responsibilities with the U.S. government lay in the area of weapons development, and he made it a point not to speak his mind on larger strategic issues outside the many committees to which he had been appointed. However, his geopolitical views have become known. They were markedly different from Wiener's in being totally supportive of a relentless escalation of the arms race. Professor Heims has dwelt on them at length in his recent study of the two men {H6}. The question he asks, of obvious interest to sociologists is: *What made these two great devotees of mathematics take so different a position in matters geopolitical?* Heims also brings the Wiener and von Neumann differences to bear on his study of *how the unlimited growth of science and technology might jeopardize the continuation of a moral social order.* He writes:

> The traditional ethics of the scientific community, dictated by the god of scientific and technological progress, needs modification so as to allow for the reemergence of suppressed human and cultural needs. {H6, p. 159}

This question continues to engage public attention. To see how the geopolitical differences of Wiener and von Neumann bear on the answer, it is necessary to find out more fully von Neumann's position. We do this as best we can, from his occasional verbal support of the "atomic diplomacy" that governed the Truman-Acheson and Eisenhower-Dulles foreign policies, and from the known bits and pieces of the advice he proffered in committee as to the type of arsenal the United States should have, and on how this arsenal was to be built and financed. The following are the more salient bits of the evidence we have:

1. Von Neumann was opposed to the international control of atomic energy and the supervised abolition of nuclear weapons. He declined a request from Norman Cousins of the *Saturday Review* (Spring 1946) to sign a letter supportive of the Acheson-Lilienthal Report of November 1945 accepting such control.

2. Von Neumann attended the atomic bomb tests at Bikini Atoll in July 1946, notwithstanding opposition to such open-air testing from the Federation of Atomic Scientists.

3. He supported building the largest ("biggest bang") atomic bombs that technology and finance would allow, at the fastest possible rate. He also strongly advocated a switch from the bomber as delivery system to the inter-continental ballistic missile as quickly as feasible. This becomes clear from his remarks to colleagues such as Professors Oppenheimer and Ulam, cf. {U2, p. 209} and {H6, pp. 236, 258–259}.

4. When the Soviets exploded their first atomic bomb in Siberia in August 1949, and fresh ideas were being sought on American policy, von Neumann advocated a quick nuclear "first strike" by the United States, while it still had a clear lead in atomic stockpiling, cf. Blair, {B12}.

5. On the question of proceeding with a large-scale program to develop and manufacture hydrogen bombs (early 1950), von Neumann's answer was an unqualified "yes". As early as 1945 he joined Dr. Edward Teller in working on the bomb design at Los Alamos (cf. Los Alamos Scientific Laboratory Report LA-574, June 12, 1946, de-classified). In 1952 he used the computer of the Institute for Advanced Study for the thermo-nuclear program.

6. When the United States exploded a hydrogen bomb in the Marshall Island region in March 1954, that not only destroyed the island but contaminated Japanese fishermen and their catch 1,000 miles away, the biochemist Dr. L. Pauling and the geneticist Dr. H.J. Muller called attention to the dangers of fallout from open-air testing. But von Neumann in committee (May 27–29, 1954) expressed the opposite view that human tolerance to fallout was much greater than the 10,000 megatons that the AEC-sponsored "Sunshine" studies had indicated. (See {H6, pp. 263–264} for details.)

7. In January 1954, after the Eisenhower "Atoms for peace" speech, Dr. James B. Fisk, Director of Research at the Bell Laboratories, and Dr. I. Rabi, the Science Advisor to President Eisenhower, backed further discussion with the Soviets on shutting down all plutonium-producing facilities. But at a meeting of the General Advisory Committee of the U.S. Energy Research and Development Administration (January 6–8, 1954) von Neumann disagreed. He felt that the Russians were "better qualified for clandestine operations" than the United States, and that the latter would be at a disadvantage. He added that an agreement with the Russians "would have the very unfortunate effect that the U.S. would not be willing to spend as much on air defenses. . . . Also it would increase the AEC's difficulty in getting money from Congress" {H6, p. 267}.

8. By 1955 the AEC had lost the confidence of a large number of American scientists in the matter of protecting public health from nuclear hazard. Many began to endorse setting up a United Nations group to study the relevance of radioactive fallout to public health. But von Neumann went on record that "It is my conviction that such a step would be contrary to the interests of the United States" (AEC memorandum, August 19, 1955).

9. Von Neumann sided with Lewis J. Strauss in having electric power supplied to the Oak Ridge Laboratory of the AEC by the (private) Dixon-Yates Co. as against the (public) Tennessee Valley Authority, although it was estimated that this would cost the AEC more money (1955), {H6, p. 277}.

10. In 1955 it became official policy to make the production of nuclear reactors attractive to the private investor. In an official letter von Neumann alerted Senator Clinton Anderson on the importance of "providing indemnity against possible liability resulting from nuclear catastrophe" and on "the possible need for legislation". This letter was an important factor in the initiation of the Price-Anderson Act according to which the U.S.

government would furnish free insurance to American corporations up to half a billion dollars per nuclear accident.

These ten pieces of evidence are indicative of a clear bellicosity in von Neumann's geopolitical position vis-à-vis Wiener's. Until those, who have had the good fortune of being the direct beneficiaries of von Neumann's wisdom and of knowing his true thought, can provide contravening evidence, we have to proceed on the basis of the factuality of this position. This writer requested Professor Abraham Taub, the editor of von Neumann's *Collected Works* {V6} and author of his biography {T1}, to shed further light on the ten pieces of evidence. But all that Professor Taub could suggest, by way of a possible refutation, was von Neumann's testimony at the Oppenheimer security hearings. It is von Neumann's geopolitical evaluations, however, that concern us, not his stance on Dr. Oppenheimer's patriotism.

Unfortunately, we have very little from von Neumann's own pen to tell us of the strategic thinking that underlay his belligerent stance. So once again we are left to seek guidance from the little that is available. A passage from a letter von Neumann wrote to Lewis Strauss in 1951 sheds some light on his thinking:

> The preliminaries of war are to some extent a mutually self-excitory process, where the actions of either side stimulate the actions of the other side. These then react back on the first side and cause him to go further than he did "one round earlier", etc. . . . Each one must systematically interpret the other's reactions to his aggression, and this, after several rounds of amplification, finally leads to "total" conflict. . . I think, in particular, that the USA-USSR conflict will probably lead to an armed "total" collision, and that a maximum rate of armament is therefore imperative. {H6, p. 287}

This passage embodies the principles 3 to 5 of General von Clausewitz (see Note on p. 317), i.e. it idealizes the USA-USSR conflict under the Clausewitzian concept of *absolute war*. Ignored in it are the general's principles 6 to 10 governing the second term of the Kantian dichotomy on war, viz. *real war:* a continuation of political commerce by other means "in which the play of probabilities and chance make it a free activity of the soul". That actual war is a self-divided activity comprising both these conflicting variables is clear from history: in most wars political claims and conditions have prevailed over military demands in the strict sense. Thus von Neumann's "war" was merely one aspect of the full-fledged cold war. But the letter suggests that he let this one aspect govern his thinking, and made it the sole basis of the stratagem to wage the full-fledged war. To this extent his analysis of the cold war was superficial. At the other extreme, Sir Arnold Toynbee in *Russia's Byzantine Heritage* gives an interesting account in which the full-fledged cold war emerges as the continuation of a nearly 2000-year contest between two civilizations {T5, pp. 164–183}.

Wiener, on the other hand, instinctively saw the importance of Clausewitz's principle 9 and also noted the strong points of the enemy's position, as his letter to Reuther shows, as indeed did Reuther himself. The last part of his letter to Wiener clearly states that the American offensive will require a strong social arm to challenge "the source of communist power" (see p. 273).

The Soviet side also took a full Clausewitzian view of the war. As the noted scholar and political scientist, W.B. Gallie, has pointed out:

> There was . . . one group of thinkers who, from the 1850s until the 1920s fully appreciated Clausewitz's contributions not only to military thought but to social thought in general-and who drew out some of their most interesting implications. These were the founding fathers of Marxism. {G1, p. 66}

Adding that Engels's writing in the military field (over 2000 closely printed pages) outnumbered those in any other field, he continues: "I find much more to praise than to condemn in the specifically military thinking of the great Marxists" {G1, p. 67}

But von Neumann did not appreciate the enemy's strong points. In his testimony before Congress in January 1955 (during confirmation hearings pending his appointment as Commissioner of Atomic Energy), he described himself as being "violently opposed to Marxism ever since I can remember". There is no evidence to indicate that his opposition to Marxism rested on a study rather than on hearsay or emotion. Von Neumann did not know, for instance, that the ideas in his very important 1945 paper on general economic equilibrium {V5} had been anticipated by Karl Marx in the 1860s. This has been pointed out by the eminent econometrist, Professor M. Morishima {M17}.

These limitations of von Neumann's appreciation of the USA-USSR conflict in disregard of the dichotomous nature of war, cast doubt on his maturity as a global strategist. Characterizations of him as an "exclusively rational" strategist (Ulam {U1, p. 6}), or "hard boiled" strategist (Blair {B12}), are thus inaccurate. In truth, his idealization of the cold war was scientifically inadequate, in stark contrast to the profundity of his earlier idealizations of quantum phenomena and of ratiocinating automata, and his astute though tentative idealizations of non-linear phenomena. The bellicosity in his stance thus stemmed from an *insufficiently scientific appreciation* of war, and not, as Professor Heims {H6} suggests, from an *exclusively scientific one*.

The truth is that the excessive inhumaneness in our technological society stems not from too much of a scientific attitude, but from too little of it. The continuation of a moral social order, far from being in conflict with the growth of science, as Heims's words (p. 311) suggest, demands an evermore vigorous and widespread application of the scientific methodology. This was Wiener's position.

As to von Neumann's strategic stance, let us remember that his official responsibilities were not in global strategy but in weapons development. He wrote no papers on global stratagem that we know of, and his thoughts on the question do not affect his greatness as a scientist. The fact that he was not an "exclusively rational" strategist is not relevant.

A reader may nevertheless ask why von Neumann, a great mathematician and scientist, instead of generalizing the theory of games into a comprehensive theory of war stratagem, adopted a less than scientific position on the matter. Why did Galileo, the father of modern mechanics, slip on the question as to why water rises only up to 34 feet in a pump? Why did H. Poincaré, universal mathematical genius, pathetically classify Cantor's Mengenlehre as a disease? Why did P. E. A. Lenard, Nobel Prize-winning physicist, vent hocus-pocus about the superiority of "Aryan science" over "Jewish science"? "The most complicated object under the sun" was von Neumann's apt description of the human nervous system (p. 243). This "complicated object" has turned man into Homo sapiens, but only to a degree; it has also allowed him to turn into something else. (See § 21 B.)

G The Black Mass, twentieth century

Wiener's great analogical mind saw a parallel between the contemporary concern over the misuse of technology and the medieval concern over the misuse of the Mass—the so-called *Black Mass*. In this Mass the orthodox dogma that "the priest performs a real miracle and that the Element of the Host becomes the very Blood and Body of Christ" remains the principle, but because of its ulterior motivation, the ceremony is severely condemned by the Church. To quote Wiener:

> The orthodox Christian and the sorcerer agree that after the miracle of the consecration of the Host is performed, the Divine Elements are capable of performing further miracles. They agree moreover that the miracle of transubstantiation can be performed only by a duly ordained priest. Furthermore, they agree that such a priest can never lose the power to perform the miracle, though if he is unfrocked he performs it at the sure peril of damnation.
>
> Under these postulates, what is more natural than that some soul, damned but ingenious, should have hit upon the idea of laying his hold on the magic Host and asking its powers for his personal advantage. It is here, and not in any ungodly orgies, that the central sin of the Black Mass consists.

The magic of the Host is intrinsically good: its perversion to other ends than the Greater Glory of God is a deadly sin. [64e, pp. 50–51] (emphasis added)

Wiener believed that the wisdom in this medieval attitude could be profitably transferred to the contemporary scene. The central theme is "the use of great power for base purposes" [64e, p. 52]. So interpret "the magic of the Host" as being the magic of automatization, and interpret "other ends than the Greater Glory of God" as being avaricious gain or nuclear holocaust. We then arrive at Wiener's injunction:

> There is a sin, which consists of using the magic of modern automatization to further personal profit or let loose the apocalyptic terrors of nuclear warfare. If this sin is to have a name, let that name be Simony or Sorcery. [64e, p. 52]

Wiener was aware of mankind's perennial apprehension of inherent evil in "the use of great power for base ends". It is the moral of the story, "Fisherman and the Jinni" in the *Arabian Nights,* and of several myths and legends of antiquity. The theme appears in Goethe's *The Sorcerer's Apprentice,* and in other fiction, among which Wiener's favorite was *The Monkey's Paw,* by W. W. Jacobs, cf. [64e, pp. 55–59]. In this way, Wiener was able to mobilize the "accumulated common sense of humanity" to warn against the light-hearted use of the great powers that modern science and technology have placed at our disposal for base ends, i.e. for catering to the new sins and idolatries of the modern secular age: materialism, gadget-worship and the transference of the power of being beyond good and evil, which in truth is ideal and belongs to God alone, to our all too human rulers.

Note: Von Clausewitz's principles of war

The Clausewitzian aspect of Wiener's military thought can be seen from a list of ten major principles of his doctrine. Von Clausewitz was influenced by Immanuel Kant, the philosopher of "unbridgeable dichotomies" {G1, p. 12}, and saw in war a self-divided activity differing from other forms of concerted action in its heavy reliance on violence. Following Kant's approach, he made *the maximal and unbridled use of violence* (absolute war) the first term of a Kantian dichotomy. But war is always a social venture subject to social pressures. He accordingly made the *subordination of war to politics* (real war) the second term of the Kantian dichotomy. His principles thus break up into two groups which bring out the underlying contradiction. Clausewitz's emphasis on the dichotomous nature of war is compelling, and his ten principles provide the embarkation point of a scientific theory of war.

The ten principles of War[5]

1. War is an act of violence intended to compel our opponent to fulfill our will.

2. In its "element" or essence, war is nothing but an extended duel, e.g. between two wrestlers, in which each tries to throw his adversary and thus render him incapable of further resistance.

3. As the use of physical violence by no means excludes the use of intelligence, it comes about that whoever uses force unsparingly, . . . finds that he has the advantage over him who uses it with less vigour.

4. Hence, as each side in war tries to dominate the other, there arises a reciprocal action which must escalate to an extreme.

5. The disarming or destruction of the enemy . . . or the threat of this . . . must always be the aim in warfare.
   (Principles 1–5 introduce the idea of *Absolute War*)

6. War is a political act . . . also an effective political instrument, a continuation of political commerce and a carrying out of this by other means.

7. Under no circumstances is war to be considered as an independent thing . . . Policy is interwoven with the whole action of war and must exercise a continuous influence upon it . . .

8. Wars must differ in character according to the motives and circumstances from which they proceed.

9. The first and greatest and most decisive act of the statesman or general is to understand the kind of war in which he is engaging, and not to take it for something else, or to wish it was something else which, in the nature of the case, it cannot possibly be.

10. War . . . is a wonderful trinity, composed of the original violence of its elements, of the play of probabilities and chance which make it a free activity of the soul, and of its subordinate nature as a political instrument, in which respect it belongs to the province of Reason . . .
    (Principles 6–10 introduce the idea of *Real War*)

---

5   The wording is that of the edition {C10} of *On War*, as adapted by Professor Gallie {G1}.

21 Wiener's Excursion into the Religious Domain

A The problem of evil

As the study of transducers in the wide sense, cybernetics cuts across the traditional boundaries separating the sciences, cf. § 18B. The attitudes engendered by this scientific eclecticism, such as the view that we can learn about the human mind by studying an electrical network, or that the imprecise term "soul" must either be redefined or abandoned, conflicts with a religion which treats man as God's favorite creature and the rest of creation as inferior, and so succumbs to its own misty conceptualizations. But the cybernetical attitude does not conflict with more cosmic religious attitudes, which seek the spiritual in the meanest of things, the attitudes exemplified for instance by St. Francis of Assisi, Mahatma Gandhi, Albert Schweitzer and others, and in the Leibnizian doctrine of monadic souls. It is in terms of the possibility of this harmony that Wiener's musings of the 1950s and 1960s on the religious bearing of cybernetical thought has to be gauged. We have also to extrapolate, for his writings are scanty and his death has left us with several gaps in his train of thought.

In his book, *The Human Use of Human Beings,* a socially oriented sequel to *Cybernetics,* published in 1950 [50j], Wiener discussed first the moral implications of the *indeterminism* inherent in contemporary physics, and *ipso facto* in all empirical science. This indeterminism has three aspects: (1) noise is all-pervading, (2) all measurement is inaccurate, (3) the laws of atomic physics are intrinsically probabilistic, cf. § 12C. Wiener felt that this recognition by modern science of the existence of randomness within nature, and the resulting limitations it imposes on our probing into the truth, brings it close to the tradition of St. Augustine:

> For this random element, this organic incompleteness, is one which without too violent a figure of speech we may consider evil; the negative evil which St. Augustine characterizes as incompleteness rather than the positive malicious evil of the Manichaean. [50j, p. 11]

Since the scientific concept of entropy is a measure of the "random element" in the universe, Wiener here drew a parallel between *entropy* and (negative) *evil*.

This analogy between the entropy concept of statistical physics and the concept of evil of Augustinian theology is credible in so far as St. Augustine regards "evil" not as a substance coexisting with the "good", but as a deprivation. Basing himself on the Pauline text, "Every creature of God is good" (1 Timothy 4:4), the Saint wrote:

> No nature, therefore as far as it is nature, is evil, but to each nature there is no evil except to be diminished in respect of the good. {A7, Ch. 17}

Hence, *evil is "nothing else than corruption, either of measure or the form or the order that belong to Nature"* {A7, Ch. 4}. On replacement of the word "evil" by Wiener's "random element", and of the expression "measure, form, order" by "the orderliness of the universe", Augustine's statement reduces to the assertion: "the random element is nothing but the corruption of the orderliness of the universe". This assertion, a paraphrase of "noise disrupts message", is in line with cybernetical thought. Moreover, the religious view that evil is never fully eradicable is analogous to the truth that noise is never fully eradicable. Thus, the replacement of "entropy" by "evil" converts certain truths of physics into Augustinean propositions.

Wiener's statement misses, however, an important element. It does not reveal that Augustine, unlike Leibniz, desisted from describing as "evil", the accidents and calamities that occur by virtue of the Entropy Principle. He had no concept of so-called "natural evil". Earthquakes and animal fights over food, and the inevitable pain and death were not evil. Behavior that may harm in one situation or level can do good in another. For instance, the venom of the snake, that may kill a child, protects the snake from attackers, and also has medicinal uses, cf. {A8, 11:22}, {A9, I, 16}. The Saint was fully aware of the indispensable role of death in the cosmic pattern:

> But to things falling away and succeeding a certain temporal beauty in its kind belongs, so that neither those things that die, or cease to be what they were, degrade or disturb the fashion and appearance and order of the universal creation: as a speech well composed is assuredly beautiful, although in it syllables and all sounds rush past as it were in being born and in dying. {A7, Ch. 8}

In short, the (negative) evil of Augustine refers only to the unnatural destruction practiced by man.

Wiener's omission, though serious, does not sever the connection he detected between entropy and evil. His great analogical mind was on the right trail, for to see a better connection, we have only to extend Wiener's own theory of *phylogenetic learning* (§ 18F) to one of *phylogenetic corruption.*

St. Augustine In His Study
Black and white reproduction of a painting by Botticelli (c. 1480) in the Church of
Ognissanti, Florence. The open volume on the shelf shows a geometrical text with
diagrams.

B Phylogenetic corruption, the Fall. Homo peccator

Consider an assembly $\mathbb{S}$ of teleological and learning machines S_1, S_2, ..., each having the objective to play and win in a strategic game. Suppose that this game, unlike both chess and dueling, involves violent moves, but has no sharply defined and legally enforceable regulations. Adaptation, i.e. ontogenetic learning, will now enhance efficiency in the arts of deception and violence. Imagine further that the units S_1, S_2, etc. can reproduce according to the Lamarckian genetic scheme or incubator discussed in § 18F, and that all belong to the posterity of a single ancestor unit S_0. We then have exactly as in § 18F, a growing genealogy of automatons, increasingly efficient in the deceptive and deadly arts. If we adopt the common parlance of calling these arts "evil", then we can say that the assembly $\mathbb{S}$ itself learns evil. We may use the term *phylogenetic corruption* for this phylogenetic process of acquiring evil.

In this we have taken account not just of the Entropy Principle and its temporal asymmetry, but of Bergsonian time and the possibilities it opens up, of which Wiener was well aware (§ 12F). To see the new fit with Augustinean ideas, we must consider, together with the Saint's thoughts on the negativity of evil, his thoughts on the phylogenetic corruption that he and other theologians call *the Fall of Man*. Here are three of his propositions:

(1) Prior to the Fall man had the potency to sin (i.e. to commit evil) as well as the potency not to; he possessed the freedom to choose;

(2) The Fall came from Adam's sin: "by desisting what was better he committed an evil act" {A7, Ch. 34};

(3) After the Fall, man lost his (natural) potency not to sin.[1]

Proposition (2) is premissed on strict monogenism, i.e. on our entire descent from a single initial pair (Adam and Eve). All evidence points to the falsity of this premise. Thus the proposition (2), which rests on this premise, is also false. We claim, however, that with the extension of Wiener's learning theory to phylogenetic corruption, and our present knowledge of man's phylogeny, the concept of the Fall can be retained, and with it also the proposition (3), with a suitable demarcation of the concept of sin.

---

1 Later propositions in this chain pertain to Divine Grace. According to St. Augustine such Grace is not (and cannot be) bestowed on the basis of meritorious behavior: it is pure charity. It follows from the Augustinean propositions that God could cut down the growth of evil by munificent distribution of Grace. It was the Saint's belief, however, that the bulk of people are not recipients of Grace. He could not answer how such impoverishment could occur under a totally benevolent God. According to him, the answer was beyond human understanding and lay in God's infinite wisdom.

We turn first to the demarcation of the concept of sin. Empirical observation shows that mankind is afflicted by a malicious form of evil, quite different from the tension exemplified in conflict, illness, pain, death, etc. that affects the rest of nature. It is the occurrence of such malicious evil that the term *sinful* is meant to denote. The most obvious manifestation of such evil is *infra-specific warfare,* marked by genocide, in distinction to inter-specific killing for food. Homo sapiens is the only mammalian species that regularly indulges in such killing. This propensity to murder is linked to man's high intelligence, for the overwhelming percentage of human-inflicted death comes from the use of intelligently constructed weapons, and the most destructive of these are the ones that embody our deepest knowledge, the work of our Einsteins and Rutherfords. Man also has a battery of "tool-free" vices, depending exclusively on his high intelligence, excellent memory and remarkable linguistic powers: dishonesty, deceit, hypocrisy, conceit, avarice and others, all selfish and all directed to the humiliation of other human beings. Thus, descriptions of mankind such as "tool-bearing species", "rational species", "Homo sapiens" or "Homo faber" are not enough, and must be supplemented by descriptions such as "Homo peccator" or "sinful species", etc. And, with our explication of the term *sinfulness,* the latter descriptions are just as scientific as the former and no less verifiable by observation and experiment.

Another verifiable difference between non-human mammals and man is the spontaneity with which the former perform their tasks, and the tension and hesitation that accompanies human performance, sometimes prompt and efficient, sometimes sluggish and derelict. This disability is called *self-alienation.* Wiener did not dwell on this important aspect of human evil. Nor was he aware that it is the center of Karl Marx's theory of the self-alienation of man in civil society, his distinction between "egoistic man" and "species being" {M7, pp. 235–237}.

We turn next to the demarcation of the Fall, replacing Augustine's monogenism by the phylogeny of man, as presently known.

Unlike pseudo "natural evil", which is of thermodynamic origin, the malicious evil of which we just spoke has its source in the slow evolution, lasting about three million years, that turned the primarily herbivorous hominids, the Australopithecines, into the more intelligent, language-oriented, tool-bearing social hunters, the Homo sapiens. The considerable increase in both individual violence and individual control over the world that came from this transition upset the balance or homeostasis in the species between phylogenetic and ontogenetic interests. It would be tempting to settle extra-tribal disputes over land and food resources by the violent ways used in the social hunting for food, rather than by a slow process of compromise. Next, the same temptation

Homo peccator (Hiroshima sometime after atomic bombing). The perversion of $E = mc^2$ to other ends than the greater Glory of God (see p. 316).

could easily intrude into the settlement of disputes within the tribe itself. In this way the range of murderousness and deception would keep widening, with increasing sorrow, despair and anxiety following in the wake.

What has emerged from this evolution, as we know, is the most self-centered and cruel of the mammalian species, a race made up of self-alienated individuals. This retrogression is what we shall call the *Fall,* for it is a scientific accounting of the kind or moral devolution of man that the sages and saints wanted to portray. Briefly, *the Fall is the transition that took the hominid from the animal-hunt to the man-hunt, and that brought in self-alienation.*

This Fall differs from the process of phylogenetic corruption of reproducing automata only in that the acquired characteristics are transmitted by social heredity and not by Wiener's Lamarckian incubator (§ 18F). The offspring acquire their latent skills in the violent arts by early schooling from their parents. This Fall differs from Augustine's only in its substitution of evolutionary anthropogenesis in place of monogenesis. This train of thought, resulting from Wiener's observation of the analogy between entropy and evil, and his learning theory has two positive aspects:

(1) It brings about a certain reconciliation between religion and science; for the propositions that man is a sinner, and that the source of his sinfulness is not within his social history, but extends to his very beginnings are absolutely fundamental to the teachings of all great religious sages, not just St. Augustine.

(2) It suggests that the intelligent attitude to religion is to view it (along with science) as a prosthesis for the human race, as man's response to the challenge of his evil—a response that is corruptible as history shows, but nevertheless basically prosthetic. Evil, like finitude, is not self-sustaining, as Whitehead has said. And as Wiener reminds us: "It (the Church) has given birth to the universities and other intellectual institutions such as academies . . ." [56g, p. 362].

Wiener's train of thought thus lends extra credence to the words of Einstein:

Science without religion is lame. Religion without science is blind. {E5, p. 26}

C Redemption: Mythic perception and faith as integral parts of man's ascent

Wiener was aware that in man's struggle for emancipation from sin, the faculties of imagination and idealization are crucial.

The importance of *imagination* for man's liberation from ignorance, avarice and injustice is obvious. Any movement away from the status quo requires some conception of what "ought to be", as P. Tillich {T4} has emphasized, and this last, since it does not exist, can only be imagined. The origins of the faculty of imagination can be traced back to animal ritual. This, as we now know, not only serves the emotional needs of the species, but also joins with the external environment in orienting its evolutionary trend.

Ritual becomes an even more vital formative agent for man. As A.N. Whitehead has remarked: "Mankind became artists in ritual" {W9, p. 21}. In the human ritual speech plays an important part. A *rite,* i.e. a ceremonial ritual with incantation, invariably accompanies the initiation of important activity such as the hunt. *Myths,* involving gods and deities, are invented to lend credence to the rite, and so justify the efficacy of the activity. *These myth-makers were our first scientists, who tried to explain why things work.* In these myths each situation is perceived as a blurred organic whole, in much the way a child or a poet perceives, incognizant of the distinctions and polarities occurring in a logical description. Man's earliest mode of perception was mythic, as Vico very clearly pointed out. There is profundity in the remark that "Poetry is the mother-tongue of humanity" {C9, p. 35}.

Wiener was profoundly aware of the creative and moral role of the myth. In the first place, he affirms the wisdom of fairy tales and the significance of Prometheus's defiance of the gods [50j, pp. 183–184]. Secondly, with regard to scientific creativity, he attached great weight to "discomfort and pain", and to "the power to operate with temporary *emotional symbols* and to organize out of them a semi-permanent recallable language" [56g, p. 86]. He shared the experience that Einstein described in the words (communicated to Hadamard):

> . . . the combinatory play (of psychic signs) seems to be the essential feature in productive thought—*before* there is any connection with logical construction in word or other kinds of signs . . . {H1, p. 142} (emphasis added).

Third, Wiener was aware of the challenging role of mythical and paradoxical concepts in science such as *Maxwell's demon,* cf. [50j, pp. 28–30]. (The exorcising of this demon helped not only in the understanding of the modern statistical theory of information, but of the lovely "demonic" role of chlorophyll particles, § 12G.)

A typically Wienerian insight on the issue of science and myth pertains to the fact that in mythology the name of the object is deemed to have the potency of the object itself. (Examples: prayer, with expressions such as "we ask in the *name* of"; slogans such as "In the *name* of Stalin, attack".)[2] Science has increasingly militated against such identification of symbol and essence, and indeed in areas such as metamathematics a sharp distinction between symbol and designatum, between mention and use, is crucial. But in Wiener's theory of regeneration of machines, the symbolic description of an object is made to serve as its operative image, somewhat as in mythology. For instance, a good diagram of an electric circuit can do the job of the circuit itself, for as Wiener tells us:

> . . . an electric circuit may fulfill a relatively complicated function, and *its image,* as reproduced by a printing press using metallic inks, *may itself function as the circuit it represents.* These printed circuits have obtained a considerable vogue in the techniques of modern electrical engineering. [64e, p. 31] (emphasis added)

Thus, the allied notions in mythology of (i) the name being a true image of the object, and (ii) the name serving as the object (of the "word" becoming "flesh" as it were), can be put to use in the designing of efficient and inexpensive engineering hardware.

2   A slogan used by the Red Army during battles in World War II.

Nowhere in Wiener's writings is there any reference to Cassirer's extensive study of mythic symbolism. But Wiener, like Cassirer, recognized the fundamental place of symbolism—linguistic, mythic, artistic and religious—in man's nature, and saw the naiveté of the hackneyed view, often labeled "progressive", that with the rise of the scientific mode of perception, the mythic mode would, should, or could disappear. Wiener's life and work both underline the truth of Cassirer's assertion:

> Reason is a very inadequate term with which to comprehend the forms of man's cultural life in all their richness and variety. Hence, instead of defining man as *animal rationale,* we should define him as *animal symbolicum.* {C8, p. 26}

With the growth of the crafts, and the accompanying development of prose and logical discourse, imagination found an intellectual arena. We have the *idealizations* of science, i.e. imaginative concepts of ideal, non-existent entities suggested by entities that exist. A typical example is Euclid's concept of a *point,* a being that has location but no volume. Such idealizations are the starting points of scientific theorizing, a fact amply illustrated in our previous chapters. Idealization, was, of course, Wiener's forte.

Besides the mythic mode, a religious idea that Wiener found to be of profound importance in science is faith:

> I have said that science is impossible without faith. What I say about the need for faith in science is equally true for a purely causative world and for one in which probability rules. No amount of purely objective and disconnected observation can show that probability is a valid notion. To put the same statement in other language, the laws of induction in logic cannot be established inductively. *Inductive logic, the logic of Bacon, is rather something on which we can act than something which we can prove, and to act on it is a supreme assertion of faith.* It is in this connection that I must say that Einstein's dictum concerning the directness of God[3] is itself a statement of faith. *Science is a way of life which can only flourish when men are free to have faith.* [50j, p. 193] (emphasis added)

In a nutshell, *the initiation of creative science rests on faith and on mythic perception, not on reasoning by inductive or deductive logic.* Logical reasoning plays a crucial role in science, but only after an initiation during which it is almost entirely disregarded. Naturally, Wiener regarded as totally inadequate, descriptions of science such as "rational activity" or "organized common-sense", a view shared by other great scientists, e.g. G.D. Birkhoff {B10}.

---

3   I.e. the words we quoted on p. 267 above.

D Redemption: Self-abnegation and duty

Wiener also had sound ideas on the social aspects of the emancipation struggle, on the intellectual aspects of which we just dealt. Imagination, and its intellectual manifestation in idealization provide the necessary wisdom and knowledge. They give man the conviction of his potential control over the course of events. The exercise of this control depends, however, on man's altruistic and self-abnegating propensities, as well as knowledge. The acquisition of these propensities is also traceable to our animal pre-history. Altruism clearly manifests itself in mammalian life, a conspicuous example being instinctive maternal care and parental protection of the young. Altruism beyond a certain point becomes self-abnegation, and this too is observable in the animal world. These propensities in man account for the concept of *responsibility* and the cognate concept of *duty*. Wiener had something to say on both.

Wiener's starting point here is the concept of *vocation,* i.e. a calling and a consecration, not just a job. This concept is taboo in current economics, but is central to the normative economics of the Hindu caste system and of medieval Christendom, as well as to Marxian economics in view of its concern over labor alienation.

In an article [60e] on this subject, Wiener pointed out that in order to merit the rewards and privileges belonging to a vocation, the practitioner is duty-bound to carry out its responsibilities, howsoever difficult and painful. The responsibilities vary from vocation to vocation. Wiener, like Plato and Aristotle, began with the soldier:

> The soldier has accepted a very special responsibility for physical bravery and moral bravery as well, and above all the officer in charge of troops. I do not think that any honest man looks forward with any pleasure at the prospect not merely of dying in action but of having to face a situation in which the only honorable thing for him to do is to die in action. Nevertheless, in becoming a soldier, and especially in becoming an officer, he must accept the contemplation of this possibility. [60e, p. 26]

Wiener spoke of the scholar as "the custodian of the intellectual development of this society, of the understanding of truths already known, and of the development of new truths and concepts". He spoke of the teacher as "the guardian of the task of passing down these new truths and discoveries already known, to a new generation" [60e, p. 27]. Convinced of the great need of the scholar, the scientist, and the teacher for phylogenetic survival, Wiener wrote:

> . . . the tampering with the truth on the part of the man normally devoted to it is a dereliction of duty quite comparable with the dereliction of an officer who runs away from his own soldiers in the face of the enemy . . . [60e, p. 27]

But Wiener stressed the fugitiveness of all anti-entropic activity, and this included dutiful action as well, as we shall see in § 21F. Nevertheless, he believed that the pursuit of vocation merited sacrifice of the same kind as that of the early Christian in the pursuit of his faith. That dutiful action, though it is itself fugitive, is paramount, is implicit in great religious thought. Thus Wiener's ideas on these matters were in essence those of the great world religions. This concordance came, however, not from any expert knowledge of religions or from any special religious commitment on his part, but rather from the tide of his thermodynamic and cybernetical wisdom reaching the shores of the religious domain.

E The long-time State. The mandate from heaven

A very important religious reflection of Wiener, based on his fine understanding of time series, is on the formulation of very long-time policies, and their execution by governments. In conformity with traditional thought, Wiener held that the feasibility of such long-time governance could only rest on faith in a benevolent God.

Institutions such as cities, churches, universities and academies have life-spans several times that of a human generation. The engineering infrastructure that ensures their survival, water supply, dams, sewage, cathedrals, mosques, buildings and roads, also have correspondingly long life spans. The sustenance of these vital long-time institutions, and this includes science, is the responsibility of what Wiener has called the *long-time State* [62c]:

> The State itself is a long-time institution. Even in those cases where the State has been subject to frequent changes of regime, the internal continuity of the State has been much greater than the external changes of regime might indicate. [62c, p. 36]

Wiener viewed this "State" as a long-life transducer taking in and giving out messages. To execute its responsibilities, this State has to adopt an approach that is markedly different from what is involved in the day-to-day operations of the political state.

Since political boundaries fluctuate, the long-time State is not confined to a particular geographic area. Parts of its responsibility may encompass an entire continent, if not the whole earth. The plethora of political regimes that come and go are transients in the life of the long-time State. Such a distinction between the short-time and the long-time State, while perhaps alien to post-Renaissance political thought, is in tune with the more classical political perception of a duality in government: *regnum* and *sacerdotium*. Indeed, as Wiener has written:

> Perhaps the best thing that has been said about the continuity of the State
> was formulated two centuries before Christ by the Chinese sage Menscius.
> Menscius said, in effect, that the rule of the emperor is from heaven, but that
> when a country has come through a long period of misrule and misfortune,
> it is a sign that the emperor and even the dynasty has lost the mandate of
> heaven, and that the country must seek elsewhere for its rulers. This view
> represents an interesting attempt to combine a certain permanency in the
> essentials of government with the transience of its details and in the selection
> of those on whom the task of government lies. [62c, p. 36][4]

Thus, when the ruling class has lost "the mandate of heaven", i.e. when
it holds on to an economic order that constricts the development of the
productive forces, the resulting social revolution destroys the ruling
class, but not the institution of the State.

Wiener brought his experiences in the prediction of time-series to
point to the great differences in the objectives, mode of inquiry and type
of prediction involved in long-time and short-time considerations. Even
in air warfare, a policy that has been viable over a six-month period
could prove disastrous if prolonged say for six years [62c, p. 32]. A
short-time type 1 policy involves type 1 feedback: the deviation of shell
from bomber (type 0 entities). For a long-time policy to succeed, it must
involve type n feedback: the deviation of the (n-1) type policy from its
goal.

In general, short-time planning may be based on reasonably
accurate forecasts, deterministic or stochastic. When it is the latter, only
the linear extrapolation of a short segment of a fairly stationary time
series is usually involved. Such prediction would be futile for long-term
purposes, say planning for a city like Athens with its 2,500-year history.
The extrapolation has now to take into account a much larger past-
segment of the time-series. The latter is non-stationary, and the prediction, to be useful, must be non-linear.

> The use of this long-time information is so different from the one of short-
> time information that it is not economical to trust them to the same instru-
> ments and the same computation. [62c, p. 32]

For instance, to finance long-time city planning by private investment-
for-profit would be about as absurd as extending a short-sprint technique to run a marathon race (cf. § 19C). Briefly, *"The support of the
long-time needs of the human race cannot be left exclusively on the basis
of the returns they make on the welfare of short-time institutions"*, [62c,
p. 36] (emphasis added).

---

4   It should be noted that a dual theory of kingship, similar to the Chinese,
    prevailed in India at about the same time, 300 B.C. {C12}, and that the
    corresponding conception emerged in Europe, under a neo-Platonic aus-
    pices, in the *two sword* governmental ideas of St. Augustine {A8}.

Another major difference is that the courses of long-time series are profoundly affected by events of very low probability but very high import: natural or man-made catastrophes—"acts of God" in insurance parlance—as well as individual luminousness that results in great discoveries, inventions, new ideas, new attitudes and new leadership—"acts of Grace", a term suggested by Wiener's friend, the political scientist Karl Deutsch {D2, p. 217}. Wiener held that the use of such quasi-religious terms to designate unforeseeable incalculable risks and also unforeseeable benevolences is not fortuitous:

> For if religion purports to deal with the eternal, its vocabulary should be very suitable when we wish to treat with those matters which, although not eternal, are of very long duration. [62c, p. 35]

Our hope in the feasibility of long-time planning can only rest on the belief that the powerful noise component, arising from natural and man-made calamities and the long-range effects of human perversity, will not entirely damp out the effects of the "acts of Grace", which constitute the message. How can such a belief be justified. To quote Wiener:

> In all undertakings which contemplate eternity, or which if they do not contemplate eternity at least contemplate a period far greater than that of human life, something like faith is necessary. In these matters we cannot live to see the consequences of our own acts, or at least what we see of them does not pertain to their final consequences but merely to other intermediate consequences which we take on faith as an indication of what their final consequences will be. For example, what the great scholars or the great artists or the saints of the present day will do that will affect the human race centuries from now, is something that we cannot see with our own eyes. But we can have some indication that these people are scholars, or artists, or saints and we can have a faith that it is good for the human race, or even necessary, to have scholars, or artists, or saints. Without such a faith, it is fairly clear that their work will fall to earth like a dead sparrow, noticed perhaps by God, unnoticed by man.
> . . . .
> This faith should not, indeed, be a blind and rigid faith, but should be based on the best sense of human values and of history which we possess. For without these we cannot begin to glimpse into the dark mystery of the future. Nor can we do our duty to the generations to come. [62c, pp. 35–36]

Briefly, a faith more religious than Einstein's is needed. God has to be not only immanently rational, but also immanently benevolent.

The doctrine of the major religions that God is benevolent and that the State rules by divine right, viewed from a cybernetical perspective, provides a theory of the State which recognizes the transiency of its short-term arms—the inevitability of their losing "the mandate of heaven" and being dethroned—while maintaining its grandeur as a regulatory force contributing to man's emancipation.

It is noteworthy that these theological reflections grew from three very practical but *long-time* questions that events forced Wiener to address: the placement of German refugee scholars in American universities [34e, 35f], cf. § 13C, the prevention of ecological disaster [60d] and city-planning [62c].

F Sin grows with doing good: Tragedy and catharsis

Wiener had the wisdom to realize that moral struggle, even of the highest order, cannot eradicate evil. For he saw an analogy between the moral conduct of a person or group of persons striving for good in the midst of evil, and the anti-entropic activity of a teleological or learning mechanism in the midst of a contingent and noisy universe. The behavior of a teleological mechanism, whether organic or metallic, is anti-entropic (even if it is designed, like a nuclear missile, to explode and increase entropy); it attempts "to control entropy through feedback" [50j, p. 34]. Thus, it is

> ... a device which locally and temporarily seems to resist the general tendency for the increase of entropy. By its ability to make decisions it can produce around it a local zone of organization in a world whose general tendency is to run down. [50j, p. 34]

This anti-entropic activity is fugitive by dint of the Second Law of Thermodynamics. Sooner or later it must cease, cf. § 12G (end). Wiener felt that moral progression shares a similar inevitability of collapse. For moral evil, like physical noise, is not eradicable below a certain threshold. While it is controllable by the pursuit of good, the control is fugitive and the good will wither away much like a local zone of physical negentropy. In moral life as in physical life, "The paradox of homeostasis is that it always breaks down in the end."[5] This thought on the persistency of evil has been expressed by religious minds throughout the ages, but nowhere perhaps as succinctly and as poignantly as in a line of T.S. Eliot, Wiener's great contemporary, with whom he had an interesting exchange in his Cambridge years (cf. § 5D), the poet T.S. Eliot: "Sin grows with doing good" {E6, p. 44}.

One aspect of the concurrence of good and evil is the close correspondence between the growth of our scientific and philosophical wisdom and that of our war technology. Hiroshima was a stark demonstration of the verity of Einstein's relativistic equation $E = mc^2$; Wiener's cybernetics was a scientific-philosophical ramification of radar-based anti-aircraft ballistics; and St. Augustine's *The City of God* was a

---

5 From Wiener's manuscript *Prolegomena to Theology 1961* (unpublished), p. 103 [MC, 877–881], Terry Lectures at Yale University, 1962.

reflection based on the destruction of Rome by Alaric. Naturally, Wiener could never share the Marxian belief in the actual removability of alienation, i.e. in the actual transformation of civil society into communism, "an organization in which the free development of each is the condition for the free development of all" {M8}. For him, communism, like the Sermon on the Mount, offers an ideal to mankind, forever unattainable but one nevertheless towards which it must forever strive.

Guided by the great tragedy writers of ancient Greece, Wiener sought dignity in this tragic situation. The Second Law of Thermodynamics tells us that all life must come to an end.

> Yet we may succeed in framing our values so that this temporary accident of living existence, and this much more temporary accident of human existence may be taken as all-important positive values, notwithstanding their fugitive character. [50j, p. 40]

Adding that even in a shipwreck "human decencies and human values do not necessarily vanish and we must make the best of them" [50j, p. 40], Wiener concludes:

> The best we can hope for the role of progress in a universe running downhill as a whole is that the vision of our attempts to progress in the face of overwhelming necessity may have the purging terror of Greek tragedy. [50j, p. 41]

He contrasted the well-balanced tragic attitude with which the Greeks perceived Prometheus, the fire-bearer, and others who defied the Gods, with the overconfident and anxiety-ridden attitude of modern man with his naive belief in unlimited progress. He pointed out the cathartic need for a well balanced tragic conception in an age of atom-splitting and automatization. By contrast,

> The simple faith in progress is not a conviction belonging to strength, but one belonging to acquiescence and hence to weakness. [50j, p. 47]

G Rational Logos, or rational and altruistic Logos?

Wiener's firm ingress in the religious domain did not lead him to a belief in a personal god. His conception of the Logos was one of immanent rationality and immanent benevolence. He thus went a step beyond Einstein who could not believe "in a God who has anything to do with the fate and affairs of man".[6] For Wiener, on the other hand, God's benevolence, operating through "acts of Grace", keeps saving mankind from thermodynamic and man-made disasters, and from the long-range effects of human perversity. It is directed, however, to the human species not the human individual.

---

6   Einstein, quoted in *The New York Times, 25* April, 1929, p. 60, column 4.

Wiener's exclusion of the individual from the range of God's benevolence had something to do with one of his childhood experiences. At the age of six or seven Wiener conceived the transcendent as being the figure of Jehovah, and to use his own words, written in 1952:

> I have never made up my childish quarrel with Jehovah, and a skeptic I have remained to the present day . . . [53h, p. 142]

Indeed, Wiener was a skeptic in that he neither observed religious ritual nor worshipped overtly. We must point out, however, that Wiener's religious faith increased with the passage of time and that his own description "childish" for the quarrel with Jehovah was accurate. Hence to replace Wiener's own delineation "skeptic" by the stronger term "atheist", and claim that he was an atheist, as Professor S. Heims and Father N.J. Faramelli have done, cf. {H6, p. 375} and {F1, p. 259}[7] is quite unwarranted.

On the contrary, it is easy to get to a personal God by completing Wiener's own thought: we have only to extend the range of his analogical probing. For, as Dr. Ashby has explained (cf. § 18C above), the ideal concept of the (non-existent) massless spring is of the "highest importance" to the engineer who wants to understand and deal with an ordinary clock or watch. Is it then so absurd to expect that for an individual to understand and deal with his neighbor or spouse or child, some ideal conception may also be of the "highest importance"? Is it not then natural and even beneficial for man, struggling against evil, to explore endowing the Logos with a personal altruistic dimension? If so, Einstein's statement: "I believe in Spinoza's God who reveals himself in the harmony of what exists"[8] which could be enlarged by adding "notwithstanding the entropy and noise that complicate my equations and also my measurements", needs an accompaniment. This should be the analogical statement:

> And I also believe in the God of religion who reveals himself in the altruism that prevails in the world, notwithstanding the ever-persisting evil that complicates my life.

Thus, in the light of Wiener's own analogy, man's quest for a personal God appears as an attempt to adapt the Pythagorean-Platonic aspect of

---

7 Father Faramelli footnotes, however, that he got this information from Wiener's family, friends and colleagues and not from Wiener's writings {F1, p. 328}. We would refer the interested reader to Father Faramelli's study {F1} for other aspects of the religious implications of Wiener's thought, carried out from a standpoint that is more ontological than ours. For more on the methodological standpoint here adopted, see {M11}.

8 Einstein, quoted in *The New York Times*, 25 April, 1929, p. 60, column 4.

mathematical physics (the "massless" spring) to the domain of human relationships.[9]

There is much in personal theism and ceremonial religion that is escapist and even evil, as all great religious thinkers have affirmed. But from the Wienerian perspective we have taken, its description as "unscientific" is untenable. Rather it is atheism which appears to be "unscientific" and even "reactionary", for it seems to bar the transcendent, the crucial element in the method of mathematical physics, from the human domain, and thereby to retard the growth of human and humane relationships. In any case, Professor Heims's description of Wiener as "a modern scientific atheist" {H6, p. 375} is totally inadequate. A requirement of atheism would ostracize from the modern scientific community not only Wiener, but Einstein, Maxwell and a host of other great scientists. The term "modern scientific atheist" must either be very carefully defined or discarded.

Wiener has to be ranked, we feel, as one of the most incisive theological minds of this century, despite his scanty theological writings and despite his "childish quarrel with Jehovah".

Note. The landscape on page 323 does little justice to the calamity it is meant to depict. Graphic pictures of the holocaust are available. See F. Greene, *Let There Be a World*, Fulton Publishing Co., Palo Alto, California 1963, or *Hibakusha*, translated by G. Sekimori, Kosei Publishing Co., Tokyo, 1984.

9 Cf. A. N. Whitehead: "Mathematics and the Good", {W10, pp. 666–681}.

22 Wiener's Literary Predilections and Initiatives. His Concept of Art

A Wiener's literary work

Wiener's literary writings ranged from short story, essay and letter to autobiography and novel. In this literary bent he was more like his mentors Leo Wiener and Bertrand Russell than his mentor G.H. Hardy or his colleague John von Neumann. But unlike some scholars who indulge in hack writing for magazines, Wiener did not regard popular concerns as matters for unscholarly or perfunctory treatment. As a cyberneticist he freely crossed conventional fences in the course of his writings, but he never abandoned high standards of scholarship and a high quality of prose.

Wiener's literary abilities had much to do with his very early exposure to good books. He began reading great literature from a very young age, and had a thorough grounding in the classics:

> The lapidary poetry of Horace is not something buried between the pages of my schoolbooks: it is engraved on the tablets of my memory. The sweep and grandeur of Homer are recollections I can never forget. [53h, p. 170]

As an adolescent he loved Rudyard Kipling. But as he matured he leaned heavily on Heinrich Heine:

> I came to love the heart-rending cries of Heine, in which not one word is missing or in excess to obscure his love and his venom. I know, as did my father, almost every word of his *Hebraische Melodien,* and there are no poems that can move the Jew in me to greater pride or agony. [53h, p. 169]

Among the British men of letters, he speaks of Thackeray and "that great, paradoxical, sincere genius" John Ruskin [18b, p. 200]. His words are "what I have read and loved, I have taken into my memory so that it became a part of me never to be discarded" [53h, p. 170].

It is not surprising that with his deeply ingrained feeling for mythology, Wiener's initial literary inclinations were toward imaginative and penetrating science fiction. In childhood he enjoyed Lewis Carroll, Jules Verne, and H.G. Wells. In college he became an admirer of Swift's *Gulliver's Travels.* Such imaginative writing, which began perhaps with the Utopias of Plato and St. Thomas More, has played a positive role in the development of science and philosophy, for as Otto Neurath has pointed out, the preparation of the intellectual environment for new scientific conceptions (be it non-Euclidean geometry, relativity, cybernetics, or socialism) is more often the work of imaginative amateurs, essayists and poets than of systematic philosophers {N3, pp. 11–14}. But while Wiener thoroughly approved of this fine tradition of imaginative writing, he thoroughly disapproved of its degradation in the bulk of current science fiction. He characterized the latter as a "pernicious article", "a vicious form of day dreaming" symptomatic of intellectual paucity, spinelessness and greed [56g, pp. 270–271].

At age eleven Wiener wrote his first philosophical essay on *The Theory of Ignorance* [53h, p. 96]. In later years, after the publication of his 15 or so philosphical papers, he began to write mystery science fiction. Some of these stories remain in manuscript form in the Wiener Archives, but two were published, first in the *Technology English News* and subsequently in anthologies: *The Brain* [50k] and *The Miracle of the Broom Closet* [52b]. His major opus in fiction is the novel *The Tempter* [59e]. This portrays the lives of scientists and engineers caught in industry, and the temptations to surrender integrity for personal success. The leading characters, Woodbury and Dominguez, are prototypes of Sir Oliver Heaviside, a Wiener favorite among the non-establishment scientists, and Dr. M.I. Pupin, the well-known Serbian-American engineer, whom Wiener felt had fallen short of the highest standards of his vocation. But the bulk of Wiener's literary effort was outside the realm of fiction. It included his autobiographical books *Ex-Prodigy* [53h] and *I Am a Mathematician* [56g], his essay (with K. Deutsch) on *The Lonely Nationalism of Rudyard Kipling* [63c], and to an extent his writings in social philosophy, for instance his books *The Human Use of Human Beings* [50j] and *God and Golem, Inc.* [64e], as well as some essays.

The response to these books from the literary public was on the whole enthusiastic. While some of his opinions were questioned (and for good reason, cf. § 1B), there were no complaints about his prose. A few readers responded by writing letters to Wiener. Among them were working people, as we saw in § 19A, as well as scholars. Reproduced are two letters from the latter group to illustrate the kind of impact Wiener's writings had on other minds. The first (in longhand) is from the noted author, and scholar on technics and civilization, Lewis Mumford [MC, 128]:

4 November 1950

Dear Dr. Wiener:

      I cannot put down *The Human Use of Human Beings* without telling you how much I admire the scope and method and purpose of your argument. The combination of intelligence, human insight, and courage you display in that book sets a high level for the rest of us. I feel as if I should thank you, not just personally, but on behalf of the human race!

With warm respect,

Faithfully yours,
Lewis Mumford

Another letter is from the distinguished astronomer Professor V. Ambartsumian, then president of the Armenian Academy of Sciences in the USSR [MC, 322]:

January 25, 1963

Dear Professor Wiener,

Many thanks for the copy of your novel "The Tempter". I have just finished reading of the book and should like to say that it interested me very much. As you know we have here almost forgotten the times when there were private companies in our country. Therefore a novel describing the activities of companies as regards to application of new scientific ideas and technical inventions and the moral conflicts arising from these activities has opened to me an unknown aspect of the life of your country. Perhaps it will be useful for our young generation also to be acquainted with these problems.

Therefore I think that it is worth while to publish here a translation of this book and I suppose to try to do this through State Publishing House. Of course, this is only my intention and I write this to you only to show how much I was impressed by your novel.

Being an astronomer almost completely devoted to my science I have still some interest in mathematical problems and therefore I take this opportunity to send you my deep appreciation of your research work.

With kindest regards.

Yours sincerely,
V. Ambartsumian

Wiener's response to this letter is of some interest since it touches on the issue of international copyright [MC, 332]:

November 13, 1963

Professor V. Ambartsumian
President
The Academy of Science
Armenian S.S.R.
Yerevan
Barekamumian No. 24
USSR

Dear Professor Ambartsumian,

Many thanks for your letter of January 25, 1963 in which you expressed your interest and appreciation of the book *The Tempter*.

It was not quite clear that this letter implied your request for permission to translate my book into Armenian, but Professor Parseghian visited me this morning and told me that you were still waiting for permission. You and I know that with the absence of a copyright on books published outside the Soviet Union, such a permission is not necessary for you to go ahead. On the other hand, I appreciate greatly your asking for such a permission. I am thoroughly sympathetic with your project and should be highly complimented if you should proceed with your plans for such a translation.

In loyalty to my publishers over here, I must register a formal protest concerning the absence of copyright for foreign books. This protest, however, is purely formal as I know that nothing can be done about it. Therefore, privately and without committing my publishers in any way, I say "go ahead" with my best wishes. And do accept my apologies for not answering sooner.

Sincerely yours,
Norbert Wiener

In passing we may note that the USSR signed the International Copyright Convention in May 1973. The free reproduction and translation of important periodicals and books by learned societies and individuals on both sides of the curtain (which was a boon to libraries and working scientists throughout the world) ended, and instead came bureaucratic entanglements involving the availability of hard currency, permission-seeking and other spurious issues. Wiener's "go-ahead, with my best wishes" was the attitude of most enlightened representatives of the scientific community on this issue.

B Wiener's theatrical initiatives

Wiener was not merely a reader of mystery novels and science fiction; he also enjoyed watching them played out in the theater and in motion picture. He especially liked plots that centered on paradoxes.[1] Not surprisingly, Wiener wanted to see some of his own literary ideas put on the stage or in motion picture. In such situations he sometimes let his wishes be known to producers of plays and movies. The following letter to Alfred Hitchcock is a case in point. After two unduly apologetic paragraphs Wiener gets down to business [MC, 146]:

---

1   He once mentioned in this regard the movie *The Bridge On the River Kwai*. In this, a British officer's actions, conforming to the best moral code of the British army, and the actions of a British commando unit, just as dutiful, combine in in a way that inexorably leads to catastrophe.

February 4, 1952

Mr. Alfred Hitchcock
Care of The Directors' Guild
Hollywood, California

Dear Mr. Hitchcock:

I am very much an admirer of your technique of the suspense and horror movie, and I appreciate the amount of careful thought and construction which have gone into your work. Like many of your audience I have from time to time believed that I have come across situations which could be employed by you. I know how much you must be annoyed by chiselers who approach you with the intention of tying you up in a plagiarism suit as well as by legitimate but stupid correspondents who do not realize the risk you run in answering fan mail and the time it takes you.

Let me tell you just who I am in order to establish a presumption of my good faith and that you may know that I stand in a position in which I should be absolutely destroyed by any breach of good faith. My name is Norbert Wiener and I am Professor of Mathematics at the Massachusetts Institute of Technology. I have recently published two books. *Cybernetics* was published by John Wiley and Sons and is concerned with communication and control in the machine and in the animal. *The Human Use of Human beings* was published by Houghton-Mifflin. You will find my personal record in *Who's Who* or in *American Men of Science*.

Now to business. I have recently been in Mexico working in a scientific laboratory where I have run into a combination of characters and even of possible situations lending themselves ideally to a suspense and horror movie of the type in which you are expert. Together with my daughter, Miss Peggy Wiener, and an American doctor, Dr. Morris E. Chafetz, we have written a synopsis of the proposed movie. Because we are without experience in screen technique, we have not attempted to go further and write it up as a scenario. This synopsis has been registered with The Authors' Guild in New York, and we should be delighted to pass it on to you for examination if and when it should be possible.

I can understand the difficulty a man in your position finds with unsolicited material, and I shall be perfectly ready to accept a letter from you indicating that you do not wish to receive the document. I should regret this, both from my own point of view and from yours as I am sincerely convinced that we have laid our hands on an original and amusing situation.

If, then, you see any way clear to submitting our document either directly to yourself or to some person who can report to yourself, I shall be most delighted. Please let me know of your decision at the earliest opportunity.

Very sincerely yours,

Norbert Wiener
Professor of Mathematics

We do not know what response, if any, Hitchcock made to this letter.[2] In a similar vein, when Wiener found something in a story he was reading that struck him as being incorrect or off balance or for which he had a better idea, he sometimes took the liberty of making this known to its author. A letter of this kind is the one he wrote in 1933 to Paul de Kruif, who wrote extensively in the *Saturday Evening Post*. This remarkable document has over six typed pages and it will suffice to quote a couple of excerpts, and briefly describe what has been left out [MC, 38].

August 3, 1933

My dear Mr. de Kruif:

> I have read several of your books on scientific biography with great interest, as well as Lewis "Arrowsmith" to which you contributed so much of the background and I have recently been following your story of Kettering in the S.E.P. As a practicing scientist myself, I naturally set a high value on the popularization of science and of the circumstances under which science is carried on. I wish to congratulate you for your pioneer work in a little-exploited field. May I then, without giving offense, submit to you a few questions and criticisms?
>
> First of all, you leave the impression in the mind of the reader that applied science, followed with an eye to the benefit of humanity, is perhaps the most worthy of all pursuits, but that pure science, the fruit of mere curiosity, is likely to be dilettantish and fruitless. This impression may not be intended by you: it may be due to the fact that you are writing of Kettering, who is primarily a practician. I feel, however, that is your own opinion, and I wish to protest against it.

Wiener goes on to describe his typical day at MIT; his courses, his contacts with engineers, young and old. He then comments on his immediate circle of colleagues and the little use they have for "organized research". After a philosophical discussion of how good research, both pure and applied, is done, he charges de Kruif with "misplaced emphasis" in his portrayal of C.F. Kettering, and points to Oliver Heaviside as the real discoverer of the distortionless telephone line. The letter ends with the words:

> Heaviside, by the way, should be rich literary material for you.
>
> In this connection, I should like to see a series of biographies of engineers from your hand. Other names could be Nicola Tesla and B.A. Behrend. I should be glad to render you any possible help in the matter.

---

2 Unfortunately, correspondence initiated by Wiener is not indexed in the Wiener Archives, and we were unable to search through all the folders for a possible response.

I trust that you will not take my remarks amiss. If they interest you in any
way, I should be delighted to hear from you.

Again we regret that we do not know what response Wiener received.

C The mathematician's credo

The letter to de Kruif has three paragraphs in which Wiener spells out
the philosophical creed of his immediate circle of colleagues:

> Mathematics is a subject worthy of the entire devotion of our lives.
> We are serving a useful place in the community by our training of engineers,
> and by our development of the tools of future science and engineering.
> Perhaps no particular discovery that we make may be used in practice;
> nevertheless, much of the great bulk of mathematical knowledge will be, and
> we are contributing to that bulk, as far as lies in us.
>
> Moreover, a clearly framed question which we cannot answer is an
> affront to the dignity of the human race, as a race of thinking beings.
> Curiosity is a good in itself. We are here but for a day; tomorrow the earth
> will not know us, and we shall be as though we never were. Let us then master
> infinity and eternity in the one way open to us; through the power of the
> understanding. Knowledge is good with a good which is above usefulness,
> and ignorance is an evil, and we have enlisted as good soldiers in the army
> whose enemy is ignorance and whose watchword is Truth. Of the many
> varieties of truth, mathematical truth does not stand the lowest.
>
> Since we have devoted our lives to Mathematics—and she is no easy
> mistress—let us serve her as effectively as we may. If we work best with an
> immediate practical problem in view, well and good. If mathematical fact
> comes to our mind, not as a chain of reasoning, built to answer a specific
> question, but as a whole body of learning, first seen as in a glass, darkly, then
> gaining substance and outline and logic, well and good also. The whole is
> greater than the parts, and in a lifetime of achievement, no one will care what
> particular question of practice was in the scholar's mind at such and such a
> moment. [MC, 38, p. 4, 5]

In the letter, Wiener claims that this philosophy is molded by the
German poets Schiller, Goethe and Heine, and "that magnificent line of
philosophers from Kant to Schopenhauer". The implied association of
modern mathematics with German Romanticism appears in an earlier
article by Wiener on *Mathematics and Art* [29h].

D Why mathematics is a fine art

To turn to Wiener's concept of art, let us juxtapose our last quotation
of the mathematician's creed with the next from the paper [29h] on
mathematics and fine art. After referring to the attempts of the mathematicians A. Speiser and G.D. Birkhoff to relate mathematics to artistic
design and artistic beauty, Wiener writes:

> Our thesis is not that the arts are an expression of mathematics through the senses, but that mathematics itself is in the strictest sense of the word, a fine art. In this the author finds himself in complete accord with the views expressed by Havelock Ellis in "The Dance of Life". [29h, p. 129]

This is amplified in the shorter theses:

> ... mathematical work may produce an emotion indistinguishable from that of aesthetic contemplation; that mathematical investigation may and often does have as its goal the production of a work capable of exciting this emotion; that the creative mathematician is limited by the requirements of rigor only as any creative artist is limited by the nature of his medium; and finally, that mathematics has participated intrinsically in all the larger movements common to the several arts. [29h, p. 129]

On the last of these theses, Wiener provided more literary flourish than substance. According to him the Greek ideal of "perfection within limitation" is mirrored in Euclid's geometry which will not countenance the use of numbers and graduated instruments; the solid unspirituality of 17th-century writing, e.g. Molière, Pope and Swift, is reflected in Newton's healthy interest "in mathematical facts rather than mathematical proofs", and so forth. Turning to the 19th-century romantic movement, he wrote:

> It may seem far fetched to refer to the new impulse toward mathematical rigor as a manifestation of the romantic movement, but romantic it is in its subjectivism. It is not the theorem which now forms the chief center of attraction: it is the proof of the theorem. [29h, p. 132]

On the earlier theses, however, Wiener's offering is more substantial, and there are in it a couple of pithy points that demand attention.

The first addresses a seemingly insurmountable objection to viewing mathematics as a fine art: whereas the artist is consciously working to reproduce in his audience the particular emotion that fills his being, it is *not* the mathematician's creative emotion but rather his thought that his work is meant to relay to his audience. To this objection, Wiener responds as follows:

> The purpose of the musician or plastic artist varies according to whether the artist is, on the one hand, freely creative or, on the other, is producing a piece of work to fit into a preassigned operative scheme or decorative ensemble. At his best we speak of the artist as a *"fine" artist* and thus distinguish him from the less free, more subservient *artisan*. It is very difficult to show just what the purpose of the "fine" artist is. He probably is more *governed by an internal, compelling force* than by any philosophical reflection as to the emotion which a particular work of art may excite. Insofar as this compelling force may be rationalized into words it says to him, "This is beautiful. Let

me display it to the world." In any case it is not a matter of the artist laboring to express the emotion to the public but rather of *the emotion endeavoring to express the artist.*

The relation between the mathematician, his public, and his emotion is quite the same. You may rationalize his thoughts into some such statement as "this is an interesting idea. Let me show my colleagues—my public—where it leads". This is quite similar to the rationalized statement of the artist's impulse, and like it, is *quite false. The mathematician does research because the research demands to be done.* [29h, pp. 130, 131] (emphasis added)

This attitude towards the aesthetic impulse comes close to that of the great scholastic thinkers, e.g. Meister Eckhart, when he wrote: "What I say, springs up in me, then I pause in the idea, and thirdly I speak it out" or Dante, when he said: "Who paints a figure, if he cannot be it, cannot draw it", cf. A.K. Coomaraswamy {C11, pp. 7, 175–178}. In his masterful treatment, Coomaraswamy has summed up the scholastic view of the aesthetic intuition in the words:

Whatever object may be the artist's chosen or appointed theme becomes for the time-being the single object of his attention and devotion; and only when the theme has thus become for him an immediate experience can it be stated authoritatively from knowledge. {C11, p. 7}.

He adds: "Here indeed European and Asiatic art meet on absolutely common ground", meaning by "European" pre-Renaissance European, {C11, p. 7}. He points out how the European view began deviating from this ideal with Leonardo da Vinci {C11, p. 177}.

In regard to the Wiener and scholastic viewpoints on art two remarks are in order:

(1) Wiener wisely speaks of the *artisan,* but does not make it plain that his art can be as great as the greatest fine art. The scholastics left no doubt on this issue. As Sister Healy states in her study of St. Bonaventura:

The philosophers of the 13th century placed the productive activity of the artisans—the maker of furniture or builder of houses, on the same level with the human creative activity which inspires epics and finds expression in gorgeous cathedrals and marble statues. The mechanical arts as well as the liberal arts were possessed of the "ratio artis", and the distinction between them did not consist in the superiority of the artistic activity as such but upon differences in the processes employed. {B15, p. 82}

(2) In his article on *Aesthetics* [18b] in the *Encyclopedia Americana,* written in 1917, Wiener deals at length with its history, starting with

Plato's *The Idea of the Good*. But he dismisses the period from Plotinus to the Renaissance in the single sentence:

> ... although we do find treatments of aesthetic problems from the time of St. Augustine to that of St. Thomas, these have been singularly arid and without fruit in modern aesthetic theory. [18b, p. 199]

It is reassuring to know that by 1929 Wiener's own understanding of the aesthetic impulse was near scholastic in the best sense of the word, Occidental or Oriental. And this evolution came from introspection of his own creative experience "when the several threads of thought in a mathematical theory are seen to gather themselves together into one perfect fabric" [29h, p. 130].[3]

The second pithy observation of Wiener in [29h] is his reference to the work of Einstein to elucidate the analogy between the "double aspect" of mathematical work and the *presentative* and *representative* aspects of art:

> Einstein developed his full gravitational theory in the beginning as a *tour de force,* as a possibility, interesting, whether true or false. That it was ultimately to be verified reflects enormous credit on his intuition, but even unverified, it already existed as a mathematical essence, and this essence, this construction, parallels closely the goal of the effort of the modern artist.
>
> This double aspect of Einstein's work, and indeed of all physics, may serve as a final link between mathematics and the arts. As is well-known, most of the arts possess both a presentative and a representative aspect. A painting has beauty not merely as a study in abstract design but as a representative of the outer world. No poem is so purely an intellectual work that it is wholly without a pictorial aspect. Thus mathematics, too, besides the beauty of inner structure, has a further beauty as a representation of reality. This is most clear in mathematical physics but even in the purest of pure mathematics, mathematical physics often serves as a valid if unconscious guide. Many a pure mathematical study is an impression of some chord of the physical world. [29h, pp. 160, 162]

Thus the mathematics, which is a fine art, includes not only the work of G.H. Hardy but also that of Einstein. Whether a mathematical inquiry is initiated to settle a point of physics (in the wide sense) or of mathematics is of little consequence. What is of considerable consequence is whether the abstraction resulting from the inquiry is aesthetically appealing and rich enough to sustain a deductive system of sufficiently wide scope.

3   Unfortunately Wiener did not revise his article on *Aesthetics,* and so the 1960 edition of the *Encyclopedia* carries the same ignorant remark on the scholastic period as the 1917 edition.

E The variance between Wiener's and Halmos's views on mathematics and art. The scholastic position

It is interesting to compare Wiener's article on *Mathematics and Art* [29h] with one written forty years later with a title which Wiener could have used for his own, viz. *Mathematics as a Creative Art,* by P.R. Halmos {H3}. In this the conclusion reached in the last paragraph is flatly denied. Halmos writes:

> As I see it the main difference between mathophysics and mathology is the *purpose* of the intellectual curiosity that motivated the work. . . {H3, p. 384}

So far so good—all would agree that Hardy and Einstein had different purposes. But then Halmos adds:

> The mathophysicist wants to know the facts, and he has, sometimes at any rate, no patience for the hair-splitting pedantry of the mathologist's rigor (which he derides as rigor mortis). The mathologist wants to understand the ideas, and he places great value on the aesthetic aspects of the understanding and the way that understanding is arrived at; he uses words such as "elegant" to describe a proof. In motivation, in purpose, frequently in method, and almost always, in taste, the mathophysicist and the mathologist differ. {H3, p. 385}

Here we run into difficulty. In the first place, "rigor" becomes "rigor mortis" not just for our Diracs and Eddingtons, but also for our Eulers, Ramanujans, Galois and other such mathologists. Second, and more important, the habit of placing "great value on the aesthetic aspects of the understanding and the way that understanding is arrived at" extends well beyond the circle of mathologists, and certainly includes the great mathophysicists. For in their mathophysical quests, Pythagoras, Archimedes, Kepler, Newton, Lagrange, Hamilton, Clerk Maxwell, Einstein, de Broglie, Schrödinger, Weyl, Wiener and von Neumann, among others, were very much alive to the aesthetic origins and aesthetic import of their ideas, some seeing in them reflections of "celestial harmony", "the order of Nature" and "Spinoza's God". And much of the most sublime mathologistic architecture is founded on conceptions unearthed by these aesthetically conscious mathophysicists.

We have therefore to reaffirm Wiener's conclusion in [29h] that from the standpoint of aesthetics the division into mathology and mathophysics is of no importance; good mathophysics is as much a creative art as good mathology. And mediocre mathology can be as unartistic as mediocre mathophysics.

But a moment's reflection suggests that these dicta must apply to the entire spectrum of human labor. For if there is an aesthetic aspect to the activity of a good mathophysicist, how can this not be the case for the activities of a good experimental physicist or a good geologist or a good carpenter, or for that matter a good manufacturer of roofing and siding? We are thus irrevocably led to the all-embracing concept of art, which Coomaraswamy expressed in the words:

> The artist is not a special kind of man, but every man is a special kind of artist. {C11, p. 64}

This, as we saw earlier, is the scholastic view of art, best enunciated in the writings of the 12th-century Franciscan, St. Bonaventura. In his *Retraction of the Arts to Theology* {B15} he lists as arts not only music, architecture, painting and sculpture, but also farming, weaving, cooking and other forms of medieval industry. This view was dominant during the Middle Ages, but began to decline after the Renaissance with increasing emphasis on individual virtuosity in the fine arts. The view has suffered a near-eclipse with the systematic destruction of handicraft manufacture and its replacement by alienated wage labor that came with the Industrial Revolution. But the all-embracing conception of art retains its authenticity, and Wiener's own views on art, pushed to their logical limits, merge with this scholastic conception.

There is another fundamental, but less articulated, difference in the Halmos and Wiener viewpoints expressed in the articles {H3} and [29h]. Halmos seems to be sure that good mathological results will prove "useless" for physics. Wiener, aware that the mathematics of a Riemann or a Minkowski can be an asset to the work of an Einstein, and the logic of a Russell or a Gödel to the physiological work of a McCulloch, was inclined towards the opposite view. Wiener was also aware, of course, that the pen of an engineer such as Oliver Heaviside could be lucrative for mathology in the long run. He thus viewed mathology and mathophysics as two parts of a dialetic movement, in which each wing strengthens the other.

This Wiener viewpoint is in complete accord with the Pythagorean-Platonic tradition, which affirms the efficacy of pure mathematics in the empirical domain. As the great modern exponent of this tradition, A. N. Whitehead put it:

> The paradox is now fully established that the utmost abstractions are the true weapons with which to control our thought of concrete facts. {W8, p. 41}

F Wiener's thought on machine creativity

Gödel's arithmetization of syntax in the late 1930s shows how to any given expression (i.e. finite sequence of symbols) or to any given finite sequence of expressions of a formal language, a unique integer 1, 2, 3, ..., can be assigned, its so-called *Gödel number*. The Turing machine has the ability to recover from the Gödel number the underlying sequence of expressions, and to test if this sequence constitutes a proof. Thus, given enough time, the machine can discover all the proofs obtained by mathematicians so far, including the beautiful ones, and discover new theorems, unknown to humans, some deep. Likewise, a Turing machine can be designed to write words and sentences indefinitely, and so eventually compose some literary masterpieces. Similarly, the machine can be designed to draw and paint limitlessly, or to compose sounds.

Does such machine discovery and production share the creative aesthetic element that Wiener and Halmos assign to good mathematical work, and which the medieval scholastics discerned in all worthy, well-executed human endeavor? Wiener addresses this issue indirectly in the discussion on long-time and short-time institutions at the end of his book [56g, pp. 360–365] (cf. § 21E). In this, Wiener brings in "the writing shop of monkeys and typewriters" to illustrate his contention that only a small part of the research put out by a bureaucratically-run science laboratory will be consequential.[4] His remarks bear, however, on the issue under consideration.

> What is the real value of the work of the monkeys and the typewriters? Sooner or later, they will have written all the works of Shakespeare. Are we then to credit this mass attack with creating the works of Shakespeare? By no means, for before writing the works of Shakespeare, they will almost certainly have created just about all the nonsense and balderdash conceivable.
>
> It is only after the non-Shakespearean has been thrown away, or at least an overwhelming part of it, that Shakespeare will stand out in any significant sense, whether theoretical or practical. To say that the monkeys' work will contain the works of Shakespeare has no other sense than to say that a block of marble will contain a statue by Michelangelo. After all, *what Michelangelo does is purely critical*, namely, to remove from his statue the unnecessary marble that hides it. Thus, at the level of the highest creation, *this highest creation is nothing but the highest criticism.* [56g, p. 364] (emphasis added)

Only a contingent and noise infested cosmos has room for art and creativity. In a strictly deterministic world these concepts disappear. Creativity is criticality, i.e. the ability to tell message from noise, and create message without producing more than minimal noise. Wiener, of course, knew that an automaton could beat the human in an intellectually difficult game such as chess, and saw in the computer a prosthesis for the human brain, cf. §§ 20D, 16E. Nevertheless, he regarded *criticality*, the ability to spot message and desist from waste, to be a very important constituent of *intelligence*, and he felt that automata were

---

4   Cf. Halmos {H3, p. 382}: "I don't think a team of little Gausses could have obtained the theorem about regular polygons under the leadership of a rear admiral any more than a team of little Shakespeares could have written Hamlet under such conditions."

deficient in this regard. By virtue of their low impedance and high speed, electronic automata avoid a lot of waste of energy and time, and so possess an elemental criticality. With their present design, however, their activity forces them to wade through heaps of balderdash; they thus fall short of Wiener's standard. Wiener felt that intelligence can be arranged in a hierarchy of types,[5] and that the type of the most intelligent of present day automata is much lower than that of the human brain. He was confident, however, that increasing knowledge of cerebration will bring under construction intelligent automata of higher and higher types. From the methodological standpoint, he maintained a scientific posture and did not slip into the quagmire of "anti-reductionism".

5   The universality of the Turing machine does not affect this issue, Wiener would have contended, since its universality does not enhance its criticality.

23　Wiener, the Man and the Teacher: Authenticity and Prejudice in his Attitudes and Writings

A　The lovable quirk

Wiener's absent-mindedness, quirkishness and idiosyncrasy, amusing and even endearing, lent themselves to easy anecdote. The stories varied in sobriety. Here is one.

Wiener used to lunch in the Faculty Club in the Sloan Building of MIT. Around noon he would walk from his office to the Club and back again. During one such walk he encountered an old friend whom he had not seen for a long time. It was a balmy day. They chatted amiably, admired the trees, the Charles and its sailboats. At last, they said good-bye. But as the friend departed, Wiener, looking bewildered, stood still. "By the way" he asked, "which way was I headed when we met?" "Why, Norbert, you were headed towards your office," the friend replied. "Thanks," said Wiener, "that means I have finished lunch".

With the substitution "Walker Memorial" for "Sloan Building", and minus the embellishments (balmy weather, etc.), this story is true. The encounter occurred in 1929 with Ivan A. Getting, then a physics freshman and an organist, whom Wiener had met previously at a demonstration of a new electric organ[1]. Dr. Getting also tells us of a tennis practice in which, after failing to connect any of nearly 100 serves from him, Wiener suggested that they might exchange rackets.

The mathematicians who knew Wiener best were his erstwhile student and colleague from 1936, Professor Norman Levinson, and his colleague after 1926, Professor Dirk Jan Struik. Both have written on Wiener's life {A8}, {S10}. However, a good "first look" is portrayed in an article by Hans Freudenthal, who barely knew him:

> In appearance and behavior, Norbert Wiener was a baroque figure, short, rotund, and myopic, combining these and many qualities in extreme degree. His conversation was a curious mixture of pomposity and wantonness. He

---

1　During World War II, Dr. Getting was appointed Director of the Radar and Fire Control Division at the MIT Radiation Laboratory, and he enlisted the services of mathematicians such as Ralph Phillips, Withold Hurewicz and others. Several allied victories in the air war are attributed to his work on radar. He retired a few years ago as vice-president of the Aerospace Corporation in Los Angeles.

was a poor listener. His self-praise was playful, convincing, and never offensive. He spoke many languages but was not easy to understand in any of them. {F5, p. 344}[2]

Levinson has described his experiences as a student in Wiener's postgraduate course ("really a seminar course") in 1933:

> He would actually carry on his research at the blackboard. As soon as I displayed a slight comprehension of what he was doing, he handed me the manuscript of Paley-Wiener for revision. I found a gap in a proof and proved a lemma to set it right. Wiener thereupon sat down at his typewriter, typed my lemma, affixed my name and sent it off to a journal. A prominent professor does not often act as secretary for a young student. He convinced me to change my course from electrical engineering to mathematics. He then went to visit my parents, unschooled immigrant working people living in a run-down ghetto community, to assure them about my future in mathematics. He came to see them a number of times during the next five years to reassure them until he finally found a permanent position for me. (In those depression years positions were very scarce). {L8, pp. 24–25}

This little story is more telling of Wiener, the man, than the earlier impression. But Levinson hastens to add:

> If this picture of extreme kindness and generosity seems at odds with Wiener's behavior on other occasions, it is because Wiener was capable of childlike egocentric immaturity on the one hand and extreme idealism and generosity on the other. Similarly his mood could shift quickly from a state of euphoria to the depths of dark despair. {L8, p. 25}

When from personality and character we turn to Wiener's mind, D.J. Struik's observations are germane:

> ... the first impression was that of an enormous scientific vitality, which the years did not seem to affect. The second was to certain extent complementary, and that was of extreme sensitivity. Complementary indeed, since a man with heart and mind so close to nature and the technique of his time must have had very fine antennae; he sees, or believes he sees, he feels, or believes he feels, where others remain unresponsive. {S10, p. 35}

My own observations of Wiener during the last eleven years of his life[3] corroborate these testimonies, although the word "wanton"

---

2   In this otherwise apt description, the ambiguous term "wantonness" is totally inappropriate, and is perhaps indicative of inadequate acquaintance with the English language.

3   In the earlier period, I knew Wiener when I was a student at Harvard (1942–1946), but my acquaintance with him was minimal. Wiener knew me enough to greet me and talk about Ramanujan or his acquaintances among Indian mathematicians or on topics such as the Moghul emperors of India, but I doubt if he could have recalled my name. Although I attended his talks at Colloquia and other meetings, the first occasion when I discussed mathematics with him was during a brief encounter at Princeton in 1947, and over

would not occur to me to use for Wiener. Although I know of no instance of great personal sacrifice on his part on another's behalf, the Wiener I knew had a warm heart that felt the suffering of another. The three testimonies address very different facets of Wiener, the man. In this biography our subject is Wiener, the message (Wiener "modulo noise"). We are interested, not in Wiener's myopia (visual or psychic) or in his emotional ups and downs, but in his service to mathematics, to science and to mankind. Since it was Wiener's mind and heart that propelled him forward in this service, our perspective so far has been nearer to Struik's. But in this section we have also to deal with certain shortcomings of the Wiener-signal caused by Wiener-noise, i.e. with how evil interfered with mind and heart.

We have already alluded to the bias and inaccuracy in some of Wiener's writings in § 1B, and there and in Chs. 3 and 4 have referred to the torment and suffering in Wiener's life, and to his refuge in egoism and hostility. We have spoken of his slow advance up the academic ladder, the absence of outside offers, and the internal tension that this created (Ch. 8). Some of his shortcomings were those of his father's. On the other hand, Wiener did inherit his father's sterling qualities. We must now narrate how the positive and negative aspects of Wiener's being blended in the Wiener-signal, and how this affected other people.

B Wiener as public figure and teacher

When Wiener's name began to appear in newspapers, he started receiving letters from strangers asking all sorts of questions. A human being reveals himself to an extent in the way he handles this situation. Mahatma Gandhi, for instance, made it a point to respond in some way to every letter he received. A good portion of the 70 volumes of his *Collected Works* {G2} comprise correspondence. One such letter, from an Attica Prison inmate under life sentence for murder, was the following [MC, 49]:

P.O. Box 149
Attica, N.Y.
September 19, 1938

Dear Doctor:

In the September 14th issue of the New York Times I read of your discovery of a new Calculus.

For the past seven years—during which time I have been in prison —I have devoted all of my time to the study of calculus, including Finite

a six-week period in Bombay in 1953–1954. This led to collaboration at the Indian Statistical Institute in Calcutta in 1955–1956 and at MIT in 1957–1958, and to intermittent exchange of ideas thereafter at MIT and at his summer home in South Tamworth, New Hampshire.

differences, Quaternions, Harmonics, etc. This new subject of which you speak interests me a great deal—especially when you mention the "Perturbations" of the Planets.

In my position it is very difficult to obtain literature, reference books and other material which is indispensable to one who is interested in studying a subject thoroughly. For that reason might I prevail upon you to send me something on this new science in the field of Mathematics that deals with "Chaos"?

Any consideration you may give this request will be deeply appreciated.

Respectfully yours,

Frank J. Scimone
#1158

Wiener replied with unusual promptness [MC, 49]:

September 23, 1938

Mr. Frank J. Scimone
#1158
P.O. Box 149
Attica, New York

Dear Mr. Scimone:

I was very much interested in your note of the 19th. I shall send you in the near future as complete a collection of my reprints as I can muster. In the meantime I should like to hear something more about the scope of your reading in Mathematics so that I may be able to suggest and possibly put at your disposal the necessary introductory material.

As to your doing scientific reading and scientific work in prison, it is by no means impossible to do very effective work under such circumstances quite apart from the personal advantage to yourself of a long-time interest as occupying as mathematical studies. One of the leading young mathematicians of France is a man by the name of Bloch who has been confined for years in the hospital for the criminal insane at Charenton, near Paris, for murdering three people in a fit of insanity. Of course this man had done valuable work before his arrest but some very important papers of his have appeared since his confinement and at least one of these was in collaboration with outside mathematicians.

While I have not the facts at hand, I have heard of several cases where war prisoners in concentration camps carried on important research under conditions of confinement, therefore, I am taking your interest in mathematics very seriously and should like to hear something about your studies and to give you any advice in my power as to profitable ways of continuing.

Sincerely yours,
Norbert Wiener

There followed two more letters. The first from the prisoner, dated September 25, 1938, begins with the words [MC, 49]:

Your letter dated September 23 was most inspiring. It is difficult to explain in writing the grand reaction I experienced when your letter arrived explaining your interest. After seven years imprisonment it is good to know that someone takes an interest in my educational development.

The second letter from Wiener informed Scimone that Levinson and he were sending him Hardy's *Pure Mathematics,* Goursat's *Cours d'analyse* and Osgood's *Lehrbuch der Funktionentheorie,* [MC, 49, September 30].

From the other end of the spectrum of strangers, Wiener received the following short letter [MC, 125]:

> My dear Dr. Wiener,
>
> I am a Victorian. I should like to know what kind of world my grandson is going to live in. Will you kindly give me the names of some of your books on the new industrial revolution—referred to in the enclosed clipping. Some nice easy ones—pleasant to read.
>
> Sincerely yours,
>
> Ella J. Meyer
> (Mrs. J. Franklin Meyer)
>
> 3727 Jocelyn St.
> Washington, D.C.
> September 21, 1950

We could not find out what reply, if any, Wiener made.[4] It is fair to surmise that in this area too, Wiener's conduct could range from extreme concern to helpless indifference.

To turn to Wiener's relationships within the academic community, his most noticeable shortcoming was as a teacher—"a famously bad lecturer" is Freudenthal's apt description {F5, p. 344}. Struik remarks how Wiener could "occasionally lull an audience to sleep", but how he was also able to hold groups of colleagues and executives "at breathless attention while he manipulated his ideas and flares of vision" {S10, p. 35}. Levinson spoke of him as "a most stimulating lecturer", but at the seminar level {L8, p. 24}. Only very rarely, however, did Wiener's lecturing attain these peaks of exultation or depression. For the most part it hovered not too far from the minimum: it was chaotic, but amusingly and not irrevocably so. A typical, off-par Wiener lecture may be described in the words used by Freudenthal in a somewhat different context:

> After proving at length a fact that would be too easy if set as an exercise for an intelligent sophomore, he would assume without proof a profound theorem that was seemingly unrelated to the preceding text, then continue with a proof containing puzzling but irrelevant terms, next interrupt it with a totally unrelated historical exposition, meanwhile quote something . . . and so on. {F5, p. 344}

---

4   A limited search of the folders in the Wiener Archives yielded nothing.

Some of this chaotic style enters now and then in Wiener's scientific writings, especially in situations where he was the sole referee. But most of his persuasive discourse, whether verbal or written, reveals a very high order of coherence when analyzed carefully. Perhaps it was for this reason that his presence at scientific meetings and discussions was eagerly cherished.

It was from his father that Wiener had learned that "scholarship is a calling and a consecration, not a job" [53h, p. 292]. On almost every issue he addressed, Wiener brought to bear his deep insight and his power to single out the fundamentals from the plethora of subordinate facts. Invariably his prose was excellent, and almost invariably his presentation was coherent and cogent. An address [60c] to medical men on the applications of physics to medicine provides a fine example of Wiener's unconventional but penetrating and fundamentally germane approach. Wiener began by contrasting Newtonian physics with its time-reversibility principle with medical science with its dependence on diagnosis, prognosis and therapeutics. He then informed the doctors that in dealing with the fire control problem he proceeded more like a doctor than a Newtonian physicist. Next he alluded to the game theoretic aspects (in the von Neumann sense) present in problems of both national defense and public health. He then cautioned the doctors against assuming that medical science and physics are irreconcilable, by emphasizing the asymmetry of time in all the experimental parts of physics, the importance of collimation, the presence of noise, and the necessity for statistical analysis. Then followed a discussion of invariant measures, the work of Gibbs and of Borel and Lebesgue, and its synthesis in Birkhoff's ergodic theorem. Finally he mentioned brain wave encephalography and the study of homeostasis as fruitful areas for the application of physical ideas in medical research, citing leukemia as an interesting case.

Another example of a good Wiener lecture was the one he gave to a group of military officers at the Industrial College of the Armed Forces on February 10, 1953; [XII]. In this Wiener described negative feedback as "continual reconnaissance" and adhered to a colloquial idiom:

> Since I am talking to a military group, I am going to present feedback in military terms. Feedback is the engineers' word for reconnaissance; for not only occasional reconnaissance, but for a process of continual reconnaissance, on the basis of which the action is adjusted or changed. In the good old days of military doctrine before the First World War, your Army marched with a point and with all advance guard, and so on. These were there for feeling-out, and were meant to be driven in by the enemy, to show what the enemy was trying to do, so that we could come to grips with him. On the basis of this continual information we were able to act intelligently.

Now notice that this is not confined to military strategy. When we drive a car we drive it by a continual reconnaissance. When we hold the wheel, we do not drive according to a pattern. Say we are going through a right angle; we shall turn the wheel so many degrees and after such and such a time turn it back. Try that, and you will be in trouble with the pavement and with a policeman before you know it. What we do if we find ourselves going to the pavement is turn into the middle of the street; if we find ourselves trying to cross the line in the middle of the street, we go back toward the pavement. In other words, we correct errors as they occur. Error itself is fed back into the machine as a correction.

I lift this piece of paper. What sort of commands do I give through my muscles? I can tell you this much—there is not an anatomist alive who could do that by specifying which muscles he would ask to act and in what order. That is not the way we do it. You want to pick the paper up; you move in such a way as to reduce the amount by which you have still to move in picking up the paper. That is a negative feedback, but negative feedbacks are quite as possible to make mechanically as they are to make humanly or militarily; and you can control machine work by negative feedback. [XII, p. 4, 5]

Later on in the lecture Wiener said:

One-shot weapons don't combine well with reconnaissance and are exceedingly dangerous for everybody. They are guns which kick almost as much with the butts as they do with the bullet. [XII, p. 6]

He was speaking of the atomic bomb.

A great deal of Wiener's speaking and writing fell in this high category, and to it Freudenthal's designation, "chaotic style" does not apply at all. His autobiographies [53h, 56g], for instance, are certainly not chaotic, and while they do occasionally display a narcissistic streak (cf. § 1B), they do not convey "an extremely egocentric view of the world" as Freudenthal seems to think {F5, p. 344}.

If more of Wiener's lectures did not come in this high category, it was partly due to factors beyond his control. He was temperamentally unsuited to teach a regular postgraduate course on harmonic analysis. Nevertheless he was asked, like any other professor, to teach such a course. The initial enrollment of about a dozen would soon drop off. In one semester, it dwindled to one, and that too an auditor. Levinson has written how in his later years, Wiener lost touch with mathematicians under the age of 35 {L8, p. 30}. But this failure must be viewed in the context of American culture: these younger mathematicians were not attuned to Wiener's mode of thought.

In his later writings, Wiener was wont to present penetrating new ideas without actually demonstrating their feasibility. A typical instance is the 2-page Note to the last chapter of his 1948 *Cybernetics* [48f, 61c]. This concerns the construction of a machine that plays good chess by following a minimax technique. The idea was quite sound, but Wiener never followed up with a paper on the designing or programming of such a machine. This was done by C.E. Shannon in 1950 {S8}. Such writing, useful though it may be, is not well received in scientific circles, especially when the author makes it a habit.

C Resignation from the National Academy of Sciences; muddled ideas on science academies

Wiener's reflections about certain institutions and individuals were biased and inaccurate. His attitude towards the National Academy of Sciences of the United States is a case in point. As we saw (§ 9E), Wiener was elected a fellow in 1934. But his relationship with several members of the Academy was shaky, and in 1941 he resigned. This event caused unhappiness in the scientific community. His avowed reasons for resigning are stated in the following letter he wrote to Dr. Frank Jewett, then president of the Academy [MC, 60], but there was more to it:

53 Cedar Road
Belmont, Mass.
September 22, 1941

Dr. Frank B. Jewett
President of the National Academy of Sciences
2101 Constitution Avenue
Washington, D.C.

Dear Dr. Jewett:
It is with great regret that I read of your prolonged ill-health, and with great pleasure that I hear that you are now coming into shape again. I am sorry to continue to disturb you on Academy matters, particularly after your very kind letter, which I sincerely appreciate. Nevertheless, I feel that I must do so.

The Academy operates in at least three distinct roles, and to my mind these roles are not compatible with one another. It is at least a quasi-official agency of the United States Government, entrusted with the advice of the Government on scientific matters. It is the custodian of certain journals and funds for research. It is a self-perpetuating society of restricted membership, considering the gift of that membership as a high honor on the recipient, among other honors and prizes which are also within its gift.

As a government agency, it is distinguished from most others by possessing a personnel concerning which no other department of the government has any say, either as to term of office, or as to appointment. The corps of officers of the Army, the Navy, and other related services share the long term of office of the Academy; but their appointments are much more definitely regulated by Congress. This is likewise true of the judiciary. I know no other important case besides the Academy in which Congress, after appointing an organization as a government agency, has completely left it to its own devices, and has conveyed a continuing authority upon a self-perpetuating and (in the strict sense of the word) irresponsible body of men.

This is somewhat glazed over by the fact that Congress has incurred no financial responsibility for the Academy, either in the matter of salaries or otherwise, and that the Academy is maintained by dues. However, the fact remains that the Academy is accustomed to regard itself as a government agency, to ask diplomatic privileges for its official representatives in their travels, and in other ways to speak as the scientific mouthpiece of the United States of America. As such, we have a government agency which bestows titles of honor; which is based on the principle of superordination and subordination, not of functions in an organization, but of personalities; and which in many other ways is in glaring contradiction with the declared principles of the United States of America as well as the actual practice in which these principles are embodied in the government at large.

With the second function of the Academy, that of the custodian of certain journals and funds for research, I have no quarrel, providing that the Academy accepts a position simply on the same level as that of other agencies with a like custodianship. I have no sympathy whatever with the Science Fund idea, which seems to me an excellent means to discourage independent gifts to science, and to stifle all work not pleasing to whatever group is at the moment running the scientific politics of the country. I say this with full respect to the personnel now in charge of the fund. When they go, the overcentralization of scientific funds will remain.

As to the third purpose of the society—the conveying of honors—I have no sympathy at all. I have always regarded exclusiveness as an attribute chiefly of use in selling unwanted junk to parvenus. I do not wish to belong to any scientific organization which has more than one grade of membership, nor to one in which that grade of membership is not available to every person with a sincere interest in the field. We all judge the ability of others, but I have no desire to see my unsolicited opinion of a man published with official sanction to injure either him or his competitors, nor will I accept such an unsolicited and officially published opinion of myself nor of anyone else. This would apply to the best available opinion, from which, either because of organized electioneering by influential colleges, government departments, and commercial laboratories, or because of the general fallibility of a group of persons none too well-informed concerning the work of one another, I have found the official judgment of the Academy to differ quite appreciably. As a young man, I have felt far too much of the weight of the unsolicited disapproval or sanction of the elders of science to wish to have any connection with a body of self-appointed judges. Every time a new member is appointed, an unnecessary gift of prestige or position is made to one man; and this gift comes from the one place from which it can come: from the pockets and reputation of someone at a more remote institution or with less influential friends. I am afraid that I can not be reconciled to injustice even by becoming its beneficiary.

As to medals, prizes, and the like, the less said of them the better. The heartbreak to the unsuccessful competitors is only equalled by the injury which their receipt can wreak on a weak or vain personality, or the irony of their reception by an aging scholar long after all good which they can do is gone. I say, justly or unjustly administered, they are an abomination, and should be abolished without exception. So long as I am a member of the American Mathematical Society, I shall work against the acceptance of a single penny or gift to be spent on medals or prizes, and for the liquidation of those prize funds already established. I can not in honor continue in an organization devoted in principle to their support.

I do not wish to speak in detail of the many faults I have had to find with the Academy—of the bad catering, of the tedious and expensive dinners, of the general atmosphere of select and costly pomposity which has hung over the meetings, of the camp-followers of the press and the camera, of the excessive age of most of the new members—first, because you have taken strong steps to improve these situations; and secondly, because they do not touch the essence of my attitude, which is, that I am profoundly suspicious of honors in science, and of select, exclusive bodies of scientists, and that I do not like to see the relations of my country to science committed to the care of such a body. With these convictions, I can only resign from the National Academy of Sciences, and rectify the error, commited under the well-meaning appeals of my friends, which I committed in accepting membership in it. I hereby do so resign.

I wish to express to you, Mr. President, my thankfulness for your considera-tion, and my willingness at any time to undertake as a private individual any work, scientific or other, which I can perform in behalf of my country.

Very respectfully yours,

Norbert Wiener.

This letter, despite the generosity of its tone and its memorable words, "I cannot be reconciled to injustice even by becoming its benefi-ciary" has to be grouped with Wiener's less mature writings. For apart from the logically fuzzy treatment of the notions of "official", "quasi-official", "responsible", etc., Wiener does not bring the full force of his tremendous understanding to bear on the important issues he raises in the letter, and does not propose amendments to the charter of the Academy; instead he offers his resignation.

Even though Wiener's cybernetical ideas had not crystallized in 1941, he surely had an inkling (of what he was to say in 1961) that the chief function of science is

> that of subserving a homeostasis in human life; that of maintaining a rapport with the environment, which will enable us to face our environment and its changes, as we may come to them.[5]

One aspect of this is finding and making the best environment for man as a physiological unit, a long-time problem in Wiener's terminology. Another is grappling with emerging social problems such as overpopula-tion, economic dislocation and war. Wiener knew that to fulfill these functions, there had to be semi-official bodies of scientists that could advise governments on scientific matters. He knew of the origins of the Accademia dei Lincei (1603), the Royal Society (1662) and the Acadé-mie Francaise (1666) and their singular service in the development of science and in the formulation of governmental policy. Wiener criticizes the system of appointment to the Academy by co-option by the acade-micians, and raises the question of accountability, but he offers no alternative arrangement. In fact, the quest for alternatives runs us into a quagmire, as is apparent from the following quotation of what

5   From p. 102 of "Prolegomena to Theology, 1961", unpublished manuscript [MC, 877–881].

Professor J.D. Bernal, FRS, a serious student of these questions, had to say in 1939:

> The simplest alternative would be the democratic one of direct election of the academicians for life or for a fixed period of years by the body of all qualified scientists.[6] It may be objected that this would expose science to the evils of vote-catching and political partisanship. Perhaps it would, but their effects are not likely to be worse than the toadying which is rife in science today. A more serious objection is that the main body of scientists would neither be competent nor interested enough to act as electors. This difficulty might be overcome by dividing the academy into sections based on subjects, but this would serve to perpetuate existing divisions. Alternatively the academy could be divided into age-groups of fixed relative numbers to which both candidates and voters should belong. Another method which has the advantage of democratic choice with the safeguard of academic competence is the one of reciprocal election suggested by Dr. Pirie. In this, election to the academy would be not by the scientists as a whole but by a body of some two thousand electors chosen for their general scientific competence by the academicians themselves. Thus the academy would tend to represent the active and responsible scientists of the day. Some such method combined with a limited time of service and the separation of the honorific from the functional aspects of the academy should serve to make it a suitable body for the general direction of scientific work. {B8, p. 283}

Briefly, we are confronted with the fact that when it comes to assessment of merit and to voting, scientists are as fickle as the rest of mankind,[7] and that there is no obvious solution to the selection problem that is clearly superior to co-option.

There is a bit of irony in Wiener's condemnation of the honorific role of the Academy and its bestowal of prizes. For speaking of his receipt of the Bocher prize of the American Mathematical Society in 1933, Wiener recounts that "it was a pleasant thing to be recognized . . ." [56g, p. 177]. And after his resignation Wiener did not decline honorary degrees from Tufts College, 1946, and Grinnell College, 1957, nor the Alverega Prize from the College of Physicians in Philadelphia, 1952, nor the Medal of the Rudolf Virchow Medical Society, 1957, nor the ASTME Research Medal, 1960, nor the National Medal of Science, 1964. We also know of the excellent research that was stimulated by the prizes offered by the Académie Française. The frenzy for virtuosity, priority and recognition is a post-Renaissance aberration, encouraged by secularism and by a social order which bungles on the hope that greed can be made to stimulate useful activity. Wiener was by no means

---

6 Bernal does not say how the class of "qualified scientists" is to be defined.
7 See C.P. Snow's novel *The Masters* {S9} for this fickleness and the tragi-comic consequences.

immune to the frenzy, as the Kellogg episode (Ch. 8) testifies. But until a substantial change occurs in the socio-economic and moral climate, science will have to cope with those drawn to its temple for the pleasures of brain sport or for reasons of ambition, money-making, or vanity.

In his autobiography Wiener describes the National Academy of Sciences and his connection with it in the following terms:

> This is the organization which was entrusted during the Civil War with the task of putting the services of the scientists of the United States at the command of the American government. In the course of the years, its governmental importance had gradually given way to the secondary function of naming those American scientists who might be considered to have arrived. There has always been a great deal of internal politics about science, and this has been distasteful to me. The building of the National Academy was for me a fit symbol of smug pretentiousness, of scholarship in shapely frock coat and striped trousers. After a brief period, during which my inquisitiveness concerning the nature of the high brass of science was amply satisfied, I got out. [56g, p. 176]

From this one might surmise that Wiener's resignation from the Academy was caused by its internal politics rather than by dissatisfaction with the more fundamental aspects of its organization that he voiced in his resignation letter. When questioned about this, Wiener mentioned that he became tired of the canvassing practiced by the "Harvard group". He cited G.D. Birkhoff's pressure on him to back the election to the Academy of the late Professor J.L. Walsh. Wiener would have probably done this on his own, for he held Professor Walsh in high regard, but he resented the pressure. This "pressure" could only have come in 1936 when Wiener was in China. In his letter to Wiener of January 8, 1936, [MC, 44], G.D. Birkhoff pointed out that Walsh had received the highest number of votes cast, and added:

> This is just enough to enable us to present Walsh's name and I hope you will join me in making a vigorous effort to see that he goes through. To this end, would you write me a letter, which may be used as seems best, concerning your estimate of Walsh's work? What you write will, of course, have much influence.

Birkhoff's letter is not very different from the kind that a church trustee might write to another who is away. And it is hard to see how any organization can function without such preliminary agreements. But Wiener disapproved of such requests, and evidently by 1941 had gotten tired of them.

In this episode Wiener-noise seems to have beaten the Wiener-message. For apart from the ambiguities in the letter, it does not give the full reasons for his dissatisfaction. Dr. Jewett's reply of September 24 (delayed by illness) ends both sensibly and sympathetically [MC, 60]:

While I still feel you are making a mistake and that you can render a better service by staying inside the Academy and using your influence to make it conform more nearly to what you think it should be, I realize that you alone must judge your own desires.

I am sorry I have not been able to dig up a problem which would show you the value I see in a body like the Academy even though it is not all I myself should like to have it. However, one cannot always produce white rabbits out of a hat on demand.

Whatever your final decision, believe me to be,

Sincerely your friend,
Frank B. Jewett
President

It should be noted that Wiener's criticism of internal politics within the Academy was not entirely groundless. As D.S. Greenberg has stated in his study of this subject in 1967 {G15}, the Academy "employs a wondrously arcane electoral process that has all the attributes of a papal election except smoke". His account of the maneuverings in the Academy's 1950 presidential election that led to the defeat of Dr. James Bryant Conant suggests that politics in the Academy can become as involuted as in the Roman Curia.

D Unobjective attitude towards Harvard

Another institution that Wiener viewed with noisy consideration was Harvard University. Harvard's positive role in Wiener's life is easy to catalog. Harvard gave his father a permanent faculty position as Professor though he lacked conventional credentials, and through Royce it gave Wiener his initiation in mathematical philosophy. Harvard awarded Wiener a doctorate and a traveling fellowship, and upon his return from Europe gave him a faculty position with the privilege to lecture on his own research. It was a Harvard professor, Osgood, who was instrumental in getting Wiener placed at MIT, and another, Kellogg, who initiated him into the field of potential theory. It was Harvard's Birkhoff who proved the ergodic theorem, which substantially contributed to Wiener's outlook on statistical physics and shaped much of his later work. It was in the Harvard Medical School that Wiener met Rosenblueth, an assistant of Professor Walter Cannon, and the scientific circle in which cybernetics was conceived. We might also add that it was Harvard that gave Wiener's distant teacher, A.N. Whitehead, a permanent position after two retirements from British academic life.

Yet in speaking of the period 1919–1920, Wiener wrote:

My sense of belonging to a group which was treated unjustly killed the last bonds of my friendship and affection for Harvard. [53h, p. 272]

The last part of this sentence is an exaggeration, for as we have seen, Wiener spent many fruitful hours at Harvard or in Harvard company after 1920. Furthermore, he was a regular participant in the Harvard Mathematical Colloquium as well as in scientific functions at its faculty club and its observatory until the 1950s. The unjust treatment to which Wiener refers is the policy of restricting Jewish student admissions to Harvard to a certain percentage to which we alluded in Ch. 2. This quota policy was indefensible. But to let this one serious breach of academic virtue color his entire attitude to the institution is indicative of the immaturity in Wiener's character. This becomes clear when his attitude is compared with that of a younger British Jew at Harvard at that time, Harold Laski, great political theorist-to-be.

Harvard was fostered by circles within the very New England democracy that Wiener loved (Ch. 2), and he should have understood that Harvard could not stray too far from the mores of these circles, which in the 1920s were those depicted in Upton Sinclair's *Boston*. Even so, Harvard endeavored to maintain its position of excellence, integrity and courage, and a person of Wiener's caliber should have realized this. For instance, when Felix Frankfurter recommended for faculty appointment the 23-year-old British Jew of Polish descent, Harold Laski, Harvard complied. This was in 1916 during the presidency of Abbot Lawrence Lowell, whom Wiener inaccurately describes as a "superficially polished and rather ordinary man of conformist loyalties . . ." [53h, p. 126]. Yet when Laski took the extraordinary step of publicly supporting the Boston Police strike of 1919, and a hue and cry went up for the termination of his untenured Harvard position, Lowell told the University's Overseers: "If you ask for Laski's resignation, you will get mine". On the other hand, by a strange quirk, Wiener was one of the "black-legs" in the strike! He was roped in, as he recounts, "in a moment of misguided patriotism". During his beat, he had to arrest "a wifebeater in a slum near the North Station" in Boston. Wiener drew his revolver, but it was "trembling like the tail of a friendly dog", and fortunately did not go off, cf. [53h, pp. 275–276].

Again it was Lowell who conceived and endowed the Society of Fellows to allow outstanding students to pursue research in total freedom and then step into the faculty unencumbered by postgraduate regulations and degrees.

What Wiener and his father had to endure during the anti-Semitic trend in postwar New England and at Harvard was mild compared to what Laski had to face after his endorsement of the police strike. Editorials in the *Harvard Lampoon* attacked him with such lines as: "In the parlance of the ghetto/He would shake a mean stiletto". But Laski requested Lowell not to punish the editors, and when he left for England in 1920 did so without bitterness toward Harvard, cf. {M5}.

E Wiener and George David Birkhoff

Wiener's estimation of G.D. Birkhoff was also uncritical. He spoke of Birkhoff's ambition to become a power in American mathematics and of his intolerance of rivals, but gave no supporting evidence [56g, pp. 27–28]. Just as groundless seem statements such as "Birkhoff . . . saw to it that many academic offers which otherwise might have been open to me were diverted elsewhere" [56g, p. 128]. The truth is that in his childish moments Wiener could be very irritating, and there is evidence that Birkhoff was generous in not holding this against him. (Some letters from Birkhoff to Oswald Veblen in the early 1920s suggest this.)

Wiener drew unwarranted conclusions from the fact that he did not receive an offer from the Harvard Mathematics Department. There is no evidence to support the allegation, implicit in Wiener's account [56g, pages cited], that Birkhoff opposed the appointment of Jews to permanent rank in the Harvard Mathematics Department. On the contrary, during Birkhoff's tenure, Gabriel Marcus Green (who, had he survived the influenza epidemic of 1918, would have become Wiener's brother-in-law) was a member of the faculty, and a professorship was offered to J. Robert Oppenheimer, a Jew. There is no evidence, however, of Birkhoff's opposition to the Jewish "numerus clausus" (Ch. 2). It is possible to be critical of many attitudes of G.D. Birkhoff and yet arrive at an objective and warm appraisal of him as a man; see for instance, D.D. Kosambi's obituary of Birkhoff {K10}.

Actually, Wiener's attitude toward Birkhoff was ambivalent. This is revealed in the sympathetic references to Birkhoff in his writings [53h, pp. 230–231; 56g, p. 281], but even more in some of his unrecorded statements. For instance, when Birkhoff died in November 1944, the Interscientific Discussion Group at Harvard, of which he had been a leading member, decided to devote its forthcoming meeting to a discussion of his life and work. The Steering Committee wanted Wiener (who had been an active participant) to be one of the speakers, and asked me to convey the request to him.[8] Wiener's response on the telephone line was clear: "Please tell the committee that I shall be delighted and that I regard their invitation as an honor". Wiener's eulogy was of a very high standard and one of the warmest I have heard. It is reasonable to surmise that Birkhoff felt the same ambivalence toward Wiener. One should not overlook their considerable philosophical concord: their common bond to the unifying principles of Plato and Leibniz.

---

8   I was then a student and served the committee in a clerical capacity.

Norbert Wiener in the Soviet Union in 1960

Norbert Wiener at the White House with President Lyndon B. Johnson and colleagues (among them Drs. Weisner and Bush) when he received the National Medal of Science in 1963.

Fortunately, Wiener did not let his prejudices interfere with his scholarly pursuits. He imbibed freely what Harvard had to offer, and made Birkhoff's theorem his own. Nor, barring one or two exceptions, did he allow his subjective attitudes to affect his scientific criticism. His prejudicial statements should therefore be overlooked as the mild aberrations of a great and fundamentally impartial and honest mind.

24 Epilogue

Wiener's creativity was marked by certain crucial traits in his intellectual makeup and modus operandi.

First was Wiener's strong dependence on good external stimulation to set his creative powers in motion. What he read or heard from great minds like Spinoza, Leibniz, Bergson, Russell, Perrin, Born, Haldane, Bush, McCulloch and Rosenblueth stimulated his own tremendously. But while the inception of Wiener's ideas depended on external stimulation, once germinated his ideas became his masters and were impervious to his intellectual gregariousness. In so leaning on the great but only initially, Wiener resembled T.S. Eliot, his Harvard contemporary, and like Eliot he could at once be profoundly original and yet close to the tradition of his mentors. Thus, Wiener continued the tradition of Russell in its emphasis on type-classification and on relation-structure, but in a way entirely original and different from that of the official followers of Russell—the school of analytic philosophers.

Second was the coupling of a very deep intuition with a very strong sense of integrity. This potent combination was at the basis of most of his profound contributions. It also helped shape his direct and penetrating mode of exposition, and unavoidably made him a critic of pseudointellectual tendencies among the public and especially among the scientific community.

A third was the engineering vision and sharp criticality that allowed Wiener to see into the future. This becomes very clear when we place him next to another humane and great scientist, Robert Andrew Millikan. In India, in 1940, to measure cosmic rays at high altitudes near the magnetic equator, Dr. Millikan was asked about the possible dangerous applications of subatomic research. He answered:

> It looks increasingly improbable that there is any appreciable amount of subatomic energy available for man to tap.[1]

Here is what Wiener wrote seven years earlier, in an interesting 1933 article on energy:

1 "An interview with Robert Andrew Millikan", *The Aryan Path*, April 1940, pp. 204–208.

> ... our one hope probably lies in atomic energy ... A pound of matter has 11,300,000,000 kilowatt hours of energy in it, while a pound of gasoline as used in an internal combustion engine has at most about one kilowatt hour. The ratio is so enormous that if even a modest part ... of the energy in matter can be made available for mechanical use, all other forms of storage will be put far in the shade.
>
> . . . .
>
> Mr. Robert J. Van de Graaff, of the Institute's Department of Physics ... is quite reasonably optimistic as to the ultimate possibility of realizing, mechanically at least, a large part of the energy which appears as a change of mass of the nucleus when one element is synthesized into others, as in the work of Cockcroft and Walton. [33f, pp. 59, 70]

Wiener then went on to surmise the social ramifications of the use of atomic energy. The pressure to replace the laborer as a source of energy by the machine "would be intensified a thousandfold" [33f, p. 70]. He ended by referring to the 1930 depression and the New Deal it engendered as a "dress rehearsal for the real show which will be put on when and if atomic energy becomes available" [33f, p. 72].

A fourth factor was Wiener's great analogical mind. In discerning and seriously evaluating analogies, he was following a valuable Thomistic tradition, cf. {P4}. But in the far-reaching conclusions he drew from the steadfast pursuit of analogies, Wiener was unique. For instance, the similarity he noted in the evolutions of mechanics and electrical engineering and the resemblance he found between the disease *tabes dorsalis* and the hunting phenomenon in servomechanisms led him to theories of remarkably rich scope. His important theological probing also falls in this analogical category.

A fifth factor was Wiener's power to fuse the transcendent and abstract with the practical and concrete. His vision of the transcendent appears a bit limited when compared with that of Whitehead. But this limitation did not affect in the least Wiener's firm understanding of abstractness in the practical. As he wrote:

> ... so often in my work, the motivation which has led me to the study of a practical problem has also induced me to go into one of the most abstract branches of pure mathematics. [56g, p. 192]

He also of course understood thoroughly the practicality of good abstraction, as is clear from the applications he made of the Lebesgue integral in Brownian motion theory and electrical engineering, and the type-classification he brought into automata theory. Unlike his distinguished contemporary E.P. Wigner, who was puzzled by what he perceived as "the unreasonable effectiveness of mathematics in the natural sciences", cf. {W13}, Wiener found the effectiveness of mathematics perfectly understandable. The investigations of the remarkable medical men close to him, W. Ross Ashby, A. Rosenblueth and W. McCulloch, showed that mathematical relation-structure is all that is preserved in the course of sense-observation and the subsequent neurological transitions that constitute cognition {A6, M13, R2}. Thus the paramountcy that Pythagoras, Plato, Roger Bacon, Galileo and Dirac accorded to mathematics was not at all misplaced, and it becomes clear why, to use Whitehead's words, "the utmost abstractions are the true weapons with which to control our thought of concrete facts" {W8, p. 41}. Moreover, the relation structures of the different sciences cohere. In the words of H. Weyl:

> The fact that in nature "all is woven into one whole", that space, matter, gravitation, the forces arising from the electromagnetic field, the animate and inanimate are all indissolubly connected, strongly supports the belief in the unity of nature and hence in the unity of scientific method. {W5, p. 214}

Wiener not only shared this belief, but, in Struik's words, he "lived the unity of science" {S10, p. 34}.

Wiener did not let the revolutions of thought that make and partition the history of science obscure his vision of its fundamental continuity. The age of science began when man first made an attempt to comprehend the world around him, and this happened when his mode of perception and language were still poetic. "Mouths of rivers", "veins of minerals", such were his first scientific utterances. Wiener knew that his own cybernetical ideas went back at least to Leibniz, and that cybernetical devices were used from very early times. We now know that Plato had barely missed the notion of feedback. The notion of fractal too has roots extending to Aristotle, as Mandelbrot has indicated {M3, p. 406}. Unlike some modern writers, Wiener did not forget that non-Euclidean geometry, Mengenlehre and fractals are inconceivable without the work of Euclid, and that there was nothing "anti-Euclidean" about them. To him, as to J.E. Littlewood, the Greeks were "dons from another college", who would have fully approved of such later activity. In his view, the stark separation in the arts between the old and the new is entirely absent in the sciences.

A similar perception of continuity marked Wiener's vision of history as a whole. An admirer of both the 16th-century Renaissance and the 18th-century French Enlightenment, he was blinded by neither. The removal of medieval teleology from post-Renaissance science was a boon, but Wiener contributed to its useful restoration in a modern scientific setting. The same applies to his restoration of the long-time State, the "Sacerdotium'. He was a revolutionary-traditionalist in the best sense of the word.

Wiener was too great a thinker to accept uncritically all the mores of post-Renaissance science. He denied Locke's classification of "rational self-interest" as a virtue, and the belief in "unlimited progress" of the French Enlightenists and the Marxists. According to him a democratic republic can prosper only with a learned populace with a large store of communal information, and this requires the entrustment of the channels of communication to long time institutions. Without this, democracy reduces to counting heads which, in the words of W.R. Inge, the former dean of St. Paul's Cathedral, is "the next silliest thing" to smashing heads. Wiener also realized that science, for all its importance, is only a part of human practice, and that the scientific mode of perception has to be integrated with (and can never supplant) the more universal mythic mode of perception. This gave him a perspective on the human predicament, on life and death, work and worship, science and art, which was deeper than that of most scientists. Among his contemporaries he was closest in spirit to Hermann Weyl, Albert Einstein and Lewis Mumford.

America lost when it allowed its most profound and versatile philosopher—its first Leibniz—Charles Sanders Peirce to die unheard, while it settled for less iconoclastic but smaller minds {P1, p. xvii}. Peirce's heir was luckier: the Massachusetts Institute of Technology gave Wiener a congenial workplace, and his contributions were appreciated. Even so, America will lose again if it lets the profound thought of this, its second Leibniz, go unattended, while it pursues the outpourings of lesser minds. Wiener, like Peirce before him, has illuminated a central problem of our age, "the humanizing of the sciences and their increasing role in dealing with the moral and educational phases of our industrial age", to use the words of Peirce's editor, Philip P. Wiener {P1, p. ix}. Few others have comprehended reality so profoundly, and so clearly seen the way out of the present spiritual impoverishment.

Wiener understood the scientific and philosophical implications of Bergsonian time and the Second Principle of Thermodynamics more fully than any other thinker. A unique and significant aspect of his writings is the underlying thought that the contingent cosmos that modern physics reveals is worthy of the Pythagorean attitude that Einstein expressed in the words "Intelligence is manifested throughout all Nature". He also realized that the entropy principle has a moral dimension, that evil is like physical noise in its persistency and noneradicability. Thereby he showed that the contemporary belief in unlimited progress has no basis in science, and that the reflections on the problem of evil of our great religious predecessors can be absorbed into a scientific framework. This is perhaps the first significant dent in the schism between science and religion that has reigned during the last 300 years—a schism which Whitehead regarded as the source of much that is "halfhearted and wavering" in modern life {W8, p. 94}. Wiener's contribution was to show that the Gibbsian conception of the universe allows the precise demarcation of the concepts of contingency, purpose, freedom and entropy in stochastic terms. We may regard him as the first scientist-philosopher of the stochastic age. His thought firmly suggests that the calculus of probabilities has to become an integral tool in the philosophy and theology of modern man.

But despite his full cognizance of contingency in the universe and its ever-increasing entropy, Wiener, like Spinoza, believed in its dignity. In this he found his mentors in Aeschylus, Euripides and Sophocles. He would have contributed more to this remarkable synthesis had he lived longer. But on March 18, 1964, in Stockholm during a visit to Holland, Wiener died at age seventy.[2]

2 Commensurate with Wiener's spirit, an ecumenical memorial service was held for him at the MIT Chapel on June 2, 1964. At Mrs. Wiener's behest, his friend and religious counsellor at MIT, Swami Sarvagatananda of the Ramaskrishna Vedanta Society of Boston, led the service. Also officiating were Rabbi Herman Pollack, Rev. Dr. Myron Bloy and Father Golden, all MIT counsellors. But Father Golden could not get into the sanctuary, so packed was the chapel.

Academic Vita of Norbert Wiener[1]

1894	Born on November 26 in Columbia, Missouri, to Bertha Kahn Wiener and Leo Wiener, a professor of foreign languages at the University of Missouri.
1895	The family moved to Cambridge, Massachusetts, where Leo Wiener became a professor of Slavic languages at Harvard.
1901	Entered the third grade at the Peabody School, but was removed shortly and taught by his father until 1903.
1903	Entered Ayer High School.
1906	Graduated from Ayer High School and entered Tufts College where he studied mathematics and biology.
1909	Received an A.B. degree, cum laude, from Tufts, and entered the Harvard Graduate School to study zoology.
1910	Entered the Sage School of Philosophy at Cornell University with a scholarship, and studied with Frank Thilly, Walter A. Hammond, and Ernest Albee.
1911	Transferred to the Harvard Graduate School to study philosophy, and studied with E.V. Huntington, Josiah Royce, G.H. Palmer, Karl Schmidt, and George Santayana.
1912	Received an M.A. degree from Harvard.
1913	Received a Ph.D. degree from Harvard; dissertation under J. Royce, but supervised by K. Schmidt of Tufts College.
	Appointed a John Thornton Kirkland Fellow by Harvard, and entered Cambridge University. Studied logic and philosophy with Bertrand Russell, G.E. Moore, and J.M.E. McTaggart, and mathematics with G.H. Hardy and J.E. Littlewood.
1914	Joined the University of Göttingen and took the courses of David Hilbert, Edmund Husserl, and Edmund Landau.
	Appointed a Frederick Sheldon Fellow by Harvard; returned to Cambridge University to study mathematics and philosophy. Received the Bowdoin Prize from Harvard.
1915	Studied philosophy under John Dewey at Columbia University.
	Appointed an assistant and a docent lecturer in Harvard's Philosophy Department for 1915–1916, and lectured on the logic of geometry.
1916	Served with Harvard's reserve regiment at the Officer's Training Camp in Plattsburg, N.Y.
	Appointed instructor of mathematics at the University of Maine in Orono for 1916–1917.

---

1 This data is extracted from the Chronology in the spiral-bound publication entitled the "Inventory of Norbert Wiener, 1894–1964", processed by Mary Jane McCavitt, September 1980.

1917   Served with the Cambridge ROTC; briefly worked as an apprentice engineer in the turbine department of the General Electric Corp. in Lynn, Massachusetts.
            Appointed a staff writer for the *Encyclopedia Americana* in Albany, N.Y.

1918   Joined the Aberdeen, Proving Grounds of the U.S. Army under O. Veblen, and worked on computations of ballistic tables. Joined the American Mathematical Society.

1919   Served as an Army private at the Aberdeen Proving Ground, Maryland.
            Worked as a journalist with *The Boston Herald.*
            Appointed instructor of mathematics at MIT.

1920   Attended the International Mathematical Congress in Strasbourg as MIT's representative and presented a paper on Brownian motion. He also visited Cambridge and Paris.

1924   Appointed assistant professor of mathematics at MIT.

1925   Attended the International Mathematical Congress in Grenoble and the British Association for the Advancement of Science meeting in Southampton; visited Göttingen University.

1926   Elected fellow of the American Academy of Arts and Sciences. Married Marguerite Engemann.
            Received a Guggenheim Fellowship to study in Göttingen and in Copenhagen during 1926–1927. Collaborated with Harald Bohr, and taught a course of general trigonometric developments at Göttingen.

1928   Addressed the Symposium on Analysis Situs of the American Mathematical Society.

1929   Appointed associate professor of mathematics at MIT.
            Lectured at Brown University as exchange professor during 1929–1930.

1931–1932   Visiting lecturer at Cambridge University; lectured on the Fourier integral and its applications at Trinity College.
            Appointed professor of mathematics at MIT.
            Attended the International Congress of Mathematics, Zurich, as MIT's representative.

1933   Awarded Bocher Prize by the American Mathematical Society.
            Elected to the National Academy of Sciences.
            Began participation in the interdisciplinary seminar at Harvard Medical School under Arturo Rosenblueth. Collaborated with REAC. Paley.

1934   Delivered the AMS Colloquium Lectures at Williamston, Massachusetts.

1935   Patented electrical network systems with Yuk Wing Lee. (Two more patents were issued in 1938.)
            Lectured at Stanford University and in Japan on his way to China. Visiting professor at Tsing Hua University in Peiping, China, during 1935–1936.

1936   Attended the International Congress of Mathematicians in Oslo, Norway, and lectured on Tauberian gap theorems.
            Collaborated with Harry Ray Pitt at MIT during 1936–1937.

1937   Delivered the Dohme lecture at Johns Hopkins University on Tauberian theorems.

1938   Lectured on analysis at the semicentennial of the AMS.

1940   Appointed chief consultant in the field of mechanical and electrical aids to computation for the National Defense Research Committee.

	Consultant with the NDRC's Office of Scientific Research and Development, Statistical Research Group and Operational Research Laboratory at Columbia University.
	Consultant to the War-Preparedness Committee of the American Mathematical Society.
	Joined a team at MIT under S.H. Caldwell to study the guidance and control of antiaircraft fire.
	Worked on the theory and design of fire control apparatus for antiaircraft guns with Julian Bigelow, under NDRC Project.
1941	Resigned from the National Academy of Sciences.
1945	Participated in a study group set up by John von Neumann, and attended a meeting on communication theory in Princeton.
	Collaborated with Arturo Rosenblueth at the Instituto National Cardiologia in Mexico, and attended the Mexican Mathematical Society's Conference held in Guadalajara.
1946–1950	With Arturo Rosenblueth received a five-year Rockefeller Foundation grant that allowed them to collaborate in Mexico and at MIT on alternating years.
1946	Received an honorary Sc.D. degree from Tufts College.
	Attended the first three Josiah Macy, Jr. Foundation Conferences and the Conference on Teleological Mechanisms sponsored by the New York Academy of Sciences.
	Lectured at the National University of Mexico.
1947	Visited England and France, and gave lectures on harmonic analysis in Nancy, France.
1948	Spoke at the AMS's Second Symposium on Applied Mathematics.
1949	Received the Lord & Taylor American Design Award.
	Delivered the AMS's Josiah Willard Gibbs Lecture at the annual meeting.
1950	Attended the seventh Macy conference.
	Lectured at the International Congress of Mathematicians at Harvard University.
1951	Lectured at the University of Paris, College de France, under a Fulbright Teaching Fellowship, and also lectured in Madrid.
	Received an honorary Sc.D. degree from the University of Mexico.
1952	Received the Alvarega Prize from the College of Physicians in Philadelphia.
	Delivered the Forbes-Hawks Lectures at the University of Miami.
1953	Lectured on the theory of prediction at the University of California at Los Angeles.
	Taught a summer school course with Claude Shannon and Robert Fano on the mathematical problems of communications theory.
1953–1954	Lectured at the Tata Institute of Fundamental Research, Bombay, attended the All-India Science Congress, and visited research centers.
1955–1956	Visiting professor at the Indian Statistical Institute in Calcutta.
1956	Lectured in Japan and gave a summer school course at UCLA.
1957	Received an honorary Sc.D. degree from Grinnell College.
	Awarded the Virchow Medal from the Rudolf Virchow Medical Society.
1959	Gave a summer school course at UCLA.
	Appointed institute professor at MIT.
1960	Lectured at the University of Naples in Italy, and visited the USSR.

Received the ASTME Research Medal.

Retired from MIT, and appointed institute professor emeritus.

1961   Gave a summer school course at UCLA.

1962   Lectured at the Institute of Theoretical Physics, University of Naples, Italy.

Delivered the Terry Lectures at Yale University, titled "Prolegomena to Theology".

1963   Gave a summer school course at UCLA.

1964   Received the National Medal of Science from President Johnson.

1964   Visiting professor and honorary head of neurocybernetics at Netherlands Central Institute for Brain Research, Amsterdam.

Lectured in Norway and Sweden.

Died on March 18 in Stockholm, Sweden.

Doctoral Students of Norbert Wiener[1]

Shikao Ikehara	Ph.D.	1930
Sebastian Littauer	Sc.D.	1930
Dorothy W. Weeks	Ph.D.	1930
James G. Estes	Ph.D.	1933
Norman Levinson	Sc.D.	1935
Henry Malin	Ph.D.	1935
Bernard Friedman	Ph.D.	1936
Brockway McMillan	Ph.D.	1939
Abe M. Gelbart	Ph.D.	1940
Donald G. Brennan	Ph.D.	1959

1 Reprinted with the kind permission of Professor Irving Ezra Segal of MIT.

The Classification of Wiener's Papers

I *Mathematical Papers*
  A Mathematical philosophy and foundations
  B Potential theory
  C Brownian movement, Wiener integrals, ergodic and chaos theories, turbulence and statistical mechanics
  D Generalized harmonic analysis and Tauberian theory
  E Classical harmonic and complex analysis (orthogonal developments, quasi-analyticity, gap theorems, and Fourier transforms in the complex domain)
  F Hopf-Wiener integral equations
  G Prediction and filtering
  H Relativity and quantum theories
  I Miscellaneous mathematical papers
II *Cybernetical and Philosophical Papers*
  A Philosophical papers
  B Cybernetical papers
III *Social, Ethical, Educational, and Literary papers*
IV *Book Reviews, Prefaces, and Obituaries*
  A Book reviews and prefaces
  B Obituaries
V *Abstracts*
VI *Books and Other Publications.*

In the accompanying Bibliography, Wiener's publications have been classified into six broad categories labeled I, . . ., VI, and these larger categories have been divided into subcategories labeled A, B, C, . . .

The publications are indexed by the official year of their appearance. The internal ordering of the publications appearing in a given year is not chronological but according to the categories mentioned in the last paragraph: a run down the list a, b, c, . . . for a particular year entails a run down the list of categories IA, IB, . . ., usually with omissions and repetitions, as will be apparent from a glance at the right-hand columns in the bibliography. For instance, [33d] means "the dth paper, according to category, which appeared in a journal marked 1933".

References in the book indicated by numbers in square brackets preceded by MC, e.g. [MC, 57], refer to the Manuscript Collection of Wiener in the MIT Archives. Roman numerals in square brackets, e.g. [IV], refer to Defense Department Documents pertaining to Wiener, listed after the Bibliography. (Symbols in braces, e.g. {K3}, which refer to other authors, are listed at the end of the book.)

Bibliography of Norbert Wiener

[13a]   *On a method of rearranging the positive integers in a series*   IA
        of ordinal numbers greater than that of any given funda-
        mental sequence of omegas, Messenger of Math. 43 (1913),
        97–105.

[14a]   *A simplification of the logic of relations,* Proc. Cambridge Philos.   IA
        Soc. 17 (1914), 387–390.

[14b]   *A contribution to the theory of relative position,* Proc. Cambridge   IA
        Philos. Soc. 17 (1914), 441–449.

[14c]   *The highest good,* J. Phil. Psych. and Sci. Method 11 (1914),   IIA
        512–520.

[14d]   *Relativism,* J. Phil. Psych. and Sci. Method 11 (1914), 561–577.   IIA

[15a]   *Studies in synthetic logic,* Proc. Cambridge Philos. Soc. 18 (1915),   IA
        14–28.

[15b]   *Is mathematical certainty absolute?,* J. Phil. Psych. and Sci.
        Method 12 (1915), 568–574.IA

[16a]   *Mr. Lewis and implication,* J. Phil. Psych. and Sci. Method 13   IA
        (1916), 656–662.

[16b]   *The shortest line dividing an area in a given ratio,* Proc. Cambridge   I, I
        Philos. Soc. 18 (1916), 56–58.

[16c]   Review of Cassius J. Keyser, *Science and Region: the Rational and*   IVA
        the Superrational, J. Phil. Psych. and Sci. Method 13 (1916),
        273–277.

[16d]   Review of A.A. Robb, *A Theory of Time and Space,* J. Phil. Psych.   IVA
        and Sci. Method 13 (1916), 611–613.

[17a]   *Certain formal invariances in Boolean algebras,* Trans. Amer.   IA
        Math. Soc. 18 (1917), 65–72.

[17b]   Review of C.J. Keyser, *The Human Worth of Rigorous Thinking,*   IVA
        J. Phil. Psych. and Sci. Method 14 (1917), 356–361.

[18a]   Review of Edward V. Huntington, *The Continuum and Other*   IVA
        Types of Serial Order, J. Phil. Psych. and Sci. Method 15 (1918),
        78–80.

[18b]   *Aesthetics,* in Encyclopedia Americana, 1918–20 edition, vol. I,   IIA
        198–203.

[18c]   *Algebra, definitions and fundamental concepts,* in Encyclopedia   I, I
        Americana, 1918–20 edition, vol. I, 381–385.

[18d]   *Alphabet,* in Encyclopedia Americana, 1918–20 edition, vol. I,   IIA
        435–438.

[18e]   *Animals, chemical sense, in,* in Encyclopedia Americana, 1918–20   IIA
        edition, vol. I, 704.

18f]   *Apperception,* in Encyclopedia Americana, 1918–20 edition,   IIA
        vol. II, 82–83.

[18g] *Category*, in Encyclopedia Americana, 1918–20 edition, vol. VI, IIA
49.

[18h] *Dualism*, in Encyclopedia Americana, 1918–20 edition, vol. IX, IIA
367.

[18i] *Duty*, in Encyclopedia Americana, 1918–20 edition, vol. IX, IIA
440–441.

[18j] *Ecstasy*, in Encyclopedia Americana, 1918–20 edition, vol. IX, IIA
570.

[19a] *Geometry, non-euclidean*, in Encyclopedia Americana, 1918–20 I, I
edition, vol. XII, 463–467.

[19b] *Induction, in logic*, in Encyclopedia Americana, 1918–20 edition, IIA
vol. XV, 70–73.

[19c] *Infinity*, in Encyclopedia Americana, 1918–20 edition, vol. XV, IIA
120–122.

[19d] *Meaning*, in Encyclopedia Americana, 1918–20 edition, vol. IIA
XVIII, 478–479.

[19e] *Mechanism and vitalism*, in Encyclopedia Americana, 1918–20 IIA
edition, vol. XVIII, 527–528.

[19f] *Metaphysics*, in Encyclopedia Americana, 1918–20 edition, vol. IIA
XVIII, 707–710.

[19g] *Pessimism*, in Encyclopedia Americana, 1918–20 edition, vol. IIA
XXI, 654.

[19h] *Postulates*, in Encyclopedia Americana, 1918–20 edition, vol. IIA
XXII, 437–438.

[20a] *Bilinear operations generating all operations rational in a domain* IA
Ω, Ann. of Math. 21 (1920), 157–165.

[20b] *A set of postulates for fields*, Trans. Amer. Math. Soc. 21 (1920), IA
237–246.

[20c] *Certain iterative characteristics of bilinear operations*, Bull. Amer. IA
Math. Soc. 27 (1920), 6–10.

[20d] *Certain iterative properties of bilinear operations*, G.R. Strasbourg IA
Math. Congress, 1920, 176–178.

[20e] *On the theory of sets of points in terms of continuous transforma-* IA
tions, G.R. Strasbourg Math. Congress, 1920, 312–315.

[20f] *The mean of a functional of arbitrary elements*, Ann. of Math. (2) IC
22 (1920), 66–72.

[20g] Review of C.I. Lewis, *A Survey of Symbolic Logic*, J. Phil. Psych. IVA
and Sci. Method 17 (1920), 78–79.

[20h] *Soul*, in Encyclopedia Americana, 1918–20 edition, vol. XXV, IIA
268–271.

[20i] *Substance*, in Encyclopedia Americana, 1918–20 edition, vol. IIA
XXV, 775–776.

[20j] *Universals*, in Encyclopedia Americana, 1918–20 edition, vol. IIA
XXVII, 572–573.

[21a] *A new theory of measurement: A study in the logic of mathematics*, IA
Proc. London Math. Soc. 19 (1921), 181–205.

[21b] *The isomorphisms of complex algebra*, Bull. Amer. Math. Soc. 27 IA
(1921), 443–445.

[21c] *The average of an analytic functional*, Proc. Nat. Acad. Sci. U.S.A. IC
7 (1921), 253–260.

[21d] *The average of an analytic functional and the Brownian movement*, IC
Proc. Nat. Acad. Sci. U.S.A. 7 (1921), 294–298.

[21e]	*A new vector method in integral equations* (with F.L. Hitchcock), J. Math. and Phys. 1 (1921), 1–20.	I, I
[22a]	*The relation of space and geometry to experience,* Monist 32 (1922), 12–60, 200–247, 364–394.	IA
[22b]	*The group of the linear continuum,* Proc. London Math. Soc. 20 (1922), 329–346.	IA
[22c]	*Limit in terms of continuous transformation,* Bull. Soc. Math. France 50 (1922), 119–134.	IA
[22d]	*The equivalence of expansions in terms of orthogonal functions* (with J.L. Walsh), J. Math. and Phys. 1 (1922), 103–122.	IE
[22e]	*A new type of integral expansion,* J. Math. and Phys. 1 (1922), 167–176.	I, I
[23a]	*On the nature of mathematical thinking,* Austral. J. Psych. and Phil. 1 (1923), 268–272.	IA
[23b]	*Nets and the Dirichlet problem* (with H.B. Phillips), J. Math. and Phys. 2 (1923), 105–124. (Reprinted in [64f].)	IB
[23c]	*Discontinuous boundary conditions and the Dirichlet problem,* Trans. Amer. Math. Soc. 25 (1923), 307–314.	IB
[23d]	*Differential-space,* J. Math. and Phys. 2 (1923), 131–174. (Reprinted in [64f].)	IC
[23e]	*O szeregach $\Sigma_1^\infty(\pm 1/n)$—Note on the series $\Sigma_1^\infty(\pm 1/n)$,* Bull. Acad. Polon. Ser. A, 13 (1923), 87–90.	IC
[23f]	*Note on a new type of summability,* Amer. J. Math. 45 (1923), 83–86.	ID
[23g]	*Note on a paper of M. Banach,* Fund. Math. 4 (1923), 136–143.	I, I
[24a]	*Certain notions in potential theory,* J. Math. and Phys. 3 (1924), 24–51.	IB
[24b]	*The Dirichlet problem,* J. Math. and Phys. 3 (1924), 127–146. (Reprinted in [64f].)	IB
[24c]	*Une condition nécessaire et suffisante de possibilité pour le problème de Dirichlet,* C.R. Acad. Sci. Paris 178 (1924), 1050–1054.	IB
[24d]	*The average value of a functional,* Proc. London Math. Soc. 22 (1924), 454–467.	IC
[24e]	*Un problème de probabilités dénombrables,* Bull. Soc. Math. France 11 (1924), 569–578.	IC
[24f]	*The quadratic variation of a function and its Fourier coefficients,* J. Math. and Phys. 3 (1924), 72–94.	ID
[24g]	Review of four books on space, Rudolf Carnap's *Der Raum: Ein Beitrag zur Wissenschaftslehre,* E. Study's *Mathematik und Physik: eine erkenntnistheoretische Untersuchung* and *Die realistische Weltansicht und die Lehre vom Raume: zweite Auflage; erster Teil,* and Hermann Weyl's *Mathematische Analyse des Raum-problems. Vorlesungen gehalten in Barcelona und Madrid,* Bull. Amer. Math. Soc. 30 (1924), 258–262.	IVA
[24h]	Review of E. Study, *Denken und Darstellung: Logik und Werte; Dingliches und Menschliches in Mathematik und Naturwissenschaften,* Bull. Amer. Math. Soc. 30 (1924), 277.	IVA
[24i]	*In memory of Joseph Lipka,* J. Math. and Phys. 3 (1924), 63–65.	IVB
[25a]	*Note on a paper of O. Perron,* J. Math. and Phys. 4 (1925), 21–32.	IB
[25b]	*The solution of a difference equation by trigonometrical integrals,* J. Math. and Phys. 4 (1925), 153–163.	ID

[25c] *On the representation of functions by trigonometrical integrals,*   ID
       Math. Z. 24 (1925), 575–616.
[25d] *Verallgemeinerte trigonometrische Entwicklungen,* Göttingen   ID
       Nachr. (1925), 151–158.
[25e] *Note on quasi-analytic function,* J. Math. and Phys. 4 (1925),   IE
       193–199.
[25f] *A contribution to the theory of interpolation,* Ann. of Math. (2) 26 I, I
       (1925), 212–216.
[26a] *The harmonic analysis of irregular motion,* J. Math. and Phys. 5   ID
       (1926), 99–121.
[26b] *The harmonic analysis of irregular motion* (Second Paper), J.   ID
       Math. and Phys. 5 (1926), 158–189.
[26c] *The operational calculus,* Math. Ann. 95 (1926), 557–584.   ID
[26d] *A new formulation of the laws of quantization of periodic and*   IH
       aperiodic phenomena (with M. Born), J. Math. and Phys. 5 (1926),
       84–98.
[26e] *Eine neue Formulierung der Quantengesetze für periodische und*   IH
       nichtperiodische Vorgänge (with M. Born), Z. Physik 36 (1926),
       174–187.
[26f] *Analytic approximations to topological transformations* (with P.   I, I
       Franklin), Trans. Amer. Math. Soc. 28 (1926), 762–785.
[27a] *The spectrum of an array and its application to the study of the*   ID
       translation properties of a simple class of arithmetical functions,
       Part I, J. Math. and Phys. 6 (1927), 145–157. (Part II: *On the*
       translation of a simple class of arithmetical functions, by
       K. Mahler, *ibid,* pp. 158–163).
[27b] *A new definition of almost periodic functions,* Ann. of Math. (2) 28 ID
       (1927), 365–367.
[27c] *On a theorem of Bochner and Hardy,* J. London Math. Soc. 2   ID
       (1927), 118–123.
[27d] *Une méthode nouvelle pour la démonstration des théorèmes de M.*   ID
       Tauber, C.R. Acad. Sci. Paris 184 (1927), 793–795.
[27e] *On the closure of certain assemblages of trigonometrical functions,* IE
       Proc. Nat. Acad. Sci. U.S.A. 13 (1927), 27–29.
[27f] *Quantum theory and gravitational relativity* (with D.J. Struik),   IH
       Nature 119 (1927), 853–854.
[27g] *A relativistic theory of quanta* (with D.J. Struik), J. Math. and   IH
       Phys. 7 (1927), 1–23.
[27h] *Sur la théorie relativiste des quanta* (with D.J. Struik), C.R. Acad. IH
       Sci. Paris 185 (1927) 42–44.
[27i] *Sur la théorie relativiste des quanta* (Note), C.R. Acad. Sci. Paris IH
       185 (1927), 184–185.
[27j] *Laplacians and continuous linear functionals,* Acta Sci. Math.   I, I
       (Szeged) 3 (1927), 7–16.
[27k] *Une généralisation des fonctions à variation bornée,* C.R. Acad.   I, I
       Sci. Paris 185 (1927), 65–67.
[28a] *The spectrum of an arbitrary function,* Proc. London Math. Soc.   ID
       (2) 27 (1928), 483–496.
[28b] *A new method in Tauberian theorems,* J. Math. and Phys. 7 (1928),   ID
       161–184.
[28c] *The fifth dimension in relativistic quantum theory* (with D.J.   IH
       Struik), Proc. Nat. Acad. Sci. U.S.A. 14 (1928), 262–268.

[28d] *Coherency matrices and quantum theory*, J. Math. and Phys. 7 (1928), 109–125.   IH

[29a] *Harmonic analysis and group theory*, J. Math. and Phys. 8 (1929), 148–154.   ID

[29b] *A type of Tauberian theorem applying to Fourier series*, Proc. London Math. Soc. (20) 30 (1929), 1–8.   ID

[29c] *Fourier analysis and asymptotic series.* Appendix to V. Bush, *Operational Circuit Analysis*, New York, John Wiley, 1929, 366–379.   ID

[29d] *Hermitian polynomials and Fourier analysis*, J. Math. and Phys. 8 (1929), 70–73.   IE

[29e] *Harmonic analysis and the quantum theory*, J. Franklin Inst. 207 (1929), 525–534.   IH

[29f] *On the spherically symmetrical statical field in Einstein's unified theory of electricity and gravitation* (with M.S. Vallarta), Proc. Nat. Acad. Sci. U.S.A. 15 (1929), 353–356.   IH

[29g] *On the spherically symmetrical statical field in Einstein's unified theory: A correction* (with M.S. Vallarta), Proc. Nat. Acad. Sci. U.S.A. 15 (1929), 802–804.   IH

[29h] *Mathematics and art (Fundamental identities in the emotional aspects of each)*, Tech. Rev. 32 (1929), 129–132, 160, 162.   IIA

[29i] *Einsteiniana (Facts and fancies about Dr. Einstein's famous theory)*, Tech. Rev. 32 (1929), 403–404.   IIIE

[29j] *Murder and mathematics*, Tech. Rev. 32 (1929), 271–272.   IVA

[30a] *Generalized harmonic analysis*, Acta Math. 55 (1930), 117–258. (Reprinted in [64f] and [66b].)   ID

[30b] Review of A. Eddington's *Science and the Unseen World*, Tech. Rev. 33 (1930), 150.   IVA

[31a] *Über eine Klasse singulärer Integralgleichungen* (with E. Hopf), Sitzber. Preuss. Akad. Wiss. Berlin, Kl. Math. Phys. Tech., 1931, pp. 696–706. (Reprinted in [64f].)   IF

[31b] *A new deduction of the Gaussian distribution*, J. Math. and Phys. 10 (1931), 284–288.   I, I

[31c] *Reports from Cambridge—1931*, Tech. Rev. 34 (1931), 82–83, 131, 218, 220.   IIIE

[32a] *Tauberian theorems*, Ann. of Math. 33 (1932), 1–100 (Reprinted in [64f] and [66b].)   ID

[32b] *A note on Tauberian theorems*, Ann. of Math. 33 (1932), 787.   ID

[32c] *Back to Leibniz! (Physics reoccupies an abandoned position)*, Tech. Rev. 34 (1932), 201–203, 222, 224.   IIA

[32d] *Reports from Cambridge—1932*, Tech. Rev. 34 (1932), 62, 74.   IIIE

[32e] Review of A.S. Besicovitch, *Almost Periodic Functions*, Math. Gaz. 16 (1932), 275–277.   IVA

[32f] *Analytic properties of the characters of infinite Abelian groups* (with R.E.A.C. Paley), Abstract, Int'l. Math. Congr., Zürich, 1932, 95.   V

[33a] *Notes on random functions* (with R.E.A.C. Paley and A. Zygmund), Math. Z. 37 (1933), 647–668.   IC

[33b] *A one-sided Tauberian theorem*, Math. Z. 36 (1933), 787–789.   ID

[33c] *Characters of Abelian groups* (with R.E.A.C. Paley), Proc. Nat. Acad. Sci. U.S.A. 19 (1933), 253–257.   ID

[33d] *The total variation of $g(x + h) - g(x)$* (with R.C. Young, Trans. Amer. Math. Soc. 35 (1933), 327–340.   ID

[33e]   *Notes on the theory and application of Fourier transforms* (with   IE
        R.E.A.C. Paley) I, II, Trans. Amer. Math. Soc. 35 (1933),
        348–355; III, IV, V, VI, VII, Trans. Amer. Math. Soc. 35 (1933),
        761–791.
[33f]   *Putting matter to work (The search for chapter power)*, Tech. Rev.   IIIA
        35 (1933), 47–49, 70, 72.
[33g]   Review of Harald Bohr, *Fastperiodische Funktionen*, Math. Gaz.   IVA
        17 (1933), 54.
[33h]   *R.E.A.C. Paley—In Memoriam, Jan. 7, 1907—Apr. 7, 1933*, Bull.   IVB
        Amer. Math. Soc. 39 (1933), 476.
[33i]   *The Fourier Integral and Certain of Its Applications*, Cambridge   VI
        University Press, New York, 1933; reprint, Dover, New York,
        1959; review by E.C. Titchmarsh in Math. Gaz. 17 (1933), 129.
[34a]   *Random functions*, J. Math. and Phys. 14 (1934), 17–23.   IC
[34b]   *A class of gap theorems*, Ann. Scuola Norm. Sup. Pisa, E   IE
        (1934–1936), 1–6.
[34c]   *Quantum mechanics, Haldane, and Leibniz*, Philos. Sci. 1 (1934),   IIA
        479–482.
[34d]   *Fourier Transforms in the Complex Domain* (with R.E.A.C.   VI
        Paley), Amer. Math. Soc. Colloq. Publ. 19, Amer. Math. Soc.,
        Providence, R.I., 1934.
[34e]   *Aid for German-refugee scholars must come from non-academic*   IIID
        sources, Jewish Advocate (December 1934).
[35a]   *Fabry's gap theorem*, Sci. Repts. of Nat'l. Tsing Hua Univ., Ser.   IE
        A, 3 (1935), 239–245.
[35b]   *Limitations of science (The holiday fallacy and a response to the*   IIIA
        suggestion that scientists become sociologists), Tech. Rev. 37
        (1935), 255–256, 268, 270, 272.
[35c]   *The student agitator (Is he accepting radicalism as an opiate?)*   IIIC
        (with Carl Bridenbaugh), Tech. Rev. 37 (1935), 310–312, 344, 346.
[35d]   *Mathematics in American secondary schools*. J. Math. Assoc.   IIIC
        Japan for Secondary Education (Tokyo) 17 (1935), 1–5.
[35e]   *The closure of Bessel functions*, Abstract 66, Bull. Amer. Math.   V
        Soc. 41 (1935), 35.
[35f]   *Once more ... the refugee problem abroad*, Jewish Advocate   IIID
        (February 5, 1935).
[36a]   *A theorem of Carleman*, Sci. Repts. of Nat'l. Tsing Hua Univ., Ser.   IE
        A, 3 (1936), 291–298.
[36b]   *Sur les séries de Fourier lacunaires. Théorèmes directs* (with S.   IE
        Mandelbrojt), C.R. Acad. Sci. Paris 203 (1936), 34–36.
[36c]   *Séries de Fourier lacunaires. Théorèmes inverses* (with S. Man-   IE
        delbrojt), C.R. Acad. Sci. Paris 203 (1936), 233–234.
[36d]   *Gap theorems*, C.R. de Congr. Int'l. des Math., 1936, 284–296.   IE
[36e]   *A Tauberian gap theorem of Hardy and Littlewood*, J. Chinese   IE
        Math. Soc. 1 (1936) 15–22.
[36f]   *Notes of the Kron theory of tensors in electrical machinery*, J.   I, I
        Electr. Engrg., China 7 (1936), 277–291.
[36g]   *The role of the observer*, Philos. Sci. 3 (1936), 307–319.   IIA
[37a]   *Taylor's series of entire functions of smooth growth* (with W.T.   ID
        Martin), Duke Math. J. 3 (1937), 213–223.
[37b]   *Random Waring's theorems*, Abstract (with N. Levinson), Science   V
        85 (1937), 439.

[38a]   *The homogeneous chaos,* Amer. J. Math. 60 (1938), 897–936.   IC
        (Reprinted in [64f].)
[38b]   *On absolutely convergent Fourier-Stieltjes transforms* (with H.R.   ID
        Pitt), Duke Math. J. 4 (1938), 420–440.
[38c]   *Fourier-Stieltjes transforms and singular infinite convolutions* (with   ID
        A. Wintner), Amer. J. Math. 60 (1938), 513–522.
[38d]   *Taylor's series of functions of smooth growth in the unit circle* (with   ID
        W.T. Martin), Duke Math. J. 4 (1938), 384–392.
[38e]   *The historical background of harmonic analysis,* Amer. Math. Soc.   ID
        Semicentennial Publications Vol. II, Semicentennial Addresses,
        Amer. Math. Soc., Providence, R.I., 1938, 513–522.
[38f]   *Remarks on the classical inversion formula for the Laplace*   IE
        integral (with D.V. Widder), Bull. Amer. Math. Soc. 44 (1938),
        573–575.
[38g]   *The decline of cookbook engineering,* Tech. Rev. 40 (1938), 23.   IIIE
[38h]   Review of L. Hogben, *Science for the Citizen,* Tech. Rev. 40   IVA
        (1938), 66–67.
[39a]   *The ergodic theorem,* Duke Math. J. 5 (1939), 1–18. (Reprinted in   IC
        [64f].)
[39b]   *The use of statistical theory in the study of turbulence,* Proc. 5th   IC
        Int'l. Congr. of Applied Mechanics, Sept. 12–16, 1938, Wiley,
        New York, 1939, 356–358.
[39c]   *On singular distributions* (with A. Wintner), J. Math. and Phys. 17   ID
        (1939), 233–246.
[39d]   *Convergence properties of analytic functions of Fourier-Stieltjes*   IE
        transforms (with R.H. Cameron), Trans. Amer. Math. Soc. 46
        (1939), 97–109; Math. Rev. 1 (1940), 13; rev. 400.
[39e]   *A generalization of Ikehara's theorem* (with H.R. Pitt), J. Math.   ID
        and Phys. 17 (1939), 247–258.
[39f]   Review of Roger Burlingame, *March of the Iron Men,* Tech. Rev.   IVA
        41 (1939), 115.
[39g]   Review of W. George, *The Scientist in Action,* Tech. Rev. 41   IVA
        (1939), 202.
[39h]   *A new method in statistical mechanics,* Abstract 133 (with B.   V
        McMillan), Bull. Amer. Math. Soc. 45 (1939), 234; Science 90
        (1939), 410–411.
[40a]   Review of M. Fukamiya, *On dominated ergodic theorems in*   IVA
        Lp(p ≥ 1), Math. Rev. 1 (1940), 148.
[40b]   Review of M. Fukamiya, *The Lipschitz condition of random func-*   IVA
        tion, Math. Rev. 1 (1940), 149.
[40c]   Review of Th. De Donder, *L'énergétique déduite de la mécanique*   IVA
        statistique générale, Math. Rev. 1 (1940), 192.
[40d]   *A canonical series for symmetric functions in statistical mechanics,*   V
        Abstract 133, Bull. Amer. Math. Soc. 46 (1940), 57.
[41a]   *Harmonic analysis and ergodic theory* (with A. Wintner), Amer. J.   IC
        Math. 63 (1941), 415–426; Math. Rev. 2 (1941), 319.
[41b]   *On the ergodic dynamics of almost periodic systems* (with A. Wint-   IC
        ner), Amer. J. Math. 63 (1941), 794–824; Math. Rev. 4 (1943), 15.
[42a]   *On the oscillation of the derivatives of a periodic function* (with   IE
        G. Pólya), Trans. Amer. Math. Soc. 52 (1942), 249–256.
[43a]   *The discrete chaos* (with A. Wintner), Amer. J. Math. 65 (1943),   IC
        279–298; Math. Rev. 4 (1943), 220.

[43b]   *Behavior, purpose, and teleology* (with A. Rosenblueth and J.   IIA
Bigelow), Philos. Sci. 10 (1943), 18–24.

[44a]   Review of Hugh Gray Lieber and Lillian R. Lieber, *The Education*   IVA
of T.C. Mits; What Modern Methematics Means to You, Tech.
Rev. 46 (1944), 390, 392.

[45a]   *La teoría de la estrapolación estadística*, Bol. Soc. Mat. Mexicana   IG
2 (1945), 37–45; Math. Rev. 7 (1946), 461.

[45b]   *The role of models in science* (with A. Rosenblueth), Philos. Sci. 12   IIA
(1945), 316–322.

[46a]   *A generalization of the Wiener-Hopf integral equation* (with A.E.   IF
Heins), Proc. Nat. Acad. Sci. U.S.A. 32 (1946), 98–101; Math.
Rev. 8 (1947), 29.

[46b]   *The mathematical formulation of the problem of conduction of im-*   IIB
*pulses in a network of connected excitable elements, specifically in
cardiac muscle* (with A. Rosenblueth), Arch. Inst. Cardiol.
Méxicana 16 (1946), 205–265; Bol. Soc. Mat. Mexicana 2 (1945),
37–42; Math. Rev.

[47a]   *Sur les fonctions indéfiniment dérivables sur une demi-droite* (with   IE
S. Mandelbrojt), C.R. Acad. Sci. Paris 225 (1947), 978–980;
Math. Rev. 9 (1948), 230.

[47b]   *A scientist rebels*, Atlantic Monthly 179 (1946), 46; Bill. Atomic   IIIB
Scientist 3 (1947), 31.

[48a]   *Time, communication and the nervous system*, Ann. New York   IIB
Acad. Sci. 50 (1948), 197–220; Math. Rev. 10 (1949), 133.

[48b]   *Cybernetics*, Scientific American 179 (1948), 14–18.   IIB

[48c]   *An account of the spike potential of axons* (with A. Rosenblueth,   IIB
W. Pitts, J. Garcia Ramos, and the assistance of F. Webster), J.
of Cellular and Comparative Physiol. 32 (1948), 275–318.

[48d]   *A rebellious scientist after two years*, Bull. Atomic Scientists 4   IIIB
(1948), 338–339.

[48e]   Review of L. Infeld, *Whom the Gods Love. The Story of Evariste*   IVA
Galois, Scripta Math. 14 (1948), 273–274.

[48f]   *Cybernetics, or Control and Communication in the Animal and the*   VI
Machine, Actualités Sci. Ind., no. 1053; Hermann et Cie., Paris;
The MIT Press, Cambridge, Mass., and Wiley, New York, 1948;
Math. Rev. 9 (1948), 598. Partly reprinted as Rigidity and learn-
ing: ants and men, in *Classics in Biology (A Course of Selected
Reading by Authorities)*, Philosophical Library, New York, 1960,
pp. 205–213.

[49a]   *Sur la théorie de la prévision statistique et du filtrage des ondes*,   IG
Analyse Harmonique, Colloques Internationaux du CNRS,
No. 15, pp. 67–74. Centre National de la Recherche Scientifique,
Paris, 1949; Math. Rev. 11 (1950), 376.

[49b]   *A statistical analysis of synaptic excitation* (with A. Rosenblueth,   IIB
W. Pitts, and J. Garcia Ramos), J. of Cellular and Comparative
Physiol. 34 (1949), 173–205.

[49c]   *A new concept of communication engineering*, Electronics 22   IIB
(1949), 74–77.

[49d]   *Sound communication with the deaf*, Philos. Sci. 16 (1949),   IIB
260–262.

[49e]   *Some problems in sensory prosynthesis* (with J. Wiesner and L.   IIB
Levine), Science 110 (1949), 512.

[49f]   *Obituary—Godfrey Harold Hardy, 1877–1947*, Bull. Amer. Math.   IVB
Soc. 55 (1949), 72–77.

[49g]   *Extrapolation, Interpolation, and Smoothing of Stationary Time*   VI
Series with Engineering Applications, The MIT Press, Cambridge,
Mass., Wiley, New York; Chapman & Hall, London, 1949; paper-
back edition with the title *Time Series*, The MIT Press, 1964;
Math. Rev. 11 (1950), 118.

[49h]   Review of Philipp Frank, *Modern Science and its Philosophy*, New   IVA
York Times, Book Review, August 14, 1949, sec. 7, p. 3.

[50a]   *Some prime-number consequences of the Ikehara theorem* (with L.   ID
Geller); Acta. Sci. Math. (Szeged) 12 (1950), 25–28, Leopoldo
Fejer et Frederico Riesz LXX annos natis dedicatus, Pars B;
Math. Rev. 11 (1950), 644; Math. Rev. 12 (1951), 1002.

[50b]   *Comprehensive view of prediction theory*, Proceedings of the Inter-   IG
national Congress of Mathematicians, Cambridge, Mass., 1950,
vol. 2, pp. 308–321; Amer. Math. Soc., Providence, R.I., 1952,
Expository lecture; Math. Rev. 13 (1952), 477.

[50c]   *Some maxims for biologists and psychologists*, Dialectica 4 (1950),   IIA
186–191.

[50d]   *Purposeful and non-purposeful behavior* (with A. Rosenblueth),   IIA
Philos. Sci. 17 (1950), 318–326.

[50e]   *Cybernetics*, Bull. Amer. Acad. Arts and Sci. 3 (1950), 2–4.   IIB

[50f]   *Speech, language, and learning*, J. Acoust. Soc. Amer. 22 (1950),   IIB
696–697.

[50g]   *Entropy and information*, Proc. Sympos. Appl. Math., vol. 2,   IIB
Amer. Math. Soc., Providence, R.I., 1950, p. 89; Math. Rev. 11
(1950), 305.

[50h]   *Too big for private enterprise*, Nation 170 (1950), 496–497.   IIIA

[50i]   *Too damn close*, Atlantic 186 (1950), 50–52.   IIIA

[50j]   *The Human Use of Human Beings*, Houghton Mifflin, Boston,   VI
1950; paperback edition by Doubleday, Anchor, Garden City,
N.Y., 1954. Chapter 3 reprinted in *Classics in Biology (A Course
of Selected Reading by Authorities)*, Philosophical Library, New
York, 1960, pp. 205–213.

[50k]   *The brain* (short story), Tech. Eng. News 31 (1950), 14–15, 33–34,   VI
44, 50. (reprinted in paperback anthology, *Cross-roads in Time*,
ed. Groff Conklin, Doubleday, Garden City, N.Y., 1953.)

[51a]   *Problems of sensory prosthesis*, Bull. Amer. Math. Soc. 57 (1951),   IIB
27–35. (Reprinted in [64f].)

[51b]   *Homeostasis in the individual and society*, J. Franklin Inst. 251   IIB
(1951), 65–68. (Reprinted in [64f].)

[51c]   *Mathematical relationships of possible significance in the study of*   IIA
human leukemia (with P.F. Hahn), Federation Proc. 10 (1951).

[52a]   *Cybernetics (Light and Maxwell's demon)*, Scientia (Italy) 87   IIB
(1952), 233–235.

[52b]   *The miracle of the broom closet* (short story), Tech. Eng. News 33   VI
(1952), 18–19, 50. (Reprinted in the Magazine of Fantasy and
Science Fiction, ed. Anthony Boucher, February 1954, pp. 59–63).

[52c]   *Cybernetics*, in Encyclopedia Americana Annual 1952 edition,   IIB
187–188.

[53a]   *Optics and the theory of stochastic processes*, J. Opt. Soc. Amer. 43   IG
(1953), 225–228; Math. Rev. 17 (1956), 33.

[53b]   *A new form for the statistical postulate of quantum mechanics* (with   IH
        A. Siegel), Phys. Rev. 9 (1953), 1551–1560; Math. Rev. 15 (1954),
        273.

[53c]   *Distributions quantiques dans l'espace différentiel pour les fonctions*   IH
        d'ondes dépendant du spin (with A. Siegel), C.R. Acad. Sci. Paris
        237 (1953), 1640–1642; Math. Rev. 15 (1954), 490.

[53d]   *Les machines à calculer et la forme (Gestalt), Les machines à*   IIB
        calculer et la pensée humaine, Colloques Internationaux du Centre
        National de la Recherche Scientifique, Paris, 1953, pp. 461–463;
        Math. Rev. 16 (1955), 529.

[53e]   *The concept of homeostasis in medicine,* Transactions and Studies   IIB
        of the College of Physicians of Philadelphia (4) 20 (1953), No. 3,
        87–93.

[53f]   *Problems of organization,* Bull. Menninger Clinic 17 (1953),   IIB
        130–138.

[53g]   *The future of automatic machinery,* Mech. Engrg. 75 (1953),   IIB
        130–132.

[53h]   *Ex-prodigy: My Childhood and Youth,* Simon and Schuster, New   VI
        York, 1953; The MIT Press, Cambridge, Mass., 1965 (paperback
        edition, The MIT Press); Math Rev. 15 (1954), 277.

[53i]   *The electronic brain and the next industrial revolution,* Cleveland   IIIA
        Athletic Club Journal (January, 1953).

[53j]   *The machine as threat and promise,* St. Louis Post-Dispatch   IIIA
        (December, 1953).

[53k]   *We can't attain truth without risk of error* (from This I Believe   IIIA
        radio show), Minneapolis Tribune (November, 1953).

[54a]   *Men, machines, abd the world about,* in *Medicine and Science,*   IIB
        New York Academy of Medicine and Science, ed. I. Galderston,
        International Universities Press, New York, 1954, pp.
        13–28.

[54b]   *Conspiracy of conformists,* Nation 178 (1954), 375.   IIIE

[54c]   *Automatization* (with Donald Campbell), St. Louis Post-Dispatch   IIIA
        (December, 1954).

[55a]   *Nonlinear prediction and dynamics,* Proc. Third Berkeley Sym-   IG
        posium on Mathematical Statistics and Probability, University of
        California Press, Berkeley, Calif., 1954/5, pp. 247–252; Math.
        Rev. 18 (1957), 949.

[55b]   *On the factorization of matrices,* Comment. Math. Helv. 29 (1955),   IG
        97–111; Math. Rev. 16 (1955), 921.

[55c]   *The differential-space theory of quantum systems* (with A. Siegel),   IH
        Nuovo Cimento (10) 2 (1955), 982–1003, No. 4, Suppl.

[55d]   Thermodynamics of the message, in Neurochemistry, ed. K.E.C.   IIB
        Elliott, Thomas, Springfield, 1955, pp. 844–849.

[55e]   *Time and organization,* Second Fawley Foundation Lecture, Uni-   IIB
        versity of Southampton, 1955, pp. 1–16.

[56a]   *On a local L^2-variant of Ikehara's theorem* (with A. Wintner), Rev.   ID
        Math. Cuyana 2 (1956), 53–59.

[56b]   *The theory of prediction,* in *Modern Mathematics for the Engineer,*   IG
        ed. E.F. Beckenbach, McGraw-Hill, New York, 1956, pp.
        165–187.

[56c]   *"Theory of Measurement" in differential-space quantum theory*   IH
        (with A. Siegel), Phys. Rev. 101 (1956), 429–432.

[56d]   Pure patterns in a natural world, in The New Landscape in Art and   IIB
         Science, ed. G. Kepes, Paul Theobald and Co., Chicago, 1956, pp.
         274–276.

[56e]   Brain waves and the interferometer, J. Phys. Soc. Japan 18 (1956),   IIB
         499–507.

[56f]   Moral reflections of a mathematician, Bull. Atomic Scientists 12   IIIB
         (1956), 53–57. (Reprinted from [56g].)

[56g]   I Am a Mathematician. The Later Life of a Prodigy, Doubleday,   VI
         Garden City, New York, 1956; paperback edition by The MIT
         Press, 1964; Math. Rev. 17 (1956), 1037.

[57a]   The definition and ergodic properties of the stochastic adjoint of a   IC
         unitary transformation (with E.J. Akutowicz), Rend. Circ. Mat.
         Palermo (2) 6 (1957), 205–217, Addendum, 349; Math. Rev. 20
         (1959), rev. 4328.

[57b]   Notes on Pólya's and Turán's hypotheses concerning Liouville's   ID
         factor (with A. Wintner), Rend. Circ. Mat. Palermo (2) 6 (1957),
         240–248; Math. Rev. 20 (1959), rev. 5759.

[57c]   On the non-vanishing of Euler products (with A. Wintner), Amer.   ID
         J. Math. 79 (1957), 801–808.

[57d]   The prediction theory of multivariate stochastic processes, Part I   IG
         (with P. Masani), Acta Math. 98 (1957), 111–150; Math. Rev. 20
         (1959), rev. 4323.

[57e]   Rythms in physiology with particular reference to encephalography,   IIB
         Proceedings of the Rudolf Virchow Medical Society in the City of
         New York, vol. 16, 1957, pp. 109–124.

[57f]   The role of the mathematician in a materialistic culture (A   IIIA
         scientist's dilemma in a materialistic world), Columbia Engineer-
         ing Quarterly, Proceedings of the Second Combined Plan Confer-
         ence, Arden House, October 6–9, 1957, pp. 22–24.

[57g]   The role of the small cultural college in education of the scientists;   IIIC
         a speech given at Wabash College, Indiana, October 10, 1957.

[57h]   Cybernetics, in The Universal Standard Encyclopedia (abridgment   IIB
         of The New Funk and Wagnall's Encyclopedia), Standard Refer-
         ence Works Publishing Co., New York, 1957, p. 180.

[58a]   Logique, probabilité et méthode des sciences physiques, in La   IH
         Méthode dans les Sciences Modernes, Editions Science et Indus-
         trie, éd. François Le Lionnais, Paris, 1958, pp. 111–112.

[58b]   The prediction theory of multivariate stochastic processes, Part II   IG
         (with P. Masani), Acta Math. 99 (1958), 93–137; Math. Rev. 20
         (1959), rev. 4325.

[58c]   Random time (with A. Wintner), Nature 181 (1958), 561–562.   IIB

[58d]   Sur la prévision linéaire des processus stochastiques vectoriels à   IG
         densité spectrale bornée. I (with P. Masani), C.R. Acad. Aci. Paris
         246 (1958), 1492–1495; Math. Rev. 20 (1959), rev. 4324a.

[58e]   Sur la prévision linéaire des processus stochastiques vectoriels à   IG
         densité spectrale bornée. II (with P. Masani), C.R. Acad. Sci. Paris
         246 (1958), 1655–1656; Math. Rev. 20 (1959), rev. 4324b.

[58f]   My connection with cybernetics. Its origins and its future, Cyber-   IIB
         netica (Belgium) 1 (1958), 1–14.

[58g]   Time and the science of organization, Part II, Scientia 93 (1958),   IIB
         225–230.

[58h]   Science: The megabuck era, New Republic, 138 (1958), 10–11.   IIIA

[58i] *Nonlinear Problems in Random Theory*, The MIT Press, Cambridge, Mass., and Wiley, New York, 1958; paperback edition, The MIT Press, 1966.   VI

[59a] *A factorization of positive Hermitian matrices* (with E.J. Akutowicz), J. Math. Mech. 8 (1959), 111–120.   IG

[59b] *Nonlinear prediction* (with P. Masani), in *Probability and Statistics*, The Harald Cramér Volume, ed. U. Grenander, Stockholm, 1959, 190–212.   IG

[59c] *On bivariate stationary processes and the factorization of matrix-valued functions* (with P. Masani), Teor, Verojatnost, i Primenen. 4 (1959), 322–331. (English transl. Theor. Probability App. 4 (1959), 300–308).   IG

[59d] *Man and the machine* (Interview with N. Wiener), Challenge (The Magazine of Economic Affairs) 7 (1959), 36–41.   IIIA

[59e] *The Tempter* (novel), Random House, New York, 1959.   VI

[60a] *The application of physics to medicine*, in *Medicine and Other Disciplines*, New York Academy of Medicine, ed. I. Galderston, International Universities Press, 1960, pp. 41–57.   IIB

[60b] *The brain and the machine* (Summary of an address), in *Dimensions of Mind*, ed. S. Hook, Collier Books, 1960, (Proceedings of Third Annual New York Univ. Institute of Philosophy held on May 15–16, 1959), pp. 113–117.   IIB

[60c] *Kybernetik*, Contribution to *Wörterbuch der Soziologie*, F. Enke Verlag, Stuttgart, 1960, pp. 620–622.   IIB

[60d] *Some moral and technical consequences of automation*, Science 131 (1960), 1355–1358.   IIIA

[60e] *The duty of the intellectual*, Tech. Rev. 62 (1960), 26–27; reprinted almost in entirety in *The grand privilege*, Sat. Rev. 43 (1960), 51–52; also in Technion 18 (1961), 86–87—"A professor tells what a professor is."   IIIB

[60f] Preface to *Cybernetics of Natural Systems*, by D. Stanley-Jones, Pergamon Press, London, 1960, pp. v–viii.   IVA

[60g] *Possibilities of the use of the interferometer in investigating macromolecular interactions*, in *Fast Fundamental Transfer Processes in Aqueous Biomolecular Systems*, ed. F.O. Schmitt, Department of Biology, MIT, Cambridge, Mass., June 1960, pp. 52–53.   IIB

[61a] *Über Informationstheoreie*, Naturwissenschaften 48 (1961), 174–176.   IIB

[61b] *Science and society*, Voprosy Filosofii (1961), No. 7, 117–122; reprinted in Estratto Rivista Methodos 13 (1961), 1–8, and in Tech. Rev. 63 (1961), 49–52. Excerpts in Science 138 (1962), 651.   IIB

[61c] *Cybernetics*, Second edition of [48f] (revisions and two additional chapters), The MIT Press and Wiley, New York, 1961; paperpack edition, The MIT Press, 1965.   VI

[62a] A verbal Contribution to Proc. of the International Symposium on the Application of Automatic Control in Prosthetics Design, August 27–31, 1962, Opatija, Yugoslavia, pp. 132–133.   IIB

[62b] *The mathematics of self-organizing systems*, in *Recent Developments in Information and Decision Processes*, Macmillan, New York, 1962, pp. 1–21.   IIB

[62c] *Short-time and long-time planning*, originally presented at 1954   IIIA
ASPO National Planning Conference. Jersey Plans, An ASPO
Anthology (1962), 29–36.

[63a] *Random theory in classical phase space and quantum mechanics*   IH
(with Giacomo Della Riccia), Proc. Internat. Conference on
Functional Analysis, Massachusetts Institute of Technology,
Cambridge, Mass., June 9–13, 1963; *Analysis in Function Space*,
The MIT Press, Cambridge, Mass., 1964, pp. 3–14.

[63b] *Introduction to neurocybernetics* (with J.P. Schadé) and *Epilogue*,   IIB
in *Progress in Brain Research*, vol. 2 of Nerve, Brain and Memory
Models, Elsevier Publishing Co., Amsterdam, 1963, pp. 1–7,
264–268.

[63c] *The lonely nationalism of Rudyard Kipling* (with K. Deutsch), Yale   IIID
Rev. 52 (1963), 499–517.

[64a] *On the oscillations of nonlinear systems*, Proc. Symposium on Sto-   IIB
chastic Models in Medicine and Biology, Mathematics Research
Center, U.S. Army, June 12–14, 1963, ed. John Gurland, University of Wisconsin Press, Madison, Wisconsin, 1964, pp.
167–177.

[64b] *Dynamical systems in physics and biology*, Contribution to series   IIB
"Fundamental Science in 1984", The New Scientist (London) 21
(1964), 211–212.

[64c] *Machines smarter than men?* (Interview with N. Wiener), U.S.   IIIA
News and World Rept. 56 (1964), 84–86; abbreviated in Reader's
Digest 84 (1964), 121–124.

[64d] *Intellectual honesty and the contemporary scientist* (Transcript of   IIIA
talk given to Hillel Group at Massachusetts Institute of Technology), Tech. Rev. 66 (1964), 17–18, 44–45, 47.

[64e] *God, Golem, Inc.—A Comment on Certain Points Where Cyber-*   VI
netics Impinges on Religion, The MIT Press, Cambridge, Mass.,
1964; paperback edition, The MIT Press 1966.

[64f] *Selected Papers of Norbert Wiener* with expository papers by   VI
Y.W. Lee, Norman Levinson, and W.T. Martin, The MIT Press,
Cambridge, Mass., 1964.

[65a] L'homme et la machine, Proc. Colloques Philosophiques Inter-   IIB
nationaux de Royaumont, July, 1962; *Le concept d'information*
dans la science contemporaine, Gauthier-Villars, Paris, 1965, pp.
99–132.

[65b] *Perspectives in cybernetics*, Progress in Brain Research 17 (1965),   IIB
399–408.

[65c] *Cybernetics* in *Collier's Encyclopedia*, U.S.A., ed. William D. Hal-   IIB
sey, The Cormwell-Collier Publishing Co., New York, 1965,
pp. 598–599.

[66a] *Wave mechanics in classical phase space, Brownian motion, and*   IH
quantum theory (with G. Della Riccia), J. Math. Phys. 7 (1966),
1372–1383.

[66b] *Differential Space, Quantum Systems and Prediction* (with A.   VI
Siegel, B. Rankin, W.T. Martin), The MIT Press, Cambridge,
Mass., 1966.

[66c] *Generalized Harmonic Analysis and Tauberian Theorems* (paper-   VI
back edition of [30a] and [32a]), The MIT Press, Cambridge,
Mass., 1966.

[75a]   *Cybernetics* (with F. Landis), in Funk and Wagnall's New Ency-   IIB
        clopedia, Funk & Wagnall, New York, 1975, p. 228.
[85a]   *Letter covering the memorandum on the scope, etc., of a suggested*   IIB
        computing machine (September 21, 1940), Coll. Works, IV,
        pp. 122–124, cf. {M10}.
[85b]   *Memorandum on mechanical solution of partial differential equa-*   IIB
        tions, Coll. Works, IV, pp. 125–134, cf. {M10}.
[85c]   *Muscular Clonus: Cybernetics and Physiology* (with A. Rosen-   IIB
        blueth and J. Garcia Ramos), Coll. Works, IV, pp. 466–510,
        cf. {M10}.

Defense Department Documents[1]

Ia. S.H. Caldwell, Proposal to Section D2, NDRC (3 p.), November 22, 1940.

Ib. N. Wiener, Principles governing the construction of prediction and compensating apparatus (8 p.) accompaniment to Ia, November 22, 1940.

II. K.T. Compton, Letter to Dr. Warren Weaver, NDRC, May 13, 1941.

III. J.H. Bigelow, Minutes of Conference held at Bell Laboratories on June 4, 1941.

IV. N. Wiener, Letter to Dr. Warren Weaver, NDRC, December 1, 1941.

V. G.R. Stibitz, Note on prediction networks a la Wiener (14 p.), February 22, 1942.

VI. Warren Weaver, Letter to Dr. J.C. Boyce, MIT, March 24, 1942.

VII. N. Wiener, A.A. Directors, Summary Report of Demonstration (17 p.), June 10, 1942.

VIII. Demonstration by Wiener and Bigelow at MIT, July 1, 1942, Diary of G.R. Stibitz, Chairman, Division D2, July 23, 1942.

IX. N. Wiener: Final report on Section D2, Project No. 6 (8 p.) submitted to Dr. Warren Weaver, NDRC, December 1, 1942.[2]

X. N. Wiener, Letter to Dr. Warren Weaver, NDRC, January 15, 1943.

XI. R.S. Phillips and P.R. Weiss, Theoretical calculation on best smoothing of position data for gunnery prediction, MIT Radiation Laboratory Report 532, February 16, 1944.

XII. N. Wiener, Automatic Control Techniques in Industry, Industrial College of the Armed Forces, Washington, D.C., 1952–1953.

---

1 NDRC stands for National Defense Research Committee.
2 This report is accompanied by a Report to the Services, No. 59, of March 27, 1945, entitled "Statistical Method of Prediction in Fire Control".

References

A1 Albertson, R., A survey of mutualistic communities in America, *Iowa J. of History and Politics*, 34 (1936), 375–444. (Reprinted AMS Press, New York, 1973).

A2 American Mathematical Society Bulletin 72 (No. 1, Part II) (1966), dedicated to Norbert Wiener.

A3 Ampère, André-Marie, *Essai sur la philosophie des sciences ou exposition analytique d'une classification naturelle de toutes les connaissances humaines,* Paris, Bachelier, 1838, première partie; 1843, seconde partie.

A4 Ashby, W.R., *Design for a Brain,* Wiley, New York, 1960.

A5 Ashby, W.R., Principles of the self-organizing system, pp. 255–278, in *Principles of Self-organization,* Edited by H. von Foerster & G.W. Zopf, Macmillan Co., New York, 1962

A6 Ashby, W.R., *An Introduction to Cybernetics,* Wiley, New York, 1963.

A7 Augustine, St., Concerning the nature of good, in *Basic Writings of St. Augustine,* Vol. 1, Edited by W.J. Oates, Random House, New York, 1948.

A8 Augustine, St., The City of God, in *Basic Writings of St. Augustine,* Vol. 2, Edited by W.J. Oates, Random House, New York, 1948.

A9 Augustine, St., On Genesis against the Manichaeans. (quoted in E. Teselle's *St. Augustine, the Theologian,* Burns & Oats, London, 1970).

B1 Barnes, J.L., Information-theroretic aspects of feedback control systems, *Automatica* 4 (1968), 165–185.

B2 Bass, J., Stationary functions and their applications to the theory of turbulence, *J. Math. Appl.* 47 (1974), 354–399, 458–503.

B3 Bass, J., *Fonctions de corrélation, fonctions pseudo-aléatoires et applications,* Masson, Paris, 1984.

B4 Bell, D., *The Cultural Contradictions of Capitalism,* Basic Books, Inc., New York, 1976.

B5 Bell, E.T., *Men of Mathematics,* Simon & Schuster, New York, 1965.

B6 Benedetto, J., *Spectral Synthesis,* Academic Press, 1975.

B7 Bergman, S., *The Kernel Function and Conformal Mapping,* Math. Surveys 5, Rev. Ed., Amer. Math. Soc., Providence, R.I., 1970.

B8 Bernal, J.D., *The Social Function of Science,* George Routledge & Sons, London, 1946.

B9 Bernstein, J., *Einstein,* Fontana/Collins, Glasgow, 1973.

B10 Birkhoff, G.D., Intuition, reason and faith in science, *Science* 88, (1938), 601–609.

B11 Birkhoff, G.D., The principle of sufficient reason, *Rice Inst. Pamphlet* Vol. 28, Jan. 1941, 24–50.

B12 Blair, C., Passing of a great mind, *Life,* February 25, 1957.

B13 Bohm D., *Causality and Chance in Modern Physics*, Harper & Row, New York, 1957.

B14 Bohr, N., *Atomic Physics and Human Knowledge*, Wiley, New York, 1958.

B15 Bonaventura, St., *De reductione artium ad theologiam*, Translated by Sister E.T. Healy, St. Bonaventure College, 1940.

B16 Born, M., *Theory and Experiment in Physics*, Cambridge University Press, Cambridge, 1944.

B17 Born, M., *The Natural Philosophy of Cause and Chance*, Clarendon Press, Oxford, 1949.

B18 Born, M. and Wolf, J., *Principles of Optics*, 5th ed., Pergamon Press, Oxford, 1975.

B19 Brelot, M., Norbert Wiener and potential theory, pp. 39–41, in Ref. A2.

B20 Bridgman, P.W., *The Nature of Physical Theory*, Princeton University Press, Princeton, New Jersey, 1936.

B21 Buchdahl, H.A., *The Concepts of Classical Thermodynamics*, Cambridge University Press, 1966.

B22 Bush, V., *Operational Circuit Theory*, Wiley, New York, 1929.

B23 Bush, V., The differential analyser—A new machine solving differential equations, *J. Franklin Institute*, 212 (1931), 447–488.

B24 Bush, V., Gage, F.D. and Stewart, H.R., A continuous integraph, *J. Franklin Institute*, 203 (1927), 63–84.

C1 Cameron, R.H. and W.T. Martin, The orthogonal development of non-linear functionals in series of Fourier-Hermite functionals, *Ann. of Mathematics*, 48 (1947), 385–392.

C2 Cannon, J.W., *Topological, Combinatorial and Geometric Fractals*. Hedrick Lectures of Math. Assoc. of Amer., 1982 (unpublished).

C3 Carnap, R., *The Logical Syntax of Language*. Kegan Paul, Trench, Trubner, London, 1937.

C4 Carnap, R., *Foundations of Logic and Mathematics*, International Encyclopedia of Unified Science, Vol., I, No. 3, University of Chicago Press, Chicago, 1939.

C5 Carnap, R., *Introduction to Semantics*, Harvard University Press, Cambridge, Massachusetts, 1942.

C6 Carnap, R., *Logical Foundations of Probability*, University of Chicago Press, Chicago, 1950.

C7 Carnap, R., *Philosophical Foundations of Physics*, Basic Books Inc., New York, 1966.

C8 Cassirer, E., *An Essay on Man*, Yale University Press, New Haven, Connecticut, 1944.

C9 Cassirer, E., *Language and Myth*, Dover Publications, New York, 1946.

C10 Clausewitz, Karl von, *On War*, Translated by Colonel J. Graham, Routledge & Kegan Paul, 1966.

C11 Coomaraswamy, A.K., *The Transformation of Nature in Art*, Harvard University Press, Cambridge, Mass., 1935.

C12 Coomaraswamy, A.K., *Spiritual Authority and Temporal Power in the Indian Theory of Government*, American Oriental Society, New Haven, Connecticut, 1942.

D1 Daniell, P.J., A general form of integral, *Ann. of Math.* (2) 19 (1917–1918), 279–304.

D2 Deutsch, K., *The Nerves of Government*, Free Press, New York, 1966.

D3   Dewey, J., *Logic, the Theory of Inquiry*, H. Holt & Co, New York, 1938.
D4   Dewey, J., *Education Today*, Greenwood Press, New York, 1940.
D5   Doob, J.L., *Stochastic Processes*, Wiley, New York; Chapman & Hall, London, 1953.
D6   Doob, J.L., Wiener's work in probability theory, pp. 69–72, in Ref. A2.
D7   Draganescu, M., *Odobleja Between Ampère and Wiener*, Romanian Accad. Sci., Bucharest, 1981.

E1   Eddington, A.S., *The Mathematical Theory of Relativity*, Cambridge University Press, Cambridge, 1937.
E2   Einstein, A., *Investigations on the Theory of the Brownian Movement*, (1905), Methuen, London, 1926.
E3   Einstein, A., Méthode pour la détermination de valeurs statistiques d'observations concernant des grandeurs soumises à des fluctuations irrégulières. *Archive des Sciences Physiques et Naturelles*, t. 37 (1914), p. 254–255.
E4   Einstein, A., *Sidelights on Relativity*, Methuen, London, 1922.
E5   Einstein, A., *Out of My Later Years*, The Citadel Press, Secaucus, New Jersey, 1950.
E6   Eliot, T.S., *Murder in the Cathedral*, Faber & Faber, London, 1935.
E7   Engels, F., *Ludwig Feuerbach and the Outcome of Classical German Philosophy*, (1888), International Publishers, New York, 1941.

F1   Faramelli, N. Fr., *Some Implications of Cybernetics for our Understanding of Man: An Appreciation of the Work of Dr. Norbert Wiener*. Ph.D Thesis, Temple University, (Department of Religion, 1968).
F2   Fisher, R.A., Indeterminism and natural selection, *Philosophy Sci.* 1 (1934), 99–117.
F3   Foures, Y. and I.E. Segal, Causality and analyticity, *Trans. Amer. Math. Soc.* 98 (1955), 384–405.
F4   Frank, P., *Einstein, His Life and Times*, Jonathan Cape, London, 1948.
F5   Freudenthal, H., Norbert Wiener, pp. 344–347 in *Dictionary of Scientific Biography*, Vol. XIV, Chas. Scribner's Sons, New York, 1976.
F6   Fuller, H.J., The Emperor's New Clothes: or Prius dementat, Scientific Monthly, LXXII (no. 1), January, 1951.

G1   Gallie, W.B., *Philosophers of Peace and War*, Cambridge University Press, 1978.
G2   Gandhi, M.K., *The Collected Works of Mahatma Gandhi*, Publications Division, Ministry of Information and Broadcasting, Delhi, 1958.
G3   Garcia Ramos, G., *Libro homanaje: Arturo Rosenblueth*, Instituto Politecnico Nacional, Mexico, 1971.
G4   Giuculescu, A. The concepts of cybernetics. An historical outline, pp. 139–204 in Ref. D7.
G5   Gleason, A.M., Measures on the closed subspaces of a Hilbert space, *J. Rat. Mech. Analysis* 6 (1957), 885–894.
G6   Glushkov, V.M., *Introduction to Cybernetics*, Academic Press, New York and London, 1966.
G7   Glushkov, V.M., Contemporary cybernetics, pp. 47–70, in *Survey of Cybernetics*, Edited by J. Rose, Gordon & Breach, New York, 1969.
G8   Gödel, K., Über formal unentscheidbare Sätze der Principia Mathematica und verwandter Systeme 1, Monatsh. für Math. u. Phys. 38 (1931), 173–198.

G9 Gödel, K., Russell's mathematical logic (1944), pp. 447–469, in *Philosophy of Mathematics*, Edited by P. Benaceraff & H. Putnam, Cambridge University Press, Cambridge, 1986.

G10 Goldstine, H.H., *The Computer from Pascal to von Neumann*, Princeton University Press, Princeton, New Jersey, 1972.

G11 Goldstine, H.H. & von Neumann, J., On the principles of large scale computing machines, 1946 (unpublished), pp. 1–32, Vol. 5, in Ref. V6.

G12 Gould, K.E., A new machine for integrating a functional product. *J. Mathematics & Physics* 17 (1929), 305–316.

G13 Grattan-Guiness, I., Wiener on the logics of Russell and Schroeder; an account of his doctoral thesis, and of his discussion of it with Russell, *Ann. Sci.* 32 (1975), 103–132.

G14 Gray, T.S., A photoelectric integraph, *J. Franklin Institute*, 212 (1931), 77–102.

G15 Greenberg, D.S., The National Academy of Sciences: profile of an institution, *Science*, April 14, 21, 28, 1967.

H1 Hadamard, J., *The Psychology of Invention in the Mathematical Field*, Dover Publications Inc., New York, 1954.

H2 Haldane, J.B.S., Quantum mechanics as a basis for philosophy, *Philosophy Sci.* 1 (1934), 78–98.

H3 Halmos, P.R., Mathematics as a creative art, *American Scientist*, 56 (1968), 375–389.

H4 Hardy, G.H., *A Mathematician's Apology*, Cambridge University Press, Cambridge, 1948.

H5 Hazen, H.L. and Brown, G.S., The cinema integraph. A machine for integrating a parametric product integral. *J. Franklin Institute*, 230 (1940), 19–44, 183–205.

H6 Heims, S.J., *John von Neumann and Norbert Wiener: From Mathematics to the Technologies of Life and Death*, MIT Press, Cambridge, Mass., 1980.

H7 Hewitt, E., and K. Ross, *Abstract Harmonic Analysis*, Vol. II, Springer, New York, 1970.

H8 Hopf, E., *Mathematical Problems of Radiative Equilibrium*, Cambridge University Press, Cambridge, 1933.

K1 Kac, M., Wiener and integration in function space, pp. 52–68 in Ref. A2.

K2 Kakutani, S., Two-dimensional Brownian motion and harmonic functions, *Proc. Imp. Acad. of Sciences*, Tokyo 20 (1944), 706–714.

K3 Kakutani, S., Determination of the spectrum of the flow of a Brownian motion, *Proceedings, National Academy of Sciences*, U.S.A. 36 (1950), 319–333.

K4 Keynes, J.M., *Laisser-faire and Communism*, New Republic, New York, 1926.

K5 Kolmogorov, A.N., *Foundations of the Theory of Probability*, (1933), Chelsea Publishing Company, New York, 1950.

K6 Kolmogorov, A.N., Sur l'interpolation et extrapolation de suites stationnaires, *C.R. Acad. Sci. Paris*, 208 (1939), 2043–2045.

K7 Kolmogorov, A.N., Wiener's spiral and certain other interesting curves in Hilbert spaces, *DAN SSSR*, 26 (1940) 115–118. (In Russian.)

K8 Kolmogorov, A.N., Stationary sequences in Hilbert space (Russian). *Bull. Math. Univ., Moscou*, 2, No. 6 (1941), 40 pp. (English translation by Natasha Artin).

K9   Kolmogorov, A.N., Interpolation and extrapolation of stationary random sequences, *Izvestiya Akad. Nauk* (Mathematical Series 5) (1941), 3–14.

K10 Kosambi, D.D., Obituary: George David Birkhoff, *Math. Stud.* 12 (1945), 116–120.

K11 Kratkii filosofskii slovar (Short philosophical dictionary), Edited by Mark Moiseyevich Rosenthal, 4th ed., Leningrad, 1954, 236–237.

L1   Lafargue, P., *Socialism and the Intellectuals,* New York Labor News, New York 1967.

L2   Lau, K.S., On the Banach space of functions with bounded upper means, *Pac. J. Math.* 91 (1980), 153–172.

L3   Lau, K.S., On some classes of Hardy spaces, *J. Functional Analysis,* (to appear in 1989).

L4   Lee Y.W., Synthesis of electric networks by means of Fourier transforms of Laguerre's functions, *J. Math. Phy* 11 (1932), 83–113.

L5   Lee Y.W., *Statistical Theory of Communication,* Wiley, New York, 1960.

L6   Lee, Y.W., Contributions of Norbert Wiener to linear theory and nonlinear theory in engineering, pp. 17–34, in *Selected Papers of Norbert Wiener,* MIT Press, Cambridge, Massachusetts, 1964.

L7   Lenin, V.I., *What Is To Be done?* (1902), Edited by S.V. and Patricia Utechin, Oxford University Press, Oxford, 1933.

L8   Levinson, N., Wiener's Life, pp. 1–32 in Ref. A2.

L9   Liddell Hart, B.H., *History of the Second World War,* Paragon Books, New York, 1970.

L10 Lynd, Albert, *Quackery in the Public Schools,* Grosset & Dunlap, New York, 1953.

M1   Mackey, G.W., *Unitary Group Representations in Physics, Probability and Number Theory,* Benjamin, Reading, Massachusetts, 1978.

M2   Maimonides, Rabbi M., *The Guide for the Perplexed,* Edited by L. Roth, Hutchinson's University Library, London.

M3   Mandelbrot, B., *Fractals,* Freeman, New York, 1977.

M4   Mandelbrot, B., *The Fractal Geometry of Nature,* Freeman, New York, 1983.

M5   Martin, K., *Harold Laski, a Biographical Memoir,* Gollancz, London, 1953.

M6   Marx, K., Theses on Feuerbach, reprinted in F. Engels, *Ludwig Feuerbach and the Outcome of Classical German Philosophy* (1888), International Pub., New York, 1941.

M7   Marx, K., On the Jewish question, reprinted in *Writings of the Young Marx on Philosophy and Society,* Edited by L.D. Easton and K.H. Guddat, Doubleday, New York, 1967, 216–248.

M8   Marx, K. and Engels, F., *The Communist Manifesto,* (1848), Edited by A.J.P. Taylor, Penguin, London, 1967.

M9   Masani, P., Wiener's contributions to generalized harmonic analysis, prediction theory and filter theory, pp. 73–125 in Ref. A2.

M10 Masani, P., *Norbert Wiener: Collected Works,* Vol. I, 1976; Vol. II, 1979; Vol. III, 1981; Vol. IV, 1985; MIT Press, Cambridge, Massachusetts.

M11 Masani, P., Humanization as de-alienation, *Alternatives* 7 (1981), 265–290.

M12 Masani, P., Einstein's contribution to generalized harmonic analysis and his intellectual kinship with Norbert Wiener, *Jahrbuch Überblicke Mathematik* 1986, Vol. 19, (1986), 191–209.

M13 McCulloch, W.S., *Embodiments of Mind,* M.I.T. Press, Cambridge, Massachusetts, 1965.

M14   McCulloch, W., and Pitts, W., A logical calculus of the ideas immanent in nervous activity, *Bull. Math. Biophys.* 5 (1943), 115–133.

M15   Mehra, J. & Rechenberg, H., *The Historical Development of Quantum Theory,* Vol. III, Springer-Verlag, New York, 1982.

M16   Monod, J., *Chance and Necessity: An Essay on the Nature Philosophy of Modern Biology,* Vintage Books, New York, 1972.

M17   Morishima, M., *Marx's Economics: A Dual Theory of Value and Growth,* Cambridge University Press, Cambridge, 1973.

M18   Mumford, Lewis, *Technics and Human Development,* Harcourt, Brace Jovanovich, New York, 1967.

N1   Nagy, B.Sz., and C. Foias, *Harmonic Analysis of Operators on Hilbert Space,* North Holland, Amsterdam, 1970.

N2   Nelson, E., *Dynamical Theories of Brownian Motion,* Princeton University Press, Princeton, New Jersey, 1967.

N3   Neurath, O., Unified science as encyclopedic integration, pp. 1–27, in *International Encyclopedia of Unified Science,* Vol. I, Chicago University Press, Chicago, 1938.

N4   Nordhoff, C., *The Communistic Societies of the United States,* (1875), Dover Publications, New York, 1966.

O1   Odobleja, S., *Psychologie Consonantiste,* Librairie Maloine, Paris, 1938, 1939.

P1   Parkinson, C. Northcote, *Parkinson's Law on the Pursuit of Happiness,* John Murray, London, 1958.

P2   Peirce, Charles S., *Selected Writings,* Edited by Philip P. Wiener, Dover Publications, New York, 1966.

P3   Perrin, J., *Atoms* (English translation), Constable, London, 1915.

P4   Phelan, G.B., *St. Thomas and Analogy,* Marquette University Press, Milwaukee, Wisconsin, 1973.

P5   Pierce, J.R., *Signals and Noise: The Nature and Process of Communication,* Harper & Row, New York, 1961.

P6   Pitts, W. & McCulloch, W.S., How we know universals: the perception of auditory and visual forms, *Bull. Math. Biophysics,* 9 (1947), 127–147.

Q1   Quine, W.V., *Mathematical Logic,* Harvard University Press, Cambridge, Massachusetts, 1947.

Q2   Quine, W.V., *From a Logical Point of View,* Harvard University Press, Cambridge, Massachusetts, 1953.

R1   Randell, B., *The Origins of Digital Computers,* Springer-Verlag, Berlin, 1973.

R2   Rosenblueth, A., *Mind and Brain: A Philosophy of Science,* MIT Press, Cambridge, Massachusetts, 1970.

R3   Rudin, W., *Real and Complex Analysis,* McGraw-Hill, New York, 1966.

R4   Russell, B., *Principles of Mathematics* (1904), W. Norton & Company, New York, 1937.

R5   Russell, B., Mathematical logic as based on the theory of types, *Amer. J. Math.* 30 (1908), 222–262.

R6   Russell, B., *Mysticism and Logic,* George Allen and Unwin, London, 1932.

R7   Russell, B., *Human Knowledge, Its Scope and Limits,* Simon & Schuster, New York, 1948.

R8   Russell, B., *The Autobiography of Bertrand Russell*, Vols. I, II, III, Little, Brown, Boston, 1968.

R9   Russell, B., *Wisdom of the West*, Crescent Books, New York, 1977.

S1   Saeks, R., Causality in Hilbert space, *SIAM Rev.* 12, (1970), 357–383.

S2   Samuel, A.L., Some studies in machine learning using the game of checkers, *I.B.M. Journal* 3 (1959), 211–223.

S3   Schetzen, M., *Wiener and Volterra Series*, Wiley-Inter-Science, New York, 1980.

S4   Schrödinger, E., *What is Life?* Cambridge University Press, Cambridge, 1946.

S5   Schuster, A.R., The periodogram of magnetic declination, *Camb. Phil. Trans.* 18 (1900), pp. 107–135.

S6   Shannon, C.E., A symbolic analysis of relay and switching circuits, *Trans. Am. Inst. Electr. Eng.* 57 (1938), 713–723.

S7   Shannon, C.E., The mathematical theory of communication, *Bell Syst. Techn. Journ.* 27 (1948), 379–423; 623–656.

S8   Shannon, C.E., Programming a computer for playing chess, Phil. Mag. 41 (1950), 246–275.

S9   Snow, Sir C.P., *The Masters*, Macmillan, London, 1951.

S10   Struik, D.J., Wiener: Colleague and Friend, *American Dialogue*, March–April, 1966.

T1   Taub, A.H., John von Neumann, pp. 46–57, in *Encyclopedia of Computer Science and Technology*, Edited by H. Belzer, A.G. Holzman and A. Kent, Vol. 14, Marcel Dekker, New York, 1980.

T2   Tawney, R.H., *Religion and the Rise of Capitalism*, (1926), Mentor Edition, New York, 1954.

T3   Taylor, G.I., Diffusion by continuous movements, Proc. Lond. Math. Soc. 20 (1920), 196–212.

T4   Tillich, P., *The Socialist Decision*, Harper & Row, New York, 1977.

T5   Toynbee, Arnold, *Civilization on Trial*, Oxford University Press, Oxford, 1948.

T6   Tran, L.T., The Hausdorff dimension of the range of the N-parameter Wiener process, *Ann. Probability*, 5, (1977), 235–242.

T7   Tran, L.T., The Hausdorff α-dimensional measures of the level sets and the graph of the N-parameter Wiener process, (1977), (unpublished).

T8   Turing, A.M., On computable numbers, Proceedings, London Mathematical Society, Series 11, 42 (1937), 230–266.

U1   Ulam, S., John von Neumann 1903–1957, *Bull. Amer. Math. Soc.*, 64 (3), Part II, May 1958, 1–49.

U2   Ulam, S., *Adventures of a Mathematician*, Scribner, New York, 1976.

V1   von Neumann, J., Eine Axiomatisierung der Mengenlehre, *J. für reine u. angew. Math.* 154 (1925), 219–240.

V2   von Neumann, J., *Mathematical Foundations of Quantum Mechanics* (1930), Princeton University Press, Princeton, N.J., 1955.

V3   von Neumann, J., Physical applications of the ergodic hypothesis, *Proc. Nat. Acad. Sci.*, 18 (1932), 263–266.

V4   von Neumann, J., Zur Operatorenmethode der klassischen Mechanik, *Ann. of Math.* (2), 33 (1932), 587–642.

V5   von Neumann, J., A model of general economic equilibrium, *Rev. Econ. Studies,* 13 (1945), 1–9.

V6   von Neumann, J., *Collected Works,* Vols. I–VI, Edited by A.H. Taub, Pergamon Press, Oxford, 1961.

V7   von Neumann, J. and Morgenstern, O., *Theory of Games and Economic Behavior,* Princeton University Press, Princeton, New Jersey, 1944.

W1   Walter, W. Grey, Neurocybernetics, pp. 93–108, in *Survey of Cybernetics,* Edited by J. Rose, Gordon & Breach, New York, 1969.

W2   Watson, J.D., *The Double Helix,* Penguin Books, Middlesex England, 1971.

W3   Watts, A.W., *The Two Hands of God,* Colliers, New York, 1972.

W4   Weyl, H., *The Theory of Groups and Quantum Mechanics,* Dover Publications, New York, 1931.

W5   Weyl, H., *Philosophy of Mathematics and Natural Science,* Princeton University Press, Princeton, New Jersey, 1949.

W6   Weyl, H., *Symmetry,* Princeton University Press, Princeton, New Jersey, 1952.

W7   Whitehead, A.N., *Introduction to Mathematics,* (1919), Home University Library, London, 1945.

W8   Whitehead, A.N., *Science and the Modern World,* (1928), Cambridge University Press, Cambridge, 1933.

W9   Whitehead, A.N., *Religion in the Making* (1928), Meridian, New York, 1960.

W10   Whitehead, A.N., Mathematics and the good, pp. 660–687, in *The Philosophy of A.N. Whitehead,* Edited by P.A. Schilpp, Tudor, New York, 1951.

W11   Whitehead, A.N. and B. Russell, *Principia Mathematica,* Vol. I, II, III (1910–1913), Cambridge University Press, Cambridge, 1925–1927.

W12   Whittaker, Sir Edmund, *A History of the Theories of Aether and Electricity.* Vol. II, Harper, New York, 1953.

W13   Wigner, E.P., The unreasonable effectiveness of mathematics in the natural sciences, *Commun. Pure Appl. Math.* 13 (1960), 1–14.

W14   Wittgenstein, L., *Tractatus Logico-Philosophicus,* Routledge & Kegan Paul, New York, 1961.

Y1   Yaglom, A.M., *An Introduction to the Theory of Stationary Random Functions,* Prentice-Hall, Englewood Cliffs, N.J., 1962.

Y2   Yaglom, A.M., Einstein's 1940 paper on the theory of irregularly fluctuating sums of observations, *Problemy Peredachi Informatsii* (Problems of Information Transmission), 1985, Vol. 21, No. 4, pp. 101–107. English translation: IEEE ASSP Magazine, October 1987 (Vol. 4, no. 4).

Name Index

Subject Index

---

* This term, occurring in Wiener's lexicon, and meaning a function on the
time-domain the values of which are
empirical data, is replaced in the book
by terms such as input signal, output
signal, signal.

Photograph Index and Credits

Norbert Wiener, 1945 (Frontispiece, p. 2). Courtesy of Mrs. Margaret Wiener.

America's first Leibniz (p. 17). Courtesy of Dr. William A. Stanley, National Oceanic and Atmospheric Administration, Washington, D.C.

Figure 1 (Chalice or face profiles?) and *Figure 2* (The wave-particle or "wavicle") (p. 27). Reprinted from: James Jeans: New Background of Science, New York 1933, Frontispiece.

Bertha and Leo Wiener, parents of Norbert Wiener (p. 31). Courtesy of The MIT Museum (Photo NW 81).

Norbert Wiener at age seven (p. 36). Courtesy of The MIT Museum (Photo NW 39).

Norbert Wiener in 1912 just before he received his doctorate from Havard (p. 44). Courtesy of the MIT Museum (Photo NW 60).

Bertrand Russell (p. 51). Reprinted from: Autobiography of Bertrand Russell, Little, Brown & Co., Boston 1968, p. 277.

Norbert Wiener in army uniform (p. 69). Courtesy of The MIT Museum (Photo NW 46).

Albert Einstein (p. 80). Reprinted from: G. Kuznecov: Einstein. Leben-Tod-Unsterblichkeit, Birkhäuser Verlag, Basel 1977, Abb. 5.

J. B. Perrin (p. 81). Courtesy of Dr. Francis Perrin.

Marguerite Engemann in 1926 (p. 95). Courtesy of The MIT Museum (Photo NW 89).

Lord Kelvin (p. 103). Courtesy of The Royal Society, London.

Lord Rayleigh (p. 103). Courtesy of The Royal Society, London.

Sir Oliver Heaviside (p. 103). Reprinted from: D.K. McClerey: Introduction to Transients, New York 1961, Frontispiece. Courtesy of Wiley & Sons Inc.

Sir Arthur Schuster (p. 103). Courtesy of The Royal Society, London.

Sir Geoffrey Taylor (p. 103), Reprinted from: The Scientific Papers of G.I. Taylor, Vol. 4, ed. by G.K. Batchelor, New York 1971. Courtesy of Cambridge University Press.

G.H. Hardy (p. 106): Courtesy of Birkhäuser Boston Inc.

Wiener with Max Born (p. 118). Courtesy of The MIT Museum (Photo NW 1).

Gofffried Wilhelm Leibniz (p. 126). Courtesy of Dr. E.A. Fellmann, Euler Archiv Basel.

J.B.S. Haldane (p. 127). Reprinted from: Living Philosophers, New York 1931, p. 324. Courtesy of Harper and Row Publishers Inc.

Eberhard Hopf (p. 133). Courtesy of The MIT Museum (Photo EH 1).

R.E.A.C. Paley (p. 137). Copyright Elwin Neame. Reprinted from: Fourier Transforms in the Complex Domain, R.E.A.C. Paley and N. Wiener, American Mathematical Society Colloquium Publications (1934), Vol. 19, by permission of the American Mathematical Society.

J.-B. Fourier (p. 144). Reprinted from: I. Grattan-Guinness: Joseph Fourier 1768–1830, MIT Press, Cambridge MA 1972 (Plate 4). Courtesy of Archives de l'Académie des Sciences de Paris.

H. Lebesgue (p. 144). Courtesy of Birkhäuser Boston Inc.

J.W. Gibbs (p. 145). Courtesy of AIP Niels Bohr Library.

G.D. Birkhoff (p. 145). Reprinted from: G.D. Birkhoff: Collected Mathematical Papers (1950), by permission of the American Mathematical Society.

C.E. Shannon (p. 155). Courtesy of The MIT Museum (Photo CES 1).